Shop Floor Control Systems

Shop Floor Control Systems

From design to implementation

A. Bauer, R. Bowden, J. Browne, J. Duggan and G. Lyons

CHAPMAN & HALL
London · Glasgow · New York · Tokyo · Melbourne · Madras

Published by Chapman & Hall, 2-6 Boundary Row London SE1 8HN, UK

Chapman & Hall, 2-6 Boundary Row, London SE1 8HN, UK

Blackie Academic & Professional, Wester Cleddens Road, Bishopbriggs, Glasgow G64 2NZ, UK

Chapman & Hall Inc., One Penn Plaza, 41st Floor, New York NY 10119, USA

Chapman & Hall Japan, Thomson Publishing Japan, Hirakawacho Nemoto Building, 6F, 1-7-11 Hirakawa-cho, Chiyoda-ku, Tokyo 102, Japan

Chapman & Hall Australia, Thomas Nelson Australia, 102 Dodds Street, South Melbourne, Victoria 3205, Australia

Chapman & Hall India, R. Seshadri, 32 Second Main Road, CIT East, Madras 600 035, India

First edition 1991
Published in paperback 1994

Printed in Great Britain by Clays Ltd, St. Ives plc, Bungay, Suffolk

ISBN 0 412 58150 7

A catalogue for this book is available from the British Library

CONTENTS

	Figures	viii
	About the authors	xi
	Acknowledgements	xiii
	Preface	xv
	Foreword	xix
	PART ONE: BACKGROUND TO SHOP FLOOR CONTROL SYSTEMS	
	Overview	**1**
1	**A background to shop floor control systems**	**3**
1.1	Introduction	3
1.2	The Just in Time approach to production management	5
1.3	An overview of requirements planning (MRP and MRP II)	12
1.4	MRP/OPT versus JIT	24
1.5	Guidelines for the development and installation of production planning and control systems	25
1.6	An architecture for production planning and control	26
1.7	Conclusions	32
	PART TWO: A FUNCTIONAL ARCHITECTURE FOR SHOP FLOOR CONTROL SYSTEMS	
	Overview	**33**
2	**An architecture for shop floor control systems**	**35**
2.1	Introduction	35
2.2	Production Activity Control	36
2.3	Factory coordination	46
2.4	Conclusions	62

3	**A structured functional model for shop floor control**	**64**
3.1	Introduction	64
3.2	Structured Analysis and Design Technique	65
3.3	Overview of SADT™ model for FC and PAC	70
3.4	A0: Coordinate the factory	74
3.5	A1: Design the production environment	75
3.6	A2: Coordinate the product flow	83
3.7	Conclusions	94
	PART THREE: AN INFORMATION TECHNOLOGY ARCHITECTURE FOR SHOP FLOOR CONTROL	
	Overview	**95**
4	**An information technology architecture for shop floor control**	**97**
4.1	Introduction	97
4.2	The concepts of a layered architecture	98
4.3	The entities and core services of the reference architecture	101
4.4	A descriptive summary of Petri nets	108
4.5	A sample protocol of the reference architecture	113
4.6	Conclusions	119
5	**Implementation technologies for shop floor control systems**	**120**
5.1	Introduction	120
5.2	An overview of information technology	121
5.3	Communication systems	123
5.4	Data management systems	129
5.5	Processing systems	137
5.6	User interfaces	141
5.7	The object oriented approach	144
5.8	Conclusions	148
	PART FOUR: STATE-OF-THE-ART REVIEW	
	Overview	**149**
6	**A review of scheduling strategies**	**151**
6.1	Introduction	151
6.2	Traditional scheduling approaches	152
6.3	Modern scheduling approaches	164
6.4	Conclusions	178
7	**A review of production environment design strategies**	**179**
7.1	Introduction	179
7.2	Product based manufacturing	180
7.3	Process planning	196
7.4	Manufacturing system analysis	203
7.5	Conclusions	207

	PART FIVE: THE IMPLEMENTATION OF SHOP FLOOR CONTROL SYSTEMS	
	Overview	**209**
8	**An approach to the implementation of factory coordination and production activity control systems**	**211**
8.1	Introduction	211
8.2	Sociotechnical design	212
8.3	The contribution of sociotechnical design to the implementation of PMS	226
8.4	The environment for Factory Coordination and Production Activity Control	231
8.5	Conclusions	242
9	**A design tool for shop floor control systems**	**244**
9.1	Introduction	244
9.2	The application generator	245
9.3	The manufacturing database	248
9.4	The rulesbase	252
9.5	The PAC simulation model	255
9.6	Using the AG in the electronics industry	269
9.7	Conclusions	274
	PART SIX: AN IMPLEMENTATION OF A PAC SYSTEM	
	Overview	**277**
10	**The environment of the case study**	**279**
10.1	Introduction	279
10.2	The business environment	280
10.3	The production management system environment	283
10.4	The technical and social sub-systems	288
10.5	The information technology environment	295
10.6	Conclusions	297
11	**Implementation of a PAC system**	**298**
11.1	Introduction	298
11.2	Description of the pilot implementation	299
11.3	Architectural mapping and implementation software	303
11.4	Lessons and guidelines	308
11.5	Conclusions	312
12	**References and further reading**	**314**
12.1	References	314
12.2	Further reading	334
	Index	**336**

FIGURES

	The structure of the book.	xvii
Figure 1.1	Breakdown of the lead time in a batch production system.	4
Figure 1.2	Components of production smoothing.	9
Figure 1.3	Basic structure of an MRP system.	14
Figure 1.4	Manufacturing resource planning.	18
Figure 1.5	NBS hierarchy.	28
Figure 1.6	Manufacturing controls systems hierarchy.	29
Figure 2.1	Production Activity Control.	37
Figure 2.2	The link between production environment design and control.	47
Figure 2.3	Process based layout vs. product based layout.	49
Figure 2.4	The production environment design task within factory coordination.	50
Figure 2.5	Data exchange between the control task of factory coordination and a number of PAC systems.	55
Figure 2.6	An overall picture of the factory coordination architecture.	63
Figure 3.1	SADT^{TM} model showing structured decomposition (Ross, 1985).	66
Figure 3.2	Parent and child relationship (Ross, 1985).	67
Figure 3.3	Input, output, control and mechanism of an SADT^{TM} box (Ross, 1985).	69
Figure 3.4	A-1: Context diagram for coordinate factory.	71
Figure 3.5	A-0: Coordinate factory.	72
Figure 3.6	A0: Coordinate the factory.	75
Figure 3.7	A1: Design the production environment.	76
Figure 3.8	A11: Develop process plan.	78
Figure 3.9	A12: Maintain product based layout.	79
Figure 3.10	A13: Analyse manufacturing system.	81

Figure 3.11 A2: Coordinate product flow. 83
Figure 3.12 A21: Schedule factory. 85
Figure 3.13 A22: Dispatch factory. 86
Figure 3.14 A24: Control cells. 88
Figure 3.15 A241: Schedule cell. 89
Figure 3.16 A242: Dispatch cell. 90
Figure 3.17 A245: Monitor cell. 91
Figure 3.18 A25: Monitor factory. 93
Figure 4.1 Architectural mapping from functional to layered. 100
Figure 4.2 Layered architecture for FC and PAC. 101
Figure 4.3 Petri net graph based on Table 4.1. 110
Figure 4.4 A marked Petri net. 111
Figure 4.5 Petri net after firing transition 1. 112
Figure 4.6 Timing of a Petri net. 112
Figure 4.7 Initial Petri net model for dispatcher and producer interaction. 116
Figure 4.8 Petri net after dispatcher commands the producer to start. 117
Figure 4.9 Petri net after the producer commences. 118
Figure 5.1 General computing model. 123
Figure 5.2 OSI reference model. 125
Figure 5.3 EDI concept. 127
Figure 5.4 Implementation scenario for communication. 129
Figure 5.5 Hierarchical data model. 130
Figure 5.6 Network data model. 131
Figure 5.7 Implementation scenario for data management. 137
Figure 5.8 Implementation scenario for processing. 140
Figure 5.9 Implementation scenario for the user interface. 143
Figure 5.10 Traditional design of software systems. 144
Figure 5.11 Object oriented approach. 145
Figure 5.12 Concept of object oriented technology. 146
Figure 6.1 Possible sequences of 4 jobs A, B, C and D (Cunningham and Brownie, 1986). 156
Figure 7.1 An example of a monocode coding system. 185
Figure 7.2 Two components of similar shape and size, but different manufacturing characteristics. 186
Figure 7.3 The formation of block diagonals in a matrix. 191
Figure 7.4 An example of a composite product. 194
Figure 7.5 Parallel approach of design and manufacturing. 203
Figure 7.6 Breakdown of the production lead time. 204
Figure 8.1 Inclusion of feedback in a system. 215
Figure 8.2 The transactional and contextual environments (Pava, 1983). 216

Figure 8.3 An approach to sociotechnical design. 224
Figure 8.4 View of an organization within its environment (Brownie *et al.*, 1988). 235
Figure 8.5 The different flows in a technical sub-system. 237
Figure 8.6 Manufacturing controls systems hierarchy. 238
Figure 8.7 A two-stage scheduling procedure. 240
Figure 8.8 A sample Gantt chart. 241
Figure 8.9 The two type of inputs used in the interactive scheduling task. 242
Figure 9.1 The PAC life cycle model. 245
Figure 9.2 Combinative operation. 251
Figure 9.3 Sequential operation. 252
Figure 9.4 Inspection operation. 252
Figure 9.5 Disjunctive operation. 253
Figure 9.6 Sample output for the interactive scheduler. 257
Figure 9.7 Petri net model of a completed producer task. 261
Figure 9.8 The command from the dispatcher reaches the Petri net. 263
Figure 9.9 Petri net fires transition 1. 264
Figure 9.10 Petri net fires transition 2. 265
Figure 9.11 Test area layout for the application generator. 270
Figure 9.12 Sample output from the rulesbase. 274
Figure 9.13 Monitor screen during the simulation run. 274
Figure 10.1 Integrated manufacturing business model. 283
Figure 10.2 The planning cycle at the Digital Equipment Corporation's plant in Clonmel. 287
Figure 10.3 An ideal layout under continuous flow manufacturing. 291
Figure 10.4 The actual layout of the PCB assembly area and unit assembly cells. 292
Figure 10.5 The four workcells within the common PCB assembly cell. 293
Figure 11.1 The hierarchy of PIM systems in Clonmel. 301
Figure 11.2 PAC in the test environment. 302
Figure 11.3 PAC implementation model in the Clonmel environment. 303
Figure 11.4 Implementation area for pilot PAC system. 305
Figure 11.5 Final PAC implementation. 306
Figure 11.6 The PAC implementation software. 307
Figure 11.7 The PAC scheduler. 311
Figure 11.8 The workcell monitor. 312

ABOUT THE AUTHORS

Alfred Bauer

Alfred Bauer is a senior consultant working within the Digital European Competency Center for the Manufacturing Industries. He provides consultancy to software houses and large customer implementing production planning and control systems. His previous activities during his eleven years with Digital include hardware engineering, development of commercial applications, computer aided software engineering (CASE), technical consulting and the management of three ESPRIT projects. Alfred has publications in the field of production management systems, and is a member of various public committees such as ACM, IEEE, EuroPace and the CIM-OSA Management Committee. Alfred has a diploma in Electronic Engineering.

Richard Bowden, PhD

Dr. Richard Bowden started work with the Knowledge Based Applications and Services group of Digital Equipment International B.V. in Galway, Ireland in 1990. This group is responsible for the design, development and application of strategic planning tools and systems. Richard's role has focused on Business Modelling, Software Development and Systems Integration. Within his Business Modelling role he has worked with Digital's Corporate Revenue Analysis Group and with a leading Wall Street firm. Prior to joining Digital, Dr. Bowden completed his Batchelor's, Master's and PhD degrees in Industrial Engineering at University College Galway, Ireland, and he is professionally qualified as a Chartered Engineer.

Professor Jim Browne

Professor Jim Browne graduated with a PhD from the University of Manchester Institute of Science and Technology for work done in the area of digital simulation of manufacturing systems. He joined UCG in 1980 and was appointed Director of the CIM Research Unit in 1986. He is chairman of the Special Interest Group on Robotics in CIM Europe, is a member of the International Federation of Information Processing Technical Committee 5 and Working Group 5.7 on 'Production Planning and Control'. He is the author or co-author of in excess of eighty scientific papers, has edited a number of books and is co-author of a book entitled *Production Management Strategies – A CIM Perspective*, published by Addison-Wesley in 1988. He was awarded the degree of D.Sc. by the University of Manchester in 1991 for work on the design and analysis of advanced manufacturing systems.

James Duggan, PhD

James Duggan is employed as a software engineer with Digital's Knowledge Based Applications and Services Group in Galway, Ireland. In this role, he has worked on the development and delivery of a decision support system to assist corporate management with the task of business modelling and long-term strategic analysis. His Ph.D. thesis, which he completed at University College Galway prior to joining digital in 1990, centred on the domain of Shop Floor Control systems. Part of this work included involvement in the implementation of the prototype described in the latter stages of this book.

Gerard J. Lyons, PhD

Gerard Lyons was appointed Director of the Information Technology Centre at UCG in 1991 and was previously an information systems manager with Digitial Equipment Corporation. At UCG he is responsible for the development of degree programmes and collaborative R&D in the I.T. area. He is also engaged in teaching, applied research, and consulting. His current research interests include the analysis and modelling of business processes, and the impact of information technology on organisation design and performance. During the research for this text, he was responsible for the pilot implementation of the production planning and control systems described. Gerard is an engineer by training, holding Batchelor's, Master's and Ph.D. degrees, and is a professionally qualified Chartered Engineer (Eur. Ing.).

ACKNOWLEDGEMENTS

We acknowledge the insights we have gained from the many formal and informal discussions with colleagues from various European industries, universities, resarch institutes and the Commission of the European Community working within the Esprit and Esprit II programmes of the Economic Community. We acknowledge the financial support of the European community for much of the work discussed in this book. This support was delivered through the CIM (Computer Integrated Manufacturing) activity within ESPRIT. The authors worked together in ESPRIT project 477, entitled COSIMA (COntrol Systems for Integrated MAnufacturing). We are very grateful for the encouragement, fruitful discussions and many insights gained from our partners in COSIMA, Comau in Turin, Italy and Renault in Paris, France. We acknowledge in particular the support of M. Actis Dato and Franco Deregibus of COMAU, and Francois Feugier of Renault.

We thank our colleagues at Digital Equipment Corporation, especially Basil Cooney, Niall Connolly, John Harhen, Declan Kennedy, Manus Harley, John Lenihan, Bill O'Gorman, and James Shivnan. Within the Digital Equipment Corporation Clonmel plant, we acknowledge the financial support of the advanced manufacturing engineering group, particularly for our work on factory scheduling. We are also grateful to many people in the operations and manufacturing engineering groups for their interest in and support of our research activities, particularly Tom Malone, Paud Barry and Colin Linanne. Some of the ideas described in this book resulted from early work of the COSIMA project team in Digital Equipment Corporation Munich. In particular, we are in great debt to Hermann Konrad, Ora Jaervinen, Peggy Isakson, David Lane and Cathal Copas who have made special contributions through their dedicated work on the COSIMA project. We thank David Stone, Ernst Wellhoener, and Don Young for the management support they have given us over time.

We thank our colleagues within the Department of Industrial Engineering at University College Galway, in particular, Dr Ivan Gibson, Professor M.E.J. O'Kelly, Dr John Roche and Dr John Shiel. We are grateful to the Computer Services Group within UCG for their technical support in preparing the original manuscript.

A special thanks goes to our colleagues at the CIM Research Unit of University College Galway, particularly David O'Sullivan, Paul Higgins, Conor Morris, Sean Jackson, Una O'Connor, Michael McLoughlin, Allen Moran and Noel Fegan for their helpful criticism of the first draft of this book.

We thank Mark Hammond from Chapman and Hall for his continuing patience and help as the manuscript developed.

Finally we acknowledge the many authors whose work we have consulted in the preparation of this book and whom we have referenced in the manuscript.

PREFACE

In recent years there has been a tremendous upsurge of interest in manufacturing systems design and analysis. Large industrial companies have realized that their manufacturing facilities can be a source of tremendous opportunity if managed well or a huge corporate liability if managed poorly. In particular industrial managers have realized the potential of well designed and installed production planning and control systems. Manufacturing, in an environment of short product life cycles and increasing product diversity, looks to techniques such as manufacturing resource planning, Just In Time (JIT) and total quality control among others to meet the challenge.

Customers are demanding high quality products and very fast turn around on orders. Manufacturing personnel are aware of the lead time from receipt of order to delivery of completed orders at the customer's premises. It is clear that this production lead time is, for the majority of manufacturing firms, greatly in excess of the actual processing or manufacturing time. There are many reasons for this, among them poor coordination between the sales and manufacturing function. Some are within the control of the manufacturing function. Others are not.

This book is concerned with manufacturing lead time, i.e the time from when a batch enters the shop floor to the time it leaves manufacturing as a finished part or product. Typically this manufacturing lead time is 10 to 20 times the actual processing time. It is our contention that this manufacturing lead time must be reduced to a level comparable to the actual processing time. Further this can be achieved by the use of sound operational planning and control systems. Hence the focus of our book is on operational level production planning and control systems; normally referred to as shop floor control systems. Our experience suggests that conventional commercially available computer based systems are very weak on this aspect of production planning and control.

In this book we present an overall approach to the development and installation of sophisticated shop floor control (SFC) systems. This is achieved by offering a functional architecture for SFC, outlining a corresponding information technology architecture, offering some ideas on how SFC systems should be installed – essentially adapting a sociotechnical approach to design and installation – and developing a set of software tools to support the development and installation of SFC systems. The overall approach is verified by an industrial case study which concludes the book.

Our book is concerned primarily with the control issues within SFC and how these can be resolved using state-of-the-art software tools. We address implementation issues in terms of the approach, which we believe should be adapted to implementation. We do not, for instance, discuss implementation in terms of the devices which might be used to capture data from the shop floor.

The structure of the book is as shown in the following figure. Initially we present a short overview of the various approaches to production planning and control, position SFC below the requirements planning stage and suggest that it involves two major activities, factory coordination and production activity control (Part One). We present functional architectures for each of these sub-systems and outline in detail the building blocks within each subsystem. The functional architecture is also documented using the SADT™ (Structural Analysis and Design) approach (Part Two). In Part Three we outline an information technology architecture which matches the functional architecture presented in Part Two. Part Four presents a short summary of the state of the art in the various techniques associated with factory coordination and production activity control. (This part breaks the flow of the book but we believe it is important to offer the reader an overview of the state of the art in important relevant topics such as group technology, scheduling etc.) We discuss the design and implementation of these systems in Part Five and in particular present some ideas for an application generator for the design and development of PAC systems. We are very conscious of the difficulty of implementing such systems in practice, and thus we emphasize the importance of an implementation approach which involves likely end users from the beginning. Finally in Part Six we present a case study of the use of the software tools and the implementation of our ideas in an industrial plant.

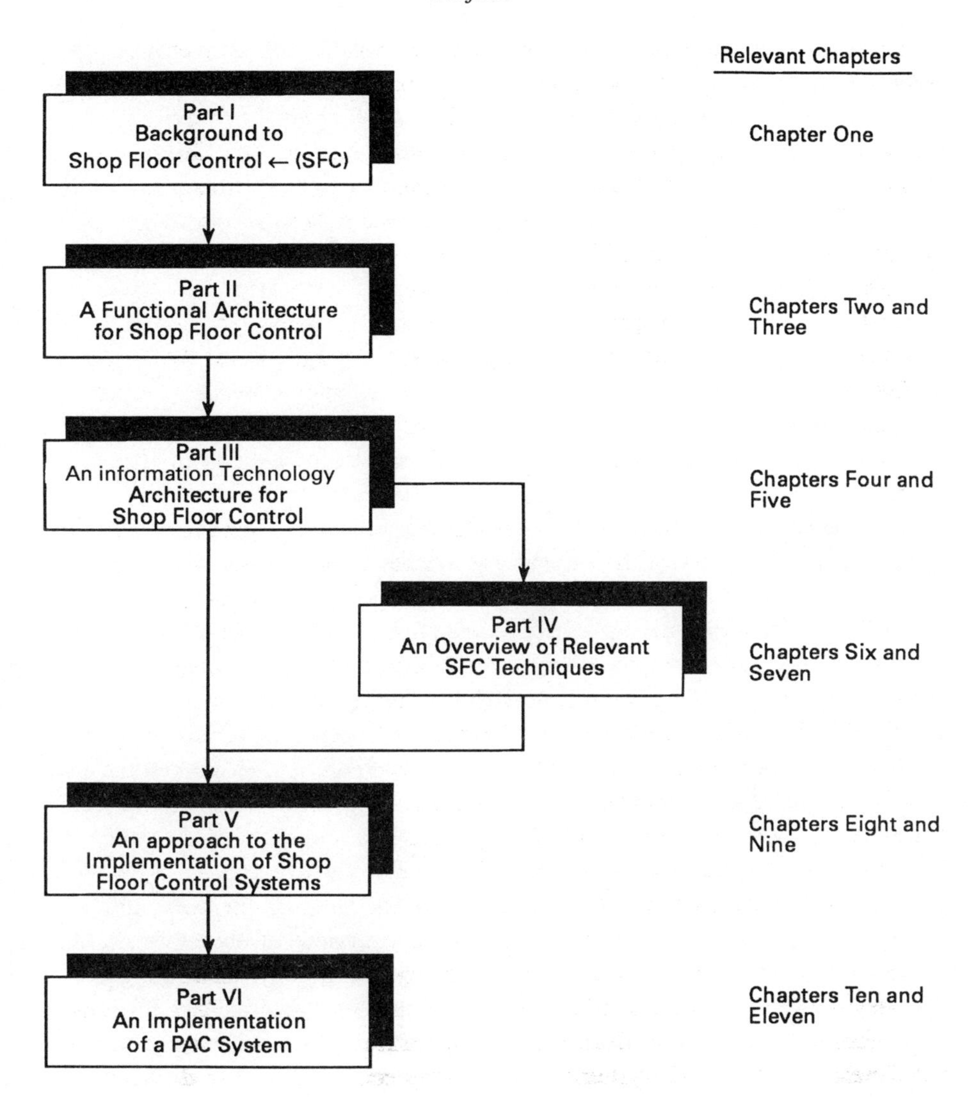

The structure of the book

FOREWORD

The modern environment of discrete parts manufacturing is sophisticated and intensely competitive. It is characterized by short product life cycles, high product diversity, and customers' demands for both excellent quality and timely delivery. If the production operation is capable of responding to these challenges, manufacturing can be a source of real competitive advantage for the business. Otherwise, the manufacturing process could become an inflexible and expensive corporate liability, and business strategists might do well to consider external sourcing of company products.

For manufacturing managers, then, the challenge is to develop a finely tuned process, capable of meeting the cost, quality, variability and time pressures imposed by the marketplace. Our primary objectives must include the reduction of manufacturing lead time to the minimum possible, and achievement of a high level of process control. The benefits accruing from such efforts should include: greater flexibility and responsiveness, better use of manufacturing resources, reduced inventory levels, and faster turn around on customer orders.

Fortunately, advanced information technology now brings the realization of such objectives within reach. It makes explicit and feasible the desire to reduce manufacturing lead time to a level approaching the actual time spent in material conversion on the shop floor. However, the application of sophisticated technology alone is unlikely to yield a durable and efficient shop floor strategy.

There is a need for a well-defined and consistent architecture which describes the production management environment within which shop floor activities take place. On the shop floor, there is a need to link and re-focus all of the discrete stages which make-up the process, so that the total manufacturing operation can be optimized. Furthermore, there is a need to understand the organizational implications of new shop floor control (SFC) technologies. Only by integrating the capabilities of manufacturing personnel, process technologies and advanced information technology can manufacturing continue to contribute to business competitiveness.

This book provides a balanced and pragmatic view of the capabilities of advanced SFC principles and their implementation in real manufacturing environments. It draws together a wealth of information technology knowledge, manufacturing experience and an appreciation of the organizational impacts of SFC designs and technologies, within a sociotechnical framework. It thus makes an invaluable contribution to the education of manufacturing professionals, who find it difficult to keep abreast of such a diverse, and yet critical, amalgam of disciplines.

As a manufacturing manager I am delighted to be associated with this book. While my contribution is limited to this Foreword, I am happy to have been involved in the development of the manufacturing operations at Digital's Clonmel facility, described in the text. Much of this book is based upon research work completed under the auspices of the European Commission sponsored ESPRIT programme. This programme not only assembled the critical mass of five authors which gave birth to this book, but also brought them into close collaboration with colleagues from other European manufacturing industries, most notably COMAU S.p.a. (of Turin, Italy) and RENAULT Automobiles (of Paris, France). I believe the breadth of experience, manufacturing strategies, technologies and corporate cultures observed in all of these industries has been distilled in this text.

Denis McPhail
Plant Manager, Clonmel Manufacturing Facility
Digital Equipment International B.V.
Clonmel, Ireland.

Part One
BACKGROUND TO SHOP FLOOR CONTROL SYSTEMS

Overview

In this introductory chapter we discuss the various production management paradigms, within which shop floor control exists. Essentially we recognize two approaches, namely the MRP (Materials Requirements Planning) approach and the JIT (Just In Time) approach. We consider OPT (Optimized Production Technology) to be an extention of the MRP approach. Based on an understanding of the basic ideas underpinning the MRP/OPT and JIT approaches we offer a hybrid production planning and control architecture, which forms the basis of our understanding of shop floor control. We try to position our proposed hybrid architecture in the context of the NBS (National Bureau of Standards) and the ISO (International Standards Organization) hierarchical models of manufacturing systems.

Our model of shop floor control is based on the recognition of two major sub-systems, namely a Factory Coordination (FC) sub-system and a Production Activity Control (PAC) sub-system. We view a factory as being composed of a series of product based manufacturing cells, the work flow through each of which is managed by a PAC system. These cells are not independent, and in general semi-finished products flow between them. Therefore it is necessary to have an overall and higher level system, whose primary function it is to coordinate the flow of work between these cells. The Factory Coordination system fulfils this role.

1

A background to shop floor control systems

1.1 Introduction

Within batch production systems the lead time or throughput time for a batch through the shop floor is typically much greater than the processing time. It is not unusual for the actual processing (including set-up time) to represent less than five per cent of the total throughput time in conventional batch production systems. The throughput time or lead time is made up of four major components, the set-up time, the process time including inspection time, the transport time and the queuing time, as illustrated in Fig 1.1. In real life this latter component is frequently the largest, often representing in excess of 80% of total throughput time.

The APICS dictionary (Wallace, 1980) defines lead time as follows:

A span of time required to perform an activity. In a production and inventory context, the activity in question is normally the procurement of materials and/or products either from an outside supplier or from one's own manufacturing facility. The individual components of any given lead time can include some or all of the following: order preparation time, queue time, move or transportation time, receiving and inspection time.

The manufacturing lead time is 'the total time required to manufacture an item. Included here are order preparation time, queue time, set-up time, run time, move time, inspection and put away time.'

In recent times the emphasis on Just In Time has focused considerable attention on all sources of waste in manufacturing, in particular on issues such as set-up time and of course queue time. It is clear that long queue times result in greatly increased inventory carrying costs, due to the resulting high Work In Progress (WIP) and reduced flexibility in terms of a manufacturing system's ability to respond to changing customer requirements. In an environment of greatly increased product diversity, reduced product life cycles and greatly

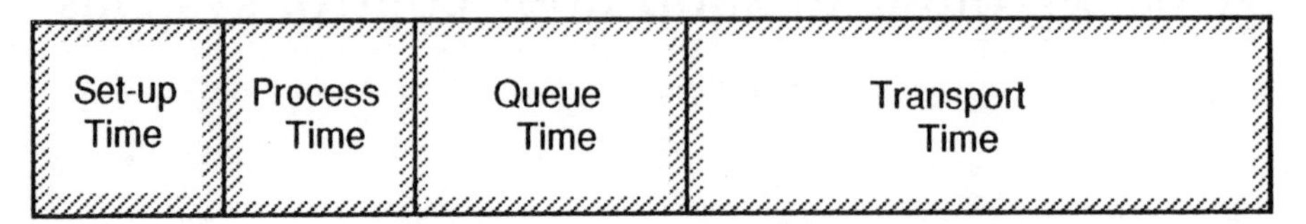

Fig. 1.1. Breakdown of the lead time in a batch production system.

enhanced customer expectations in terms of quality and delivery lead time, it is not surprising that factory managers are looking anew at product lead times and seeking to reduce them to the absolute minimum.

1.1.1 The motivation for this book

Our intention in this book is to consider modern approaches to operational level planning and control systems or shop floor control systems * approaches which will lead to greatly reduced manufacturing lead times. Our contention is that the ratio of actual process time to total lead time is a measure of the success of the operational level production planning and control system. We argue that it is possible to develop manufacturing systems, which are capable of producing a range of products in relatively modest quantities, and are able to do so in a responsive manner i.e. with low lead times. We will outline the architecture of shop floor control systems which are designed to meet this need.

Our understanding of operational level planning and control systems, and the architecture of our proposed solution arises from a synthesis of ideas developed from the three main approaches to production management, namely Materials Requirements Planning (MRP), Just In Time (JIT) and Optimized Production Technology (OPT). We will present a hierarchical operational planning and control system, which we believe is suitable for batch oriented discrete parts production systems and within that overall system we will describe in detail our ideas on operational level production planning and control.

The structure of this chapter is as follows. Firstly we will consider the JIT approach to manufacturing systems planning and control and the insights it offers in terms of creating an effective manufacturing environment (section 1.2). Next we will look at the MRP/OPT approach and highlight what we consider to be important there. We present MRP and OPT together as we

* The term shop floor control refers to both factory coordination and production activity control.

consider OPT to be an extension of MRP thinking (section 1.3). Based on the lessons and insights gained from this study of JIT and MRP/OPT (sections 1.4 and 1.5) we will present an outline architecture for a hybrid production planning and control system and in particular, emphasise the operational or shop floor level aspects (section 1.6). This is in line with our stated intention to focus on systems which reduce manufacturing lead times.

1.2 The Just in Time approach to production management

Just in Time (JIT) is an approach to manufacturing which concentrates on a simple goal, namely to produce the required items at the required quality and in the required quantities at the precise time they are required. JIT started in the Toyota automotive plants in Japan in the early 1960s but is now very widely implemented across a variety of industrial sectors. Essentially it involves a continuous commitment to the pursuit of excellence in all phases of manufacturing systems design and operation.

As Browne, Harhen and Shivnan (1988) point out, Just in Time should be seen from three perspectives, namely:

1. The JIT philosophy or overall approach to manufacturing;
2. The techniques used within JIT to design and operate the manufacturing system;
3. The shop floor control system of JIT, i.e. kanban.

The Kanban system is the most visible manifestation of the JIT approach, because of its use of Kanban cards. The Kanban technique controls the initiation of production and the flow of material with the aim of getting exactly the right quantity of items (components or sub-assemblies or purchased parts) at exactly the right place at precisely the right time. However in terms of the application of JIT thinking in the various industrial sectors, it is clear that the JIT approach and manufacturing techniques are more important. This is so because underlying the use of Kanban, is the prior application of an array of techniques to the products and the manufacturing process, in order to ensure that the application of Kanban is feasible. The techniques involve the design of the manufacturing system in its broadest sense, addressing issues of marketing, sales, product design, process engineering, quality engineering, plant layout and production management, so as to facilitate JIT production using the Kanban system.

The JIT philosophy of manufacturing, upon which JIT execution (Kanban) and the design and planning of the JIT manufacturing system are premised, is frequently the least understood aspect of JIT, but in many ways it is the most important. The JIT philosophy is a set of fundamental manufacturing strategies which when implemented, provide the basis for the JIT system and facilitate the use of the Kanban system.

In our review we will say very little about the Kanban system. Rather we will focus on the other aspects of JIT. In section 1.2.1 we will review the JIT approach or philosophy, in section 1.2.2, manufacturing systems design and planning in JIT and conclude in section 1.2.3 with a short overview of the Kanban system.

1.2.1 The JIT approach to manufacturing

In the view of Browne, Harhen and Shivnan, (1988) there are three fundamental ideas in the JIT philosophy or approach to manufacturing:

An intelligent match of product design with market demand The modern manufacturing business is faced with a tremendous challenge. Change is endemic, and the pace of change increases constantly. Product life cycles have been greatly reduced. In former times manufacturers could look forward to relatively long product life cycles where product demand was high and relatively constant over a reasonably long period. Customers were willing to tolerate reasonably long lead times from placement of an order to receipt of the product and to compromise on relatively standardised, i.e. non customised products. Furthermore customers did not get unduly upset if a proportion of the delivered order was slightly imperfect. Nowadays customers expect to have a choice of products available and will not tolerate long delays in delivery or poor quality. In this competitive market it is important for manufacturing companies to have a product range which meets the market demand in terms of variety, delivery, quality and of course cost.

Manufacturing companies must examine their market carefully and determine the appropriate manufacturing strategy in terms of product range, product quality and product cost. Clearly it is possible to offer a very wide product range to the market but at what cost and price? Products must be conceived and designed which meet the market requirements in terms of variety, quality and cost but are possible to manufacture and deliver to the market at a profit. Clearly considerable effort must be devoted to the development of a product range which meets the markets requirements and can be manufactured at a reasonable cost.

To achieve this objective, it is necessary to design products in a modular fashion. A large product range and a wide variety of product styles can result in high manufacturing and assembly costs. In general terms, it is true that the greater the flexibility required, the more expensive is the manufacturing system, and therefore the products of that system. Modular product designs are achieved by rationalizing the product range where possible and by examining the commonalty of components and sub-assemblies across the product range with a view to increasing it to the maximum level possible. Rationalization of the product range results in reduced production costs through less manufacturing set-ups, fewer items in stock, fewer component drawings etc.

The definition of product families JIT seeks to promote product based factory layout. The basis for product based layout is the definition of families of products which share common design and manufacturing attributes and consequently can be manufactured in product oriented cells.

A common approach to the identification of product families and the subsequent development of flow based manufacturing systems is Group Technology (GT) (see Chapter 7). The use of GT in JIT systems to define product families is important for a number of reasons. Firstly, Group Technology is used to aid the design process and to reduce unnecessary duplication in product design. Secondly, Group Technology is used to define families of products and components which can be manufactured in well defined manufacturing cells. Manufacturing cells generate simplified material flow patterns in a plant, and allow responsibility and ownership for a component or group of components to rest with one group of operators and their supervisor.

The establishment of relationships with suppliers to achieve Just In Time deliveries of raw materials and purchased components JIT argues strongly that manufacturing companies should establish strong and enduring relationships with qualified suppliers. The manufacturing company should share information on likely future order patterns, design changes and other relevant information with this small number of qualified suppliers; in essence supporting them in their efforts to be effective and trusted suppliers.

JIT execution, (i.e. the Kanban system) applied to purchasing, gives rise to frequent orders and frequent deliveries. On the one hand, the buyer places great demands on the supplier in terms of frequent or Just In Time deliveries of components. On the other hand, by providing the supplier with commitments for capacity over a long period and by ensuring that the supplier is aware of modifications to the company's master schedule as soon as is practicable, the company helps the supplier to meet the exacting demands of JIT.

It is clear that the JIT approach to manufacturing incorporates a business perspective as distinct from a narrow or strictly manufacturing, (i.e. inside the four walls of the factory) perspective.

1.2.2 Manufacturing systems design and planning in JIT

A primary focus of JIT is the reduction of the production lead time. Short lead times reduce the manufacturing plant's dependence on forecasts and allow the plant to have shorter planning horizons and consequently more accurate master schedules. Reduced lead times also have important consequences for a plant's ability to respond to short term unexpected changes in the market place. There are many approaches which improve manufacturing performance based on the reduced lead time. We shall group these approaches and techniques within JIT manufacturing under five identifiable headings which we will then outline in some detail:

1 Product design for ease of manufacture and assembly;
2 Manufacturing planning techniques;
3 Techniques to facilitate the use of simple but refined manufacturing control systems, (e.g. Kanban);
4 An approach to the use of manufacturing resources;
5 Quality control and quality assurance procedures;

Product design for ease of manufacture and assembly When discussing the JIT approach to manufacturing earlier, we highlighted the importance of an intelligent match of the product design with the perceived market demand. The product design team must interpret the wishes of the market place and if possible lead the market by developing a product range which allows the production system to respond effectively to market needs. This involves designing products which both anticipate the market requirement and include sufficient variety to meet consumers expectations, while being manufactured at a price which the market is willing to pay. This is achieved by using techniques such as modular design, design for simplification and design for ease of manufacture and assembly.

Manufacturing planning techniques To help production to respond effectively to short term variations in market demand, JIT attempts to match the expected demand pattern to the capabilities of the manufacturing process and to organize the manufacturing system so that short term and relatively small variations can be accommodated without major overhaul of the system. The technique used to help achieve this is known as production smoothing. Through production smoothing, single lines can produce many product varieties each day in response to market demand. Production smoothing utilises the short production lead times to mould the market demand to match the capabilities of the production process. It involves two distinct phases as illustrated in Fig. 1.2.

The first phase adapts to monthly market demand changes during the year, the second to daily demand changes within each month. The possibility of sudden large changes in market demand and seasonal changes is greatly reduced by detailed analysis of annual or even longer term projections and well thought out decisions on sales volumes in so far as this is possible. Monthly

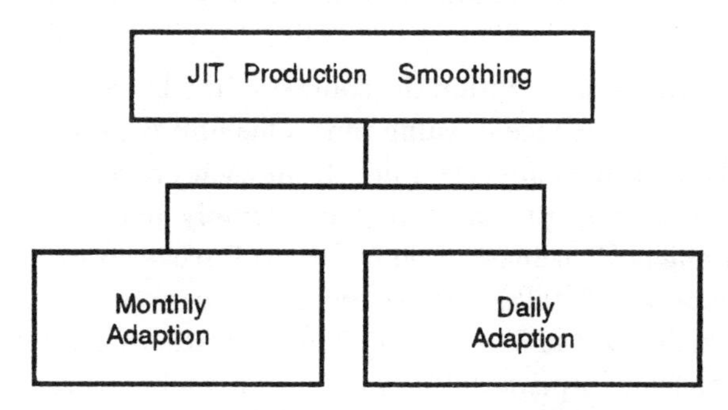

Fig. 1.2. Components of production smoothing.

adaptation is achieved though a monthly production planning process i.e. the preparation of a Master Production Schedule (MPS). This MPS gives the averaged daily production level of each process, and is typically based on an aggregate three month and a monthly demand forecast. The precise planning horizon depends very much on the industry in question – in the automotive industry where JIT originated, three months is typical. Thus the product mix and related product quantities are suggested two months in advance and a detailed plan is fixed one month in advance. This information is also transmitted to suppliers to facilitate them in providing raw materials when required. Daily schedules are then determined from the master production schedule. In fact the concept of production smoothing extends along two dimensions: firstly by spreading the production of products evenly over each day within a month; secondly by spreading the quantities of each product evenly over each day within a month. This frequently results in a mixed model final assembly line. (See Chapter 12 of Browne, Harhen and Shivnan, 1988 for details of how production smoothing is achieved in practice)

Techniques to simplify the manufacturing process and reduce lead times As we have already indicated the throughput time or lead time for a product is composed of four major components, the actual process time including inspection time, the set-up time, the transport time and the queuing time. In real life the queuing component is frequently the largest. At the factory or plant level JIT encourages product based plant layouts which greatly reduce throughput times for individual batches by facilitating the easy flow of batches between operations and work centres thus reducing queuing time. At a line or work centre level, JIT reduces throughput time by using what are termed U-shaped layouts. The U-shaped layout allows unit production and transport, since machines are close together and may be connected with chutes or conveyers. Synchronization is achieved since one unit entering the layout means one unit leaving and going on to the next work centre.

Reduction in queuing time Within the context of the U-shaped layouts various techniques are used to reduce queuing time. One unit is produced within every cycle and at the end of each cycle a unit from each process is sent to the next process. This is already prevalent in the assembly line systems of virtually all companies engaged in mass production. JIT seeks to extend the concept of unit production and transport to processes such as machining, welding, pressing etc. which feed the final assembly lines. Furthermore JIT, in separating production lots from transport lots in situations where production lots are large, is moving away from batch based production systems and towards flow based systems. Line balancing seeks to reduce the waiting time caused by unbalanced production times between individual work centres. Variances in operators' skills and capabilities are minimized through the use of carefully documented standard operations.

Reduction in operation times JIT seeks to ensure that the best possible manufacturing methods are practised by a skilled and trained workforce. So often in the conventional approach to manufacturing operations, manufacturing analysts and engineers forget the importance of good practice at the sharp end of manufacturing, namely the shop floor and come to accept unnecessary deviation in operator performance as natural. Furthermore operators in a JIT environment tend to be very versatile and are trained to operate many different machines and carry out many operations within their particular work centre. As the plant layout is product oriented, advantage can be taken of multiskilled operators.

Reduction in transport times JIT seeks to reduce transport times by intelligent layout of the production processes and faster methods of transport between production process. The U-shaped layout minimizes transport needs between individual operations on a component or assembly. Clearly unit production and transport may increase the transport frequency i.e. the number of transports of partially completed units between operations. To overcome this quick transport methods must be adopted. Belt conveyers, chutes and fork-lifts may be used.

Reduction of set-up times A major barrier to the reduction of the processing time and the ability to smooth production is the problem of large set-up times. JIT argues that machine set-up time is a major source of waste. In order to shorten the set-up time, JIT offers four major approaches (Monden, 1983):

1. Separate the internal set-up from the external set-up. Internal set-up refers to that element of the set-up process which requires that the machine be inoperative in order to undertake it.
2. Convert as much as possible of the internal set-up to the external set-up.
3. Eliminate the adjustment process within set-up. Typically adjustment accounts for a large percentage of the internal set-up time.
4. Abolish the set-up where feasible.

The reader interested in a more detailed discussion on this very important topic is referred to Shingo (1985). He discusses the SMED System. SMED is an acronym for Single Minute Exchange of Dies, which connotes a group of techniques used to facilitate set-up operations of ten minutes and under.

The use of manufacturing resources The JIT approach to manufacturing resources can be summarised in a single phrase do not confuse busy with productive. This philosophy is particularly applied to the use of labour resource and , as we will indicate in section 1.3.7 also fundamental in the OPT (Optimized Production Technique) approach to manufacturing.

In JIT, those minor changes in demand which cannot be accommodated though the use of increased kanbans are dealt with by redeployment of the workforce. Assuming multiskilled and multifunction operators, one operator may tend to a number of different machines simultaneously to meet small increases in demand. Temporary operators may be hired. Each operator may then be required to tend fewer machines thus taking up the equipment capacity slack. Adapting to decreases in demand is clearly more difficult especially when one considers that many large Japanese companies offer life-time employment. However, the major approaches are to decrease overtime, release temporary operators and increase the number of machines handled by one operator. This causes an increase in the cycle time thus reducing the number of units produced.

The important objective is to have a manufacturing system which is able to meet demand and to accommodate small, short term fluctuations in demand with the minimum level of labour. This does not imply the minimum number of machines. Companies operating JIT usually have some extra capacity in equipment, allowing for temporary operators when increased production is required.

Quality and JIT In conventional production systems the notion of Acceptable Quality Levels (AQL) is very common. In JIT systems the emphasis is on Total Quality Control (TQC) where the objective is to eliminate all possible sources of defects from the manufacturing process and from the products manufactured by that process. JIT seeks to achieve Zero Defects. Inspection is carried out to prevent defects rather than simply detect them. Machines in are designed where possible with an in-built capability to check the parts they produce as they are produced. The term Autonomation was coined to describe this condition. Autonomation suggests automatic control of defects. It implies the incorporation of two new pieces of functionality into a machine namely, a mechanism to detect abnormalities or defects, and a facility to stop the machine or line when defects or abnormalities occur.

There are other factors which assist in attaining extremely high quality levels. Small lot sizes result in quality problems being highlighted very quickly as individual items are rapidly passed to the next process and any defects are quickly detected. Similarly, a good approach to housekeeping is encouraged and is considered important as a clean well maintained working area leads to better working practices, better productivity and better personnel safety. Preventative maintenance is emphasised. Machines are checked regularly and repairs/replacements where necessary, are scheduled to take place outside working time.

1.2.3 Kanban

Kanban, the Japanese word for card, is the shop floor control or production activity control element of JIT. The Kanban technique can only be used in a repetitive manufacturing environment, where it is used to synchronize the rate at which materials are used on the final assembly line. Further Kanban requires a level schedule, normally the result of the application of production smoothing techniques, and very strict discipline to operate. It is relatively inflexible in that it is unable to accommodate all by the smallest of schedule variations. A detailed explanation of the use of Kanban cards is beyond the scope of this chapter. Thc interested reader is referred to Chapter 13 of Browne, Harhen and Shivnan, (1988) which includes a detailcd explanation of the working of Kanban, together with some simple worked examples.

In our view Kanban, which has received great exposure in the literature is the least important aspect of JIT. In some, relatively few cases, the application of JIT will create conditions where Kanban can be used. In the majority of cases this will not be so. Nevertheless implementing the ideas of the JIT approach and JIT manufacturing system design and planning will likely lead to good results..

1.3 An overview of requirements planning (MRP and MRP II)

Material Requirements Planning (MRP) originated in the early 1960s in the United States as a computerized approach to the planning of materials acquisition and production. The early MRP systems were implemented on large main-frame computers and consequently were only used by very large industrial companies. In recent years MRP applications have become available at lower cost on mini and micro-computers and are thus within the reach of small and medium size companies. Today MRP is almost certainly the most widely used computerised production planning and control system in the industrial world. The definitive book on the technique is the publication by Orlicky (1975).

1.3.1 Introduction to the MRP procedure

MRP systems are built around a Bill of Material Processor (BOMP) which converts a discrete plan of production for a product or so called parent item into a plan of production or purchasing its component items. This is achieved by exploding the requirements for the top level product, through the Bill of Material (BOM), to generate component demand, and then comparing the projected gross demand with available inventory and open orders, over the planning time horizon and at each level in the BOM.

Various operational functions have been added to extend the range of tasks that MRP software systems support. The extensions include Master Production Scheduling (MPS), Rough Cut Capacity Planning (RCCP), Capacity Requirements Planning (CRP), Purchasing and Production Activity Control (PAC). When financial modules were added and the functionality of the master production scheduling module was extended to deal with the full range of tasks in master planning and some tasks in business planning, the resultant system offered an integrated approach to the management of manufacturing resources. This extended MRP system was labelled Manufacturing Resource Planning or MRP II.

Orlicky (1975) offered several important insights to production management theory, namely;

1. Manufacturing inventory, unlike finished goods, or service parts inventory, cannot usefully be treated as independent items. The demand for component items is dependent on the demand for the assemblies of which they are part.
2. When a time phased schedule of requirements for top level assemblies is available (master schedule), the demand for the dependent components can be calculated. Consequently it makes no sense to forecast them.
3. The assumptions underlying the classical inventory control models usually suggest a uniform or at least a well defined demand pattern. However, the dependency of component demand on the demand for their parents, gives rise to a phenomenon of discontinuous demand at the component level. Orlicky termed this lumpy demand. Lumpy demand occurs even if the master scheduled parts face uniform demand, because of the effects

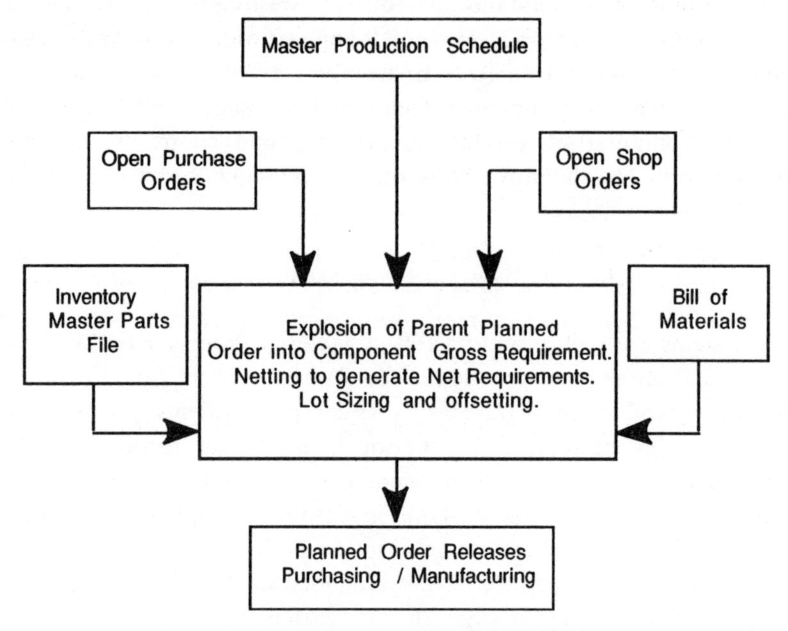

Fig. 1.3. Basic structure of an MRP system.

of lot sizing and the fact that demand for an item often arises from a number of product sources. The implication of the lumpy demand phenomenon is that order point techniques are inappropriate for managing manufacturing inventories.

4 Computers provide the data processing capability to perform the necessary calculations efficiently.

A Bill of Materials (BOM) describes the parent/child relationship between an assembly and its component parts or raw materials. Bills of material may have an arbitrary number of levels and will typically have purchased items at the bottom level of each branch in the hierarchy. An MRP system is driven by the Master Production Schedule (MPS), which records the independent demand for top level items. It is derived from evaluating forecasts and customer orders. MRP uses this requirements information, together with information on product structure from the Bill of Materials file, current inventory status from the inventory file, and component lead times data from the master parts file to produce a time-phased schedule of planned order releases on lower level items. This time phased schedule is known as the materials requirements plan. This flow of information is illustrated in Fig. 1.3.

In MRP, time is assumed to be discrete and typically is represented as a series of weekly intervals, though systems which operate on a daily planning periods are readily available. Demand for a component can derive from any of the products in which it is used, as well as independent spares parts orders. The system starts with the master production schedule as input and applies a set of procedures to generate a schedule of net requirements, and planned coverage of such requirements, for each component needed to implement the master production schedule. The system works down level by level and component by component of the BOM until all parts are planned. For each component, it applies the following procedure.

1 Netting off the gross requirement against projected inventory and taking into account any open orders scheduled for receipt, as well as material already allocated from current inventory, thus yielding net requirements.
2 Conversion of the net requirement to a planned order quantity using a lot size.
3 Placing a planned order in the appropriate period by backward scheduling from the required date by the lead time to fulfil the order for that component.
4 Generating appropriate action and exception messages to guide the users attention.
5 Explosion of parent item planned production to gross requirements for all components, using the Bill of Materials relationships.

The interested reader is referred to Chapter 5 of Browne, Harhen and Shivnan (1988) where there is a worked example of the MRP calculation based on a simple four product manufacturing system.

MRP systems are categorized into bucketed and bucketless systems. Bucketed systems limit the time horizon that may be considered, and the granularity of timing that may be ascribed to an order. Bucketless systems enable daily visibility to an order's date of requirement. Weekly time buckets are considered to be the granularity necessary for near and medium term planning. However, the normal bucket of one week may be too coarse to facilitate detailed short term planning. Further out in the planning horizon, monthly or perhaps quarterly time buckets are acceptable. In the non-bucketed approach, each element of time phased data has associated with it a specific time label, and is not accumulated into buckets. What this means is that there is the provision for daily visibility on requirements timing.

1.3.2 Features of MRP systems

Change is endemic in manufacturing. Customer orders change; delivery dates are brought forward; suppliers are unable to supply on time; or perhaps due to some unforeseen quality or manufacturing problem the manufacturing plant cannot deliver components or assemblies in the quantities required on the due date specified and agreed. MRP systems include facilities for dealing with these situations. These facilities can be categorized into top down planning and bottom up replanning systems.

Top down planning in MRP There are two basic styles of top down planning which are termed the Regenerative approach and the Net Change approach. Regenerative MRP starts with the Master Production Schedule and totally re-explodes it down through all the bills of materials to generate valid priorities. Net requirements and planned orders are completely regenerated at that time. The regenerative approach thus involves a complete re-analysis of each and every item identified in the master schedule and for all but the simplest of master schedules involves extensive data processing. Because of this, regenerative systems are typically operated in weekly and occasionally monthly replanning cycles.

In the net change MRP approach, the materials requirements plan is continuously stored in the computer. Whenever there is an unplanned event, such as a new order in the master schedule, a partial explosion is initiated for those parts affected by the change. Net change MRP can operate in two ways. One mode is to have an on-line net change system, by which the system reacts instantaneously to unplanned changes as they occur. In most cases however, change transactions are batched (typically by day) and replanning happens over night. A potential difficulty with net change systems, is that there is a reduced self purging capability. Errors may creep into the requirements plan, perhaps because of planner actions in bottom up replanning. Since the master schedule is not completely re-exploded, as in the regenerative approach, any errors in the old plan tend to remain. To counteract this problem, firms using net change MRP tend occasionally to do a complete regeneration so as to purge the system of these errors.

Bottom Up Replanning In top down planning, the MRP system itself does the planning. An alternative approach is for the planner to manage the replanning process. This is termed bottom up replanning, and makes use of two techniques namely, pegged requirements and firm planned orders.

Pegging allows the user to identify the sources of demand for a particular component's gross requirements. These gross requirements originate typically either from its parent assemblies, or else from independent demand in the MPS or from the demand for spare parts. The technique of pegging is useful, in that it allows the user to retrace the planning steps in the event of an unexpected event, such as a supplier being unable to deliver in the planning lead time. By retracing the original calculations the user can detect what orders are likely to be affected and perhaps identify appropriate remedial action. The remedial action often involves overriding the normal planning procedures of MRP. The is done by using the firm planned order technique.

The firm planned order allows the materials planner to force the MRP system to plan in a particular way, thus overriding lot size or lead time rules. A firm planned order differs from an ordinary planned order, in that the MRP explosion procedure will not change it in any way.

1.3.3 Manufacturing resource planning (MRP II)

Manufacturing Resource Planning represents an extension of MRP to support other manufacturing functions beyond material planning, inventory control and BOM control. It evolved from MRP by a gradual series of extensions to MRP system functionality. These extensions included the addition of transaction processing software to support the purchasing, inventory and financial functions of the firm. Thus MRP was extended to support Master Planning, Rough Cut Capacity Planning (RCCP), Capacity Requirements Planning (CRP) and Production Activity Control.

The term Closed Loop MRP denotes an MRP system wherein the planning functions of Master Scheduling, MRP and Capacity Requirements Planning (CRP) are linked with the execution functions of Production Activity Control (PAC) and purchasing. Closed Loop signifies that, not only are the execution modules part of the overall system, but also that there is feedback from the execution functions so that plans remain valid at all times.

With the extension of master production scheduling to deal with all master planning to support business planning, and through the addition of certain financial features to the closed loop system, so that outputs such as the purchase commitment report, shipping budget and inventory projection could be produced, Manufacturing Resource Planning or MRP II was made available. The MRP II system, is thus a closed loop MRP system, with additional features to cover business and financial planning. MRP II nominally includes an extensive what if capability. The modular structure of a typical MRP II system is shown in Fig. 1.4.

1.3.4 MRP/MRP II in practice

There is a significant divergence between what is available in state of the art MRP software systems and what is typically used in practice. Wight, (1981) developed a classification scheme to rate how well companies operate their MRP systems. The scheme involves a set of 25 questions which relate to the technical capability of the MRP software package, the accuracy of supporting data, the volume of education that has been provided to the employees and the results achieved with the system. MRP system use is rated between Class A, which represents excellence, and Class D which represents a situation where the

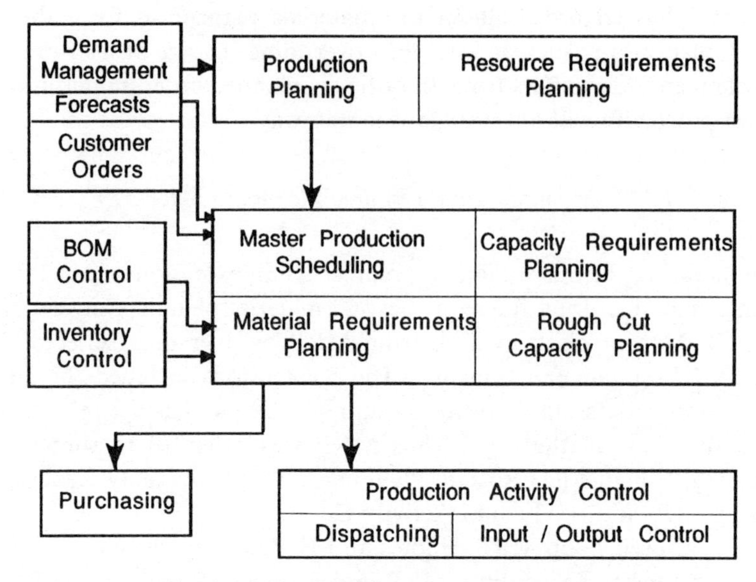

Fig. 1.4. Manufacturing resource planning.

only people using the system are those in the MIS (Manufacturing Information Systems) department.

Among the criteria that consultants and authors such as Wight (1981) used to measure effective use of MRP are the following:

1 MRP should use planning buckets no larger than a week;
2 the frequency of replanning should be weekly or more frequent;
3 if people are effectively using the system to plan, then the shortage list should have been eliminated;
4 delivery performance is 95% or better for vendors, the manufacturing shop and the MPS (Master Production Schedule);
5 performance in business goals such as reduced inventory, higher productivity and increased customer service has improved.

There have been many disappointing MRP installations. For example, a major UK based study carried out by Lawrence (1986) showed that of 33 companies analysed only 16 claimed to have successfully implemented systems. It seems to be generally agreed that failure of an MRP installation can be traced to problems such as:

1 Lack of top management commitment to the project.
2 Lack of education in MRP for those who will have to use the system.
3 Unrealistic master production schedules.
4 Inaccurate data, particularly BOM data and inventory data.

We shall discuss these problems in more detail while looking at suitable ways of implementing production management systems in Chapter 8.

1.3.5 Criticisms of the MRP approach

Burbidge (1985) points to the long planning horizon normally associated with master scheduling and the consequent errors due to our inability to make accurate forecasts of demand towards the latter end of the planning horizon. The long planning horizon arises because of what Burbidge considers the inflated lead times associated with the MRP approach 'MRP systems break the Bill of Materials into main, sub, sub-sub, lots etc. assembly and fabrication stages, estimate lead times for each stage, and add them together to establish lead times for ordering. This inflates lead times and stocks.' In fact the lead time used in MRP offsetting is the planned or expected lead time and this represents no more than an estimate of the time it takes for an individual batch to go through the system. The actual lead time will depend on the load on the manufacturing shop floor and the priority assigned to a given batch. As we have indicated above advanced MRP systems, in particular MRP II or Manufacturing Resource Planning systems, include Shop Floor Control (or Production Activity Control) subsystems, one of whose functions is to close the loop between the MRP planning system and the manufacturing shop floor. One element of this closing the loop process is the feedback of actual lead times to the MRP system. This data can then be used to establish the validity of the lead times in use and to signal the need for a change if necessary.

A second important criticism of MRP relates to the fact that the MRP approach simply accepts set up times, lead times, batch sizes etc. as given and fails to draw attention to potential improvement. This point is often made when one compares the MRP and JIT approaches to production planning and control. As we seen in our earlier discussion on JIT, JIT seeks to design the best possible production environment and then plan and control the flow of products through it. MRP simply accepts the environment within which it is placed.

1.3.6 Optimized Production Technology

We will view OPT (Optimized Production Technology) as an extension of the MRP approach to manufacturing. OPT originated in the late 70s and early 80s and attracted tremendous interest. It claimed to be based on a new understanding of manufacturing and to offer an optimized manufacturing schedule. It is based on ten rules which it is claimed are designed to meet the goal of the manufacturing system.

From the OPT perspective there is one goal for a manufacturing company to make money. All activities in the business are but means to achieve this goal. (Fox, 1982). Progress towards the achievement of this goal is measured in terms of net profit, return on investment, and cash flow. At the operational level OPT identifies three important criteria that are useful in evaluating manufacturing progress namely, throughput, inventory and operating expenses.

Throughput is the rate at which the manufacturing business generates money through selling finished goods. Inventory is defined as the raw materials, components and finished goods that have been paid for by the business but have not as yet been sold. Operating expenses are the cost of converting inventory into throughput. Operating expenses include the cost of direct and indirect labour, heat, light etc. Changes in any of these three elements, such as increasing the throughput or reducing the inventory level, result in changes in the financial measurements listed above. The goal of manufacturing, as understood by OPT, is to increase throughput while simultaneously decreasing inventory and operating expenses.

OPT is designed to achieve this goal of increasing throughput and decreasing inventory and operating expenses through realistic and optimized schedules. An analytical technique based on the ten rules of OPT is used to generate this schedule. These rules and the analytical technique have been computerized to give a software product called OPT. The ten rules of OPT may be applied to the manufacturing organization without recourse to the software system. We shall now outline these so-called rules of OPT.

1.3.7 The rules of OPT

Eight of the ten rules rules relate to the development of correct schedules, while the other two are necessary to prevent traditional performance measurement procedures from interfering with the execution of these schedules. We will now review each of the rules, starting with those relating to manufacturing bottlenecks.

Bottlenecks The resources in a manufacturing facility can be classified into bottleneck and non-bottleneck resources. A bottleneck can be defined as 'a point or storage in the manufacturing process that holds down the amount of product that a factory can produce. It is where the flow of materials being worked on, narrows to a thin stream.' (Bylinsky, 1983). OPT argues that that the non-bottleneck should work at a reduced level of utilisation, sufficient to support the bottleneck while at the same time preventing a build-up of WIP (Work in Progress) at the bottleneck station, and the bottleneck should work at 100% utilization. The strategy suggested ensures that the bottleneck resources are fully utilized at all times. With regard to non-bottleneck resources, not all of their time can be used effectively and some of their time is therefore considered as enforced idle time. The first rule of OPT derives from this insight and is as follows. The level of utilization of a non-bottleneck is determined not by its own potential, but by some other constraint in the system.

The authors of OPT suggest a second rule which makes an important distinction between doing the required work (what we should do, activation), and performing work not needed at a particular time (what we can do, utilization): Utilization and activation of a resource are not synonymous.

Set-up times The available time at any resource is divided between processing time and set-up time. However, there is a difference between the set-up times on a bottleneck and those of a non-bottleneck resource. If we can save an hour of set-up time on a bottleneck resource, we gain an hour of processing time. Relating this to the fact that bottlenecks are a limiting constraint on other resources and on the system as a whole, an hour of production gained at a bottleneck has far reaching implications. It can be equivalent to an increased hour of production and throughput for the total system.

At a non-bottleneck resource however, we have three elements, namely processing, set-up, and idle time. Clearly, if we can save an hour of set-up time, we gain an hour of idle time as the bottlenecks still constrain the capability of the non-bottleneck. Consequently, an hour saved at an non-bottleneck is likely to be of no real value. There is however one advantage to reducing set-up times at non-bottlenecks machines. Due to a lower set-up time more set-ups can be used and the batch or lot sizes can be reduced. While a smaller lot size of itself does not increase throughput, it tends to reduce inventory levels and some operating expenses.

The authors of OPT suggest three further rules:

1 An hour lost at a bottleneck is an hour lost for the total system;
2 An hour saved at an non-bottleneck is just a mirage;
3 Bottlenecks govern both throughput and inventory in the system.

Batch or lot sizes The OPT approach identifies two types of lot or batch, namely the transfer batch and the process batch. The transfer batch defines the number of items transferred together as a batch between operations. The process batch defines the number of items processed at a station before a new set-up is necessary. OPT argues that these should not necessarily be the same size, viz. The transfer batch may not, and many times should not, be equal to the process batch. Further OPT argues that it is impossible to determine from the outset a single lot size which is correct for all operations. This concept is encapsulated in another OPT rule. The process batch should be variable, not fixed. The lot size is established dynamically for each operation and balances inventory cost, set-up costs, component flow requirements and the needs for managerial control and flexibility. In addition, some operations may be bottlenecks and thus require large process batches, while non-bottlenecks may require small process batches so as to reduce lead time and the resulting inventory.

Lead times and priorities Earlier in this chapter we showed how MRP uses lead times to offset from a defined due date to calculate the time to start production or to release purchase orders. Essentially MRP uses the estimated lead time to determine the order in which jobs are processed. Priorities are assigned to jobs and those with the higher priorities are processed first. The estimated lead time is in turn dependent on the estimated queuing time for each operation. Once priorities have been established, the capacity of the production process is examined to see if the plan can be met. However, the important interaction between priority and capacity is not examined. Priority and capacity are essentially considered sequentially, not simultaneously. OPT however argues that lead times are not fixed and further that lead times are not known a priori, but depend on the sequencing at the limited capacity or bottleneck resources. Exact lead times and hence priorities cannot be determined in a capacity bound situation unless capacity is considered. Hence OPT suggests, capacity and priority should be considered simultaneously, not sequentially.

Cost Accounting and Performance Evaluation The rules which we have outlined above relate primarily to the development of correct schedules. OPT also identifies a number of barriers to the implementation of correct schedules. According to the proponents of OPT these are:

1 Conventional methods of measuring efficiency;
2 The expectation of balanced plant loads;
3 The so called hockey stick phenomenon.

Conventional cost accounting practice does not differentiate between work at a bottleneck and work at a non-bottleneck resource. Supervisors of all resources, whether they be bottlenecks or not, are encouraged to seek 100% resource efficiency since in general, they are measured by their production rates and not by how well their output impacts the output of the total manufacturing organization. However a non-bottleneck resource operating at full capacity will produce a costly build-up of inventory. The performance of the supervisor managing non-bottleneck machines should be measured, not by the amount of WIP (Work in Progress) he is responsible for creating, rather by the volume of usable product his/her area produced.

OPT argues that conventional cost accounting attempts to measure the efficiency of resources, not their effectiveness. From the perspective of the total manufacturing organization, it is the effectiveness of each resource in terms of total systems goals which is important. According to OPT thinking a manufacturing plant should produce only that which is ordered by customers, or what can reasonably be expected to be ordered by customers. Each work centre should only produce what is required at the next work centre and so on until the plant only produces what is required overall by customers. This reduces costly inventory, saves on operating expenses and facilitates maximum throughput. One should attempt to balance the flow of products through the plant rather than the plant capacity. There is an OPT rule to this effect: Balance flow not capacity.

The developers of OPT identify what they term the hockey stick phenomenon and argue that it is caused by the conflict between two measurement systems, cost accounting and financial performance, and is visible in most plants at the end of each financial reporting period. At the beginning of each period, the plant is driven by cost accounting performance measurements, which have a local focus. Measurements focus on machine and operator efficiency, standard times or costs to produce a part at a particular operation. To be efficient, large batch sizes are run through operations regardless of whether they are bottleneck or non-bottleneck, usually resulting in the build up of unnecessary inventory. As we near the end of the financial reporting period, management becomes concerned with a global measurement, the performance of the total system. There is an enormous effort to ship products, to make more money. Efficiencies, the number of set-ups etc. are no longer a consideration, and in their place, expedited lots are split, overtime is allowed, inefficient machines are put working again, anything to increase shipments! The appropriate OPT rule is: The sum of local optima is not equal to the optimum of the whole.

These ten rules of OPT were used to create a software package which it was claimed would generate an optimum schedule for a given manufacturing system. At the heart of the software was an algorithm/heuristic (?) which was referred to as the Brain of OPT. The details of this procedure were never made public and thus the claim of optimality cannot be verified. What is clear is that the software generates schedules by progressively recognizing bottlenecks in the manufacturing systems and ensuring that these bottlenecks are kept busy at all times.

1.3.8 Views on OPT

When OPT first became available it was presented as a competitor for MRP (and MRP II) and JIT. In recent years analysts, both academic and practitioner alike, seem to be moving towards the view that the two are not incompatible. We should also point out that when OPT first became available, it attracted considerable criticism and indeed continues to be criticized because of the claim, implicit in its name, that it offers an optimal schedule and also because of the fact that the scheduling algorithm on which it is based has never been revealed in the literature.

Lundrigan (1986) suggests that OPT brings together the best of JIT and MRP II into a 'kind of westernized Just in Time'. OPT has some similarities with JIT at the operational level of production management, e.g. the use of small batches, the identification of transport and process batches etc. However OPT concerns itself with scheduling to the virtual exclusion of all else.

1.4 MRP/OPT versus JIT.

It is clear even from our brief discussion that MRP is concerned primarily with the logistics of the manufacturing process. It takes the customer requirements for products and breaks these into time-phased requirements for sub-assemblies, components and raw materials. It also seeks to schedule activities within the manufacturing process and the availability of material from vendors to produce the end product on time for the customer. OPT is a scheduling tool and like MRP, is mainly concerned with the logistics of when. JIT however takes a somewhat wider view and is concerned with what the product is, how the product is manufactured as well as the logistics of delivering it on time to the customer.

Up to now JIT has been associated with mass production and repetitive manufacturing systems. As we pointed out earlier it originated in the final assembly plants of the major Japanese automotive manufacturing companies. We have distinguished between the JIT approach or philosophy, JIT techniques for manufacturing process design and planning and Kanban. Kanban is essentially a production activity control (PAC) system which functions well in a mass production or repetitive manufacturing environment. However the JIT philosophy and indeed, JIT manufacturing and planning techniques are applicable to all discrete parts manufacturing environments.

MRP and indeed OPT have been associated with batch production systems. Each is concerned with situations where there are a relatively large number of products, associated numbers of Bills of Materials (BOM) and demand which involves a combination of actual orders and forecasts. MRP, particularly in the development towards MRP II, has perhaps become over sophisticated and complex. MRP has never addressed the design of the production environment. For example it has accepted set-up times as given; nor has it challenged the unnecessarily long lead times associated with so many components in batch production systems. On the other hand JIT has emphasized the development of an efficient and simplified production environment through the use of product based layouts, U-shaped layouts, labour flexibility, set-up time reduction techniques. JIT seeks to design the production environment while MRP simply seeks to plan and control within the environment in which it finds itself.

However there is a limit to the extent that JIT can be usefully pursued in some environments. JIT has worked well in repetitive manufacturing situations. If the manufacturing system is discontinuous, in that demand is impossible to predict accurately and product variety cannot be easily constrained, then developing a JIT solution is extremely difficult. Moreover, it is not possible for all manufacturers to attain a position where their suppliers are local and captive.

1.5 Guidelines for the development and installation of production planning and control systems

As we indicated earlier our motivation in this book is to develop an understanding of shop floor planning and control systems in order to reduce the throughput times of products through manufacturing systems and to increase the overall effectiveness of manufacturing operations. Clearly there are rich insights to be gained from a study of the ideas associated with the JIT, OPT, MRP and MRP II approaches to production planning and control and these ideas can be used to help to create more effective manufacturing systems. We have tried to emphasize these points throughout our discussion and we will now gather them together before presenting an outline architecture for production planning and control systems.

1 Before installing a production management system, whether manual or computer based, it is important to evaluate the complete manufacturing system and to create an environment which facilitates sound production management practice. It is clear from our study of JIT that product design, manufacturing system design and layout are important in the creation of an environment which facilitate sound production management. Among the issues to be analysed are the following:

   (a) Is the product set designed to minimize unnecessary product variation?
   (b) Are the products designed with ease of manufacture and assembly in mind?
   (c) Is the plant layout flow based? Can it be made flow based?
   (d) How does the plant deal with vendors? Have we worked with suppliers to create enduring relationships? Is this style of relationship appropriate?

2 A factory should, where possible, be broken down into a group of cells or mini factories where each cell has been designed using Group Technology thinking to deal with a family of products. The term product is used rather loosely here to imply a group of identifiable assemblies or subassemblies of components. Within these cells groups of operators can develop responsibility and ownership for groups of products, and can manage the flow of these products through the system.

3 In designing and installing systems it is important to adopt a design methodology which gives due consideration to the social as well as the technical aspects of the system. The relative failure of many MRP installations can often be traced back to lack of involvement of the ultimate users in the development and installation of the system. Training is particularly important in this context. All those likely to come in contact with the system – and very many people will in many different ways, given the central role of production management systems should be given the appropriate level of education and training.
4 U-shaped layouts should be used where possible in the manufacturing plant.
5 Set-up times should be reduced and the techniques of SMED (Single Minute Exchange of Dies) applied as widely as possible.
6 Batch sizes should be reduced. Reduced batch sizes decrease inventory levels and increase flexibility.
7 Where large production lots are unavoidable, transfer batches should be used to facilitate flow production.
8 Lead times and schedules are intimately related. As OPT indicates very strongly, lead times are a consequence of a particular schedule and cannot be used a priori to generate accurate schedules. MRP uses planned lead times to support tactical planning. However lead times should not be used for scheduling activities at the shop floor planning and control level.
9 If they exist production bottlenecks should be identified. The plant should be scheduled to ensure that these bottlenecks never suffer from enforced idleness.

1.6 An architecture for production planning and control

An architecture establishes a framework or a set of rules and guidelines for managing the development and operation of complex systems. Clearly the planning and control of production in a batch oriented discrete parts manufacturing environment is complex and we need an overall framework within which to carry out this task. We will present our outline architecture in the form of a hierarchical planning and control model. In later chapters of this book, and in particular in Chapters 2 and 3 we will greatly expand on the shop floor planning and control aspects of this architecture. As we will see it represents a hybrid approach, that is to say it a system which is based on a synthesis of ideas drawn from the MRP (Materials Requirements Planning), Optimized Production Technology (OPT) and Just In Time (JIT) approaches to production planning and control. Before going into detail on our proposed production planning and control architecture, we will briefly review the two well known and widely used architectures.

1.6.1 An overview of the NBS and ISO models of manufacturing

The National Bureau of Standards (NBS) in the United States of America has established a hierarchical control model for automated systems. The NBS systems architecture is based on a classical hierarchical or tree-shaped structure, which is typical of many complex organizational models. The NBS model recognizes five hierarchical levels namely; facility, shop, cell, workstation and equipment (Fig. 1.5). The facility is the highest level in the structure and comprises of three major sub-systems; manufacturing engineering, information management and production management. The shop is responsible for the real time management of jobs and resources on the shop floor and achieves this through two major modules, namely those of task management and resource management. The cell level manages the sequencing of batches and materials handling facilities. The workstation directs and coordinates a set of equipment on the shop floor. A workstation consists of a set of equipment set up to realize a particular task; a typical workstation might consist of a machine tool, a robot, a material handling buffer and a control computer. The equipment controllers are linked directly to individual pieces of equipment within a workstation. Within this overall NBS model goals or tasks at the highest level are decomposed into sequences of subtasks which are passed down to the next lower level in the hierarchy. This procedure is repeated at each level until, at the bottom of the hierarchy a sequence of primitive tasks, which can be executed with simple actions, is generated. For more detailed information on the NBS model the interested reader is referred to Jones and McLean (1986).

The International Standards Organization (ISO) has developed a Factory Automation Model composed of a six-level hierarchical structure. This structure recognized six levels; namely enterprise, facility/plant, section/ area, cell, station, equipment. A number of basic functions are realized at each level including, the decomposition of higher level goals and tasks into simpler tasks, the assignment of subtasks and resources to subordinates, the analysis of feedback from subordinates and the completion of assigned tasks. In simple terms the enterprise level is responsible for the achievement of the mission of the enterprise and clearly its planning horizon is measured in years and months. The facility/plant level is responsible for the implementation of the enterprise functions and the reporting of status information to the enterprise level. It includes functions such as manufacturing engineering, information management, production management and scheduling and production engineering. The section/area level is responsible for the provision and allocation of resources and the coordination of production on the shop floor. Typically this level operates within a planning horizon of several days or weeks. The cell level is responsible for the sequencing of jobs through the various stations. Its functions include resource analysis and assignment, making decisions on job

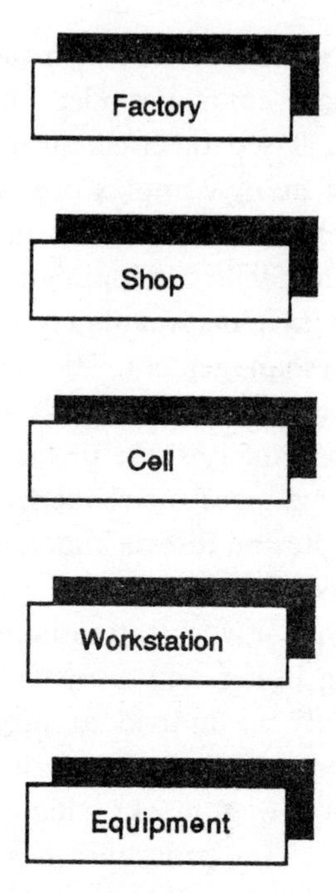

Fig. 1.5. NBS hierarchy.

routings, dispatching jobs to individual stations and the monitoring of task and station status. the station level is responsible for the direction and coordination of relatively small integrated workstations. It operates in virtual real time with planning horizons of milliseconds to hours. In a sense a station is a virtual machine. The equipment level realizes the physical execution of tasks on machines and it has a planning horizon ranging from several milliseconds to several minutes.

Readers interested in a more thorough review of the ISO model should consult ISO (1986).

1.6.2 A hybrid model for production planning and control

We see production planning and control in terms of three hierarchical levels of activity, strategic activities, tactical activities and operational activities (Fig. 1.6). In this book our interests are in the operational or shop floor planning and control issues. Nevertheless we will present a short outline of our ideas on strategic and tactical issues, since clearly decisions made at the higher levels in the hierarchy establish the environment for the operational decisions. We will now discuss each level in turn.

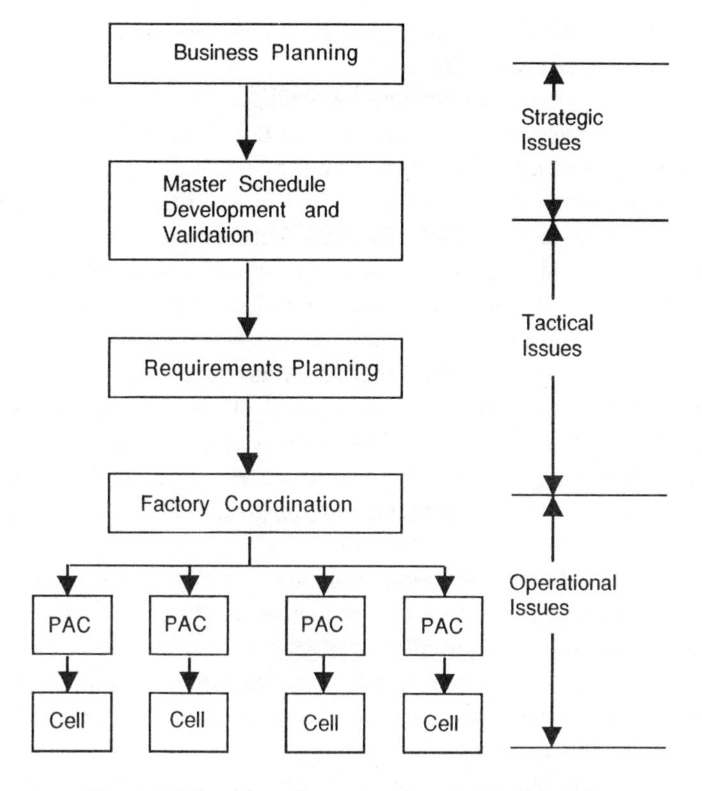

Fig. 1.6. Manufacturing controls systems hierarchy.

Strategic issues Strategic production management system issues relate to the determination of the products to be manufactured, the matching of products to markets and customers expectations and the design of the manufacturing system to ensure short production lead times, and sufficient flexibility to facilitate the production of the required variety and mix of products for the market. In an environment of greatly increased product variety, considerably reduced product life cycles and greatly enhanced customer expectations these are clearly important issues. A manufacturing firm must identify very clearly the market within which it is operating and any particular competitive advantages in terms of that market. It must ensure that the products it is producing meet the expectations of customers in terms of price, quality, functionality, level of customization, delivery lead time, and in the case of new products the time to market.

It is clear that JIT thinking has a lot to offer here. It supports the strategic layer through concentrating on matching the product to the market, the design of product to facilitate manufacturing and the use of group technology concepts to design and define families of products, assemblies and components to facilitate flow based manufacturing.

Ultimately one of the outputs from the business planning activity is a master schedule to authorize production over a medium term horizon. The objective is to verify the proposed master schedule and produce an authorized master production schedule which is likely to be realizable. If the manufacturing environment in the broadest sense of products, markets, etc. is repetitive then the development and verification of the master production schedule is greatly simplified. In such cases, JIT planning techniques discussed are clearly appropriate. However, it is unlikely this will be the case for the majority of manufacturing firms. What is required for non-repetitive manufacturing are a set of tools to facilitate verification from a capacity point of view. Such tools should be developed by incorporating many of the insights of OPT finite scheduling, such as having a planning bill of materials and process which includes the critical components, and also is aware of the likely critical i.e. bottleneck resources.

Tactical issues Tactical production management system issues relate to the generation of detailed plans to meet the demands imposed by the the master production schedule. It essentially involves the breakdown of the products in the master production schedule into their assemblies, sub-assemblies and components and the creation of a time phased plan of requirements, which is realistic in terms of capacity.

Just in Time manufacturing techniques support the requirements planning process by developing a manufacturing system which facilitates flow based manufacturing, incorporating simplicity and excellence at all stages of the manufacturing process and developing close relationships with external suppliers. This results in a manufacturing system whose performance is highly predictable and which therefore facilitates good requirements planning. In the ultimate case, this is a repetitive manufacturing system in which tactical planning can easily be achieved using production smoothing, monthly and daily adaptation as described earlier in this chapter. MRP supports the requirements planning process by planning the availability of material and of manufactured and assembled parts to meet customer requirements. MRP functions well in situations where the variety and complexity of products is such that the JIT planning techniques are inappropriate. MRP's use of planned lead times is appropriate in the context of such relatively complex manufacturing systems and the degree to which these lead times are accurate depends on the extent to which the manufacturing system has been simplified and streamlined. The important point is that planned lead times are useful at a tactical planning level, but we must understand that the output from this process is simply a plan, which should be interpreted as a guide for operational control purposes, and not a detailed schedule. Operation scheduling, in our view, is a separate issue.

Operational issues Fig. 1.6 shows that we consider the Factory of the Future to be composed of a series of mini-focused factories or product based manufacturing cells. We believe that the degree to which such cells are autonomous and the factory can be so decomposed, will depend primarily on the product/process market situation and the degree to which Just in Time thinking can be successfully applied. In later chapters of this book we will outline techniques which can be used to help define appropriate product based cells.

Operational production management system issues essentially involve taking the output from the tactical planning phase, e.g. the planned orders from an MRP system, and managing the manufacturing system in quasi real-time to meet these requirements. Our view is that it is necessary to have a production activity control system for each cell and a higher level controller to coordinate the activities of the various manufacturing cells.

We recognize that many readers may feel that it is not possible to decompose a typical factory into reasonably autonomous product based cells and further that the introduction of two layers of functionality (Factory Coordination and Production Activity Control) between requirements planning and the execution layer is unnecessary. In fact if the individual cells were autonomous, coordination i.e. factory coordination would be unnecessary. In typical manufacturing plants there will be interaction between the various product based cells. Perhaps three of the cells are manufacturing components which are assembled into a finished product in the fourth cell. Clearly in such a case it is necessary to coordinate the flow of finished components from the manufacturing cells into the final assembly cell. Perhaps certain manufacturing processes are very capital intensive and cannot be duplicated across cells i.e one of the cells may contain production facilities which must be shared by one or more of the other cells. Here again coordination is necessary.

In our view finite scheduling is appropriate at the PAC level. In particular we believe that OPT type scheduling will find application in future PAC systems. The higher level controller which coordinates the activities of the various cells will, in our view, have a similar structure to a PAC system. In a sense we see it as an upward recursion of the PAC module in that it will consist of a scheduler, perhaps using finite scheduling techniques to schedule the planned orders, a monitor which provides feedback on cell activities and a dispatcher which manages inter-cell activity. This higher level controller we term a Factory Coordination system. In terms of the NBS and ISO models discussed above we believe that the Factory Coordination system sits at the shop/area level (ISO) or the shop level (NBS). The PAC systems are at the cell level.

In Chapter 2 we will present a detailed functional architecture for the Factory Coordination and Production Activity Control systems.

1.7 Conclusions

In this chapter we have presented an overview of the various approaches to production planning and control systems, including JIT, MRP and OPT. Based on this overview we have outlined some important lessons to be learned from these diverse approaches and have outlined the architecture of a hybrid production planning and control system for discrete parts manufacturing systems. Our emphasis in this book is on the operational or shop floor level level of this hierarchy. In Chapters 2 and 3 we will present a detailed architecture for operational level production planning and control.

Part Two
A FUNCTIONAL ARCHITECTURE FOR SHOP FLOOR CONTROL SYSTEMS

Overview

Having positioned Production Activity Control (PAC) and Factory Coordination (FC) within the production management systems hierarchy, we now present a detailed statement of the functionality of these two systems.

In Chapter 2 we offer a functional architecture for PAC and FC. We argue that the PAC system can be broken down into five identifiable building blocks; namely a scheduler, a dispatcher, a monitor, producers and movers. The *scheduler* creates a short term plan for the product based manufacturing cell. This plan in turn is passed to the *dispatcher*, which can be termed a real time scheduler, in that it seeks to implement the plan created by the scheduler. The dispatcher can do this because it has access to a *monitor* which produces quasi real time information on the condition of the shop floor, e.g identifying which resources are available for work, which batches or jobs have recently completed operations thus making them available for subsequent operations etc. The monitor tracks the flow of work through the shop floor by capturing data on the condition of individual *producers* and *movers*.

The Factory Coordination system is responsible for coordinating the flow of work between the various cells. We consider FC to be, in one sense a higher level recursion of a PAC system, i.e. it also has a scheduler (factory wide), dispatcher (factory wide), monitor (factory wide), a mover to effect the flow of work between individual product cells and the PAC systems constitute its 'producers'. However FC also incorporates a second set of functionality, which we term a Production Environment Design (PED) module. The primary role of the PED module is to ensure that as new products are introduced into the factory, they are located within the correct cell. The PED module also supports process planning and facilitates analysis of the manufacturing system in order to identify elements of the process which might be improved.

In Chapter 3 we offer a formal model of the FC and PAC systems using the Structured Analysis and Design Technique (SADT™). For readers not familiar with SADT™ we offer a short overview of the technique in the early part of Chapter 3. We do not introduce any new material in Chapter 3; rather we present the ideas discussed in Chapter 2 in a more formal way.

2

An architecture for shop floor control systems

2.1 Introduction

In this chapter we present an architecture for the activities which occur at the operational level of the PMS hierarchy, namely, the day-to-day tasks involved in planning and controlling production on a shop floor. Production Activity Control (PAC) and Factory Coordination (FC) provide a framework which integrates the requirements planning functions of MRP type systems and the planning and control activities on a shop floor, and in doing so close the loop between the tactical and operational layers of the PMS hierarchy.

In Chapter 1, we reviewed two well known and widely used hierarchical control architectures: the National Bureau of Standards (NBS) and the International Standards Organization (ISO) architectures. When describing the functional architecture of factory coordination and production activity control, we shall use the ISO model. As we have already described in Chapter 1, there are six levels in the ISO model: enterprise, facility/plant, section/area, cell, station and equipment. The two FC tasks, involving control and production environment design, are situated at the facility/plant level of the hierarchy with access to the cell level, where the PAC tasks are situated.

It is important to emphasize that our aim in this chapter is to describe what tasks should be realized within FC and PAC, rather than how these tasks should be realized. The possible ways of realizing PAC and FC are the subjects of Chapters 6 and 7 respectively. The overall structure of this chapter is as follows:

1. Firstly, we discuss an architecture for PAC, and describe each of the individual functions, or building blocks, which are part of PAC. Through this discussion, we illustrate the the inter-relationships of the building blocks, the control hierarchies, and the data requirements.
2. Secondly, we describe an architecture for FC, with particular reference to

the combination of the control features and the production environment design tasks.

3 Finally, we explain both the production environment design task and the control task, by detailing each of the individual functions associated with each task. As with the PAC description, we shall discuss the interrelationships and the data requirements of the various functions within FC.

2.2 Production Activity Control

We shall now consider the typical activities used to plan and control the flow of products on any particular shop floor. Each supervisor has waiting on his/her desk a list of customer requirements, which have to be fulfilled for the forthcoming week. The main task facing the supervisor at this moment in time is to plan production over the following working week to ensure that the customer orders will be fulfilled. Factors which will influence the content of this plan include the likely availability of resources (operators, machines) and the capacity of the manufacturing system. This plan will then act as a reference point for production, and almost certainly will have to be changed due to any number of unpredictable events arising (e.g. raw material shortage, operator problems, or machine breakdown). Therefore, the three main elements for shop floor control are:

1 To develop a plan based on timely knowledge and data which will ensure all the production requirements are fulfilled. This is termed *scheduling*.
2 To implement that plan taking into account the current status of the production system. This is termed as *dispatching*.
3 To *monitor* the status of vital components in the system during the dispatching activity, either with the naked eye or by using technology based methods.

We believe that activities of scheduling, dispatching and monitoring are in fact carried out, perhaps informally, by every competent shop floor manager or supervisor. Our approach is to outline formally each of these separate tasks and show how they interact to control a manufacturing system. Essentially, this formalisation is the application of the scientific method to shop floor control. It was Thomas H. Huxley, the eminent British scientist of the nineteenth century, who defined science as 'nothing but trained and organized common sense', and architecture presented in this chapter may be viewed as a trained and organized common sense approach to the day-to-day activities of controlling a manufacturing plant.

The basis of our approach is to map the various shop floor activities onto an architecture which recognizes each individual component. The advantage

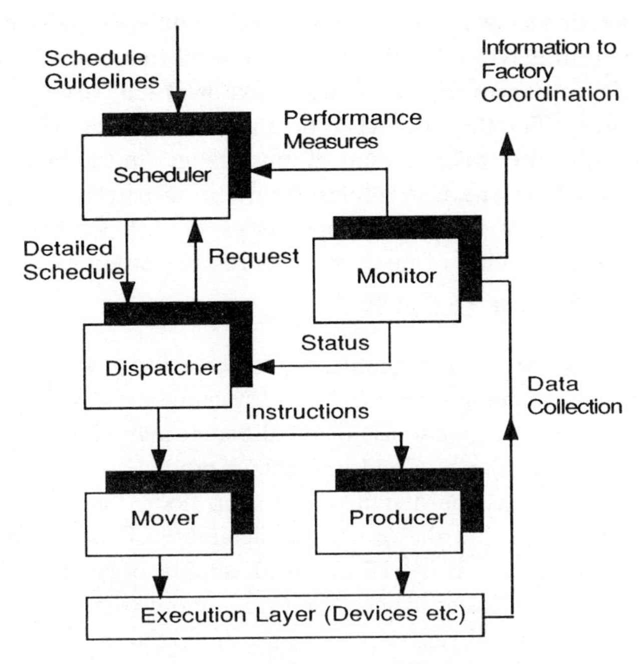

Fig. 2.1. Production Activity Control.

of having an architecture is that it both formalizes and simplifies the understanding of what occurs during production, by establishing clear and separate functions which combine into a shop floor control system.

The activities of Production Activity Control can be defined as follows:

Production activity control describes the principles and techniques used by management to plan in the short term, control and evaluate the production activities of the manufacturing organisation (Browne 1988).

As stated before, PAC realizes the lowest level of the PMS hierarchy, and the COSIMA project recognized the need for a generic and flexible architecture which identified and separated the different functions of PAC (COSIMA 1987). This separation, or *modularization* of PAC, is important, because a system of interacting well defined components can easily be changed by replacing or modifying these components on an individual basis, without effecting the other modules of the system.

The architecture is illustrated in Fig. 2.1, and the five basic building blocks of the PAC system are the *scheduler, dispatcher, monitor, mover* and *producer*. The *scheduler* develops a plan over a specified time period, based on the manufacturing data and the schedule guidelines from FC. This plan is then implemented by the remaining four modules. The *dispatcher* takes the schedule and issues relevant commands to the movers and producers, which carry out the required operation steps necessary to produce the different products. Given our inability to accurately predict the future, the need to modify the plan due to unforeseen circumstances (e.g. machine breakdown) may arise, and in this case the *monitor* notifies the dispatcher of any disturbances. The schedule may then be revamped to take account of any changes in the manufacturing environment. Based on the instructions from the dispatcher, the *producer* controls the execution of the various operations at each workstation. The *mover* organizes the handling of materials between workstations within a cell by following the dispatcher's commands.

The scheduler The task of the scheduler is to accept the production requirements from a higher planning system (i.e. a FC system), and develop a detailed plan which determines the precise use of the different manufacturing facilities within a specified time frame. Good scheduling practice is dependent on a number of factors such as the design of the shop floor, the degree of complexity of the operations and the overall predictability of the manufacturing process. It is recognized that a well designed, simply organized and stable manufacturing process is easier to schedule than a more complex and volatile system. This point will be discussed in more detail in section 2.3.

Much work has been reported on production scheduling (Graves, 1981), highlighting the fact that scheduling is a complex task, the technical difficulty being that of combinatorial explosiveness (Rickel, 1988). For instance sequencing twelve orders through six operations generates $(12!)^6$ or more than 10^{52} possible schedules in a simple job shop. This magnitude of possibilities make the goal of schedule optimality an unattainable ideal, and according to Baker (1974), there are 'relatively few situations in which general optimal solutions are known'. Regarding the ideal of optimality, it is important to note that 'the optimal solution of a model is *not* an optimal solution of a problem unless the model is a perfect representation of the problem' (Ackoff, 1977). Even in cases where optimal solutions are available, the scheduling model is somewhat removed from reality. This point is dealt with in greater detail in the review of scheduling presented in Chapter 6.

Within manufacturing there are many diverse scheduling problems, be it in mass production, job-shop or batch production, and even within these categories the scheduling problems may also differ greatly. No two scheduling problems are totally alike, and the production environment plays an important role in defining the requirements of an appropriate scheduling strategy. However, there are fundamental similarities between different scheduling problems, and we will examine these later on in Chapter 6. The scheduling task within PAC is simplified due to the nature of the PMS hierarchy presented in the previous chapter. Essentially, FC and PAC perform the scheduling tasks based on a hierarchical decentralized production control model. The advantage of this type of hierarchy was recognized by Hoefer (1985), who acknowledged that 'the creation of *self-contained tasks* aids in reducing complexity and uncertainty in task execution and it also eases the coordination of decision making.'

PAC is a *self-contained task* which controls a specific cell within the factory, and the scheduling function of PAC takes as its primary stimulus the schedule guidelines from the FC system. These guidelines specify the time constraints within which a series of job orders are to be completed, and the role of the PAC scheduler is to take these guidelines and develop a plan which can then be released to the shop floor via the PAC dispatcher. The actual development of the schedule may be based on any one of a number of techniques, algorithms or computer simulation packages.

The scheduling function includes three activities which are carried out in order to develop a realistic schedule for the shop floor.

- Firstly, a check on the system capacity is required, the objective being to calculate whether or not the schedule guidelines specified by the FC system are realistic. The method of doing the capacity analysis depends on the type of manufacturing environment, and the results of the capacity analysis will have two possible outcomes: either the guidelines are feasible or they are not. If the guidelines are feasible, the capacity constraints are then included in the procedure for developing a schedule.
- Secondly, a schedule must be generated. If there is a major problem with the available capacity, the scheduler may need to inform the FC system, and the overall guidelines for scheduling that particular cell may have to be modified.
- Finally, the schedule is released to the dispatcher so that it may be implemented on the shop floor. In a CIM environment, this release will be achieved by means of a distributed software system, which will pass the schedule between the scheduling and dispatching functions.

Scheduling represents one aspect of PAC, that is, the planning aspect. The schedule is developed taking different constraints and variables into account. When it is released to the shop floor it becomes susceptible to the reality of shop floor activity, and in particular, the *unexpected event*. The unexpected event is a true test of any system's flexibility and adaptability, and it is the role of the dispatcher to deal with inevitable unplanned occurrences which threaten to disrupt the proposed schedule.

The dispatcher Just as bus delays, traffic jams and weather conditions may spoil our well intentioned business and social plans; events such as machine breakdown, or quality problems, can have a serious effect on the production plans supplied by the scheduler. In life, people dispatch in many different situations, for instance, if when driving we see a major traffic jam, we search for an alternative route through a side street. The same principle applies to the dispatcher in a PAC system. Its main purpose is to react to the current state of the production environment, and select the best possible course of action.

In order to function correctly, the dispatcher requires the following important information:

1. The schedule, which details the timing of the different operations to be performed.
2. The static manufacturing data describing how tasks are to be performed.
3. The current shop floor status.

Thus, access to the latest shop floor information is essential, so that the dispatcher can perform intelligent and informed decision making. In fact, one of the greatest obstacles to effective shop floor control is the lack of accurate and timely data. As Long (1984) pointed out, the possibility of a good decision is directly related to the integrity of the manufacturing data. The dispatcher may use different algorithms and procedures to ensure that the schedule is followed in the most effective way. When a decision has been made as to the next step to be taken in the production process, the dispatcher will send instructions to the mover and the producer so that these steps are carried out.

The three main activities of the dispatcher involve *receiving information*, *analysing alternatives* and *broadcasting decision*. The information received is the scheduling information, as well as both static and dynamic manufacturing data. The static data may be obtained from the manufacturing database while the dynamic information describing the current status of the shop floor is received from the monitor. When received, this data is collated and manipulated into a format suitable for analysis. This analysis may be carried out using a range of software tools, or performed manually by a supervisor, based on experience and intuition. The analysis will most likely take place keeping the overall dispatching goals in mind, and the end effect of this analysis is to broadcast an instruction to the relevant building block, perhaps using a distributed software system.

The implementation of a dispatcher varies depending on the technological and manufacturing constraints of a production system. The dispatching task may be carried out manually, semi-automatically or automatically. Manual dispatching involves a human decision on what the next task should be in the system. Examples of this might be an operator deciding to select a job according to some preference. This preference might be generated using a *dispatching rule*, which prioritizes jobs in work queues according to a particular parameter (e.g. earliest due date, or shortest processing time). A semi-automatic dispatcher may be computerised application which selects jobs, but this selection can be modified by an operator. An automatic dispatcher is used in a automated environment, and it assumes responsibility for controlling the flow of jobs through the system.

To summarize, the dispatcher is the *controlling* element of the PAC architecture, and it ensures that the schedule is adhered to in so far as possible. It works in real-time by receiving information from the monitor on the current state of the system, and it issues instructions to the moving and producing devices so that the required work is performed.

The monitor Within the different levels of manufacturing, from strategic planning down to PAC, informed and accurate decision making relies on consistent, precise and timely information. Within PAC, the monitor function supplies the necessary information to the scheduler and dispatcher, so that they can carry out their respective tasks of planning and control. Thus, the role of the monitor is to make sense of the multitude of data emanating from the shop floor, and to organise that data into concise, relevant and understandable information for the scheduler and dispatcher. Put simply, the monitor can be seen as a translator of *data* into *information*, for the purpose of providing sensible decision support for the scheduling and dispatching functions.

Higgins (1988) identified three main activities of the monitor as: *data capture, data analysis* and *decision support*. The *data capture* system collects data from the shop floor. This is then translated into information by the *data analysis* system, and can then be used as *decision support* for appropriate PAC activities.

Data Capture A vital part of the monitor is the data capture system, which makes the manufacturing data available in an accurate and timely format. This data capture function should perform reliably, quickly and accurately without detracting from the normal day-to-day tasks which are carried out by humans and machines on the shop floor. Ideally, data capture should be in real time with real time updating, and the data should be collected at source. Automatic or semi-automatic collection of data is often necessary for reasons of accuracy and speed of collection.

Monitoring the shop floor and the process of data capture are very much interlinked, as information cannot be accumulated by the monitor unless the data has been collected. Therefore, data capture may be viewed as a subset of the monitor, the major difference being that data capture is only concerned with data transactions and making data available for other functions. However, the monitor analyses the data collected and either makes a decision by providing real-time feedback to the other applications within the PAC architecture, or provides a support tool with which management can make decisions at a later date. Data captured which may eventually be used for informed decision making at a higher level in the PAC architecture includes: process times, job and part status, inspection data, failure data, rework data, and workstation data.

Work in Progress Status	Job number Part name Current location Current operation Due date Number of remaining operations
Workstation Status	Workstation name Current status Current job number Utilization Percentage time in set-up Percentage time processing Percentage time down

Table 2.1. *Typical information from the data analysis of the production monitor*

Data Analysis The data analysis function of the monitor seeks to understand the data emanating from the data capture system. It is a very important component of the monitor, because it takes time and effort to filter important information from a large quantity of shop floor data. Thus, this data analysis function effectively divides the monitor into different sub-monitors, which then keep track of different important aspects of the manufacturing system. Joyce (1986) identified three main classes of monitor:

1 Production monitor
2 Materials monitor
3 Quality monitor

We will now discuss each of the main classes of monitor in turn:

Production monitor The production monitor is responsible for monitoring work in progress status and resource status on the shop floor. Table 2.1 illustrates the type of information produced as a result of the data analysis performed by the production monitor. At a glance, production personnel can see important manufacturing information regarding the progress of the schedule. This information can then be used as the basis for informed decision making. An important feature of the production monitor is the ability to recognize the point when the schedule is not implementable, and then request a new, more realistic schedule from the scheduler

Raw materials monitor The raw materials monitor keeps track of the consumption of basic raw materials at each workstation in the process. Table 2.2 shows the type of information generated by this particular monitor, and the main purpose of such a monitor is to ensure that there are no shortages in raw materials at a particular location. This is achieved by comparing current

Raw Materials Status	Material name Workstation name Buffer name Current quantity Reorder point Rate of usage

Table 2.2. *Information from the data analysis of the materials monitor*

levels of a particular raw material with the recommended reorder point, and indicating when raw materials need to be reordered.

Quality Monitor As the name suggests, the quality monitor is concerned with quality related data, and aims to detect any potential problems. Quality problems may arise from internal or external sources. Quality problems arising from external sources originate in the supply of raw material purchased from vendors which can cause problems in later stages in the production process. Problems originating in the production process which affect the quality of products are classified as internal sources. These types of problems can be indicated by the yield of the cell or of a process within the cell. If the yield fall below a defined level, an investigation into the cause may be warranted. Possible causes of a fall in yield status can range from faulty equipment to poor quality raw materials.

Thus, the data analysis module makes information available so that accurate and informed decision making can take place. It also prepares a historical reporting file so that a complete record of important manufacturing events can be kept for future reference. This type of historical reporting is particularly important if a company is required to track individual items or lots as a result of customer reports of a faulty product.

Decision Support The main function of the monitor's decision support element is to provide intelligent advice and information to both the scheduling and dispatching functions within PAC in quasi-real time. In effect, it is a form of expert analysis of output from the *data analysis* module of the monitor, sifting through a large quantity of information to detect trends which have a significant bearing on the shop floor control process. Examples of types of decision support provided by the monitor in each of the categories defined within the data analysis module are:

Overall Decision Support An overall decision support function might take aggregated information from each individual monitor and present these for the purposes of analysis to higher planning systems such as FC. This may illustrate how a particular cell is performing, and may highlight:

1 the total number of jobs produced over a certain period within the cell;
2 the overall figures on workstation utilisations and work in progress levels;
3 the relationship between what was planned and what was produced;
4 the overall figures on raw material usage, and the number of shortages that occurred in the cell;
5 information for the quality of the products and the actual process itself.

Decision support for the production monitor This decision support function can be used to assess how the current schedule is performing, and if it is likely that the schedule passed down from the FC system can be met. This feature of the monitor is important, as it ascertains whether or not a rescheduling activity may have to take place. Bottleneck workstations can be identified as well as under-utilised ones, and this information can then be used by the dispatcher and the scheduler to control the flow of work more effectively.

Decision support for the raw materials monitor If the quantity of a certain raw material has fallen below a certain pre-defined level, the raw materials monitor is in a position to tell the dispatcher that more raw materials need to be ordered.

Decision support for the quality monitor The monitor can use an early warning facility, such a flashing signal on a terminal to alert the operator. This warning can be based on statistical process control data, which may indicate that the specified tolerance limits on a particular operation have been exceeded.

It is important to have a good decision support system within the monitoring function. This facility should operate on a *need-to-know* basis (i.e. only the most important and relevant information should be presented), and the existence of this function forces production personnel to identify the information requirements of the PAC system. One of the problems identified by Ackoff (1967) within management information systems (of which monitoring may be seen as a sub-function) is the *over abundance of irrelevant information*, which can only lead to confusion and inefficiency amongst decision makers at all levels of the PMS hierarchy. Hence the importance of an intelligent monitoring function to provide an efficient, informed and intelligent support service to the planning and control activities within PAC.

The mover The mover coordinates the material handling function and interfaces between the dispatcher and the physical transportation and storage mechanisms on the shop floor. It supervises the progress of batches through a sequence of individual transportation steps. The physical realization of a mover depends on the type of manufacturing environment. It can range from a

Automatically Guided Vehicle system (AGV) to a simple hand operated trolley. The selection of the items to be moved is predetermined by the dispatcher and this decision is then transferred as a command to the mover, which then carries out the instruction. The mover monitors the states of the transporter and the storage points, translates the commands from the dispatcher to specifically selected moving devices, and also issues messages to the dispatcher signalling commencement and completion of an operation. An automated mover system might use collision avoidance algorithms, to ensure that no individual device will cross the path of another when parts are being moved to their destinations.

The producer The producer is the process control system within PAC which contains (or has access to) all the information required to execute the various operations at that workstation. The producer may be an automated function or a human. The main stimulus for a producer comes in the form of specific instructions from the dispatcher building block, and these instructions specify which batch to process. The producer accesses the relevant part programs (detailed instructions of the operation that has to be performed), and also the configuration data which specifies the necessary set-up steps that are needed before an operation can commence. The producer translates the data into specific device instructions and informs the monitor when certain stages of activity have been completed (e.g. set-up completed, job started, job finished, producer failed, etc.)

Overview of PAC Thus, a PAC system provides the necessary functions to control the flow of products *within* a cell, through the interaction of these five distinct building blocks:

1 A scheduler, which develops a schedule based on the guidelines contained in a factory level schedule.

2 A dispatcher, which controls the flow of work within the cell on a real time basis.

3 A monitor, which observes the status of the cell and passes any relevant information back to the scheduler, dispatcher and the higher level factory coordination system.

4 A mover, which organises the movement of materials between workstations

5 A producer, which controls the the sequence of operations at each workstation.

We now consider the next level of the PMS hierarchy, namely factory coordination, which coordinates a group of cells on a shop floor.

2.3 Factory coordination

So far in this chapter, we have discussed some of the main issues involved in shop floor control and the use of production activity control (PAC) to control the flow of products *within* a cell. As indicated earlier, we visualise a typical manufacturing environment as consisting of a number of Group Technology cells, each controlled by a PAC system. This layout should be as close as possible to a product based layout. This vision of a typical manufacturing environment is completed with the inclusion of a factory coordination system, which organizes the flow of products throughout a factory and ensures production in each cell is synchronized with the overall production goals of the factory.

As FC is managing a diversity of products and processes throughout a typical manufacturing environment, it requires a variety of information. Referring to the amount of information needed to complete a task and the variety involved in a task, Galbraith (1977) argued that they are dependent on:

1 The variety of outputs provided by the system such as products, services or customers.
2 The range of different input resources which are required to produce the output (e.g. operators, production equipment, etc.).
3 The level of performance as measured by some efficiency criterion, such as resource utilization, quality levels or product throughput.

In order to manage this information, we suggest the use of two principles: namely, *goal setting* and *distributed responsibility,* as proposed by Beckhard (1969) and two tasks, *integration* and *differentiation* as identified by Lawrence and Lorsch (1967). In our opinion these two principles are closely connected with the two main tasks involved in factory coordination. The *differentiation* task of FC involves maintaining a product based production environment with well defined product families. The *differentiation* task facilitates *distributed responsibility,* whereby autonomous work groups are organized around each product family and are given responsibility for managing the production of each product family. Within FC, the differentiation task is termed *production environment design.* The *integration* aspect of FC relates to the coordination of the production activities across the various cells and ensuring that order due dates are satisfied. The integration aspect is known as the *control* task. This involves the principle of *goal setting,* whereby each PAC system is given goals or guidelines within which it manages its own activities (i.e. distributed responsibility).

Galbraith (1973) argued that the most suitable method for achieving integration depends on the degree of differentiation and the rate of introduction of new products. In the context of FC, the differences in the degree of differentiation relate to the level of autonomy of each work group and the variety

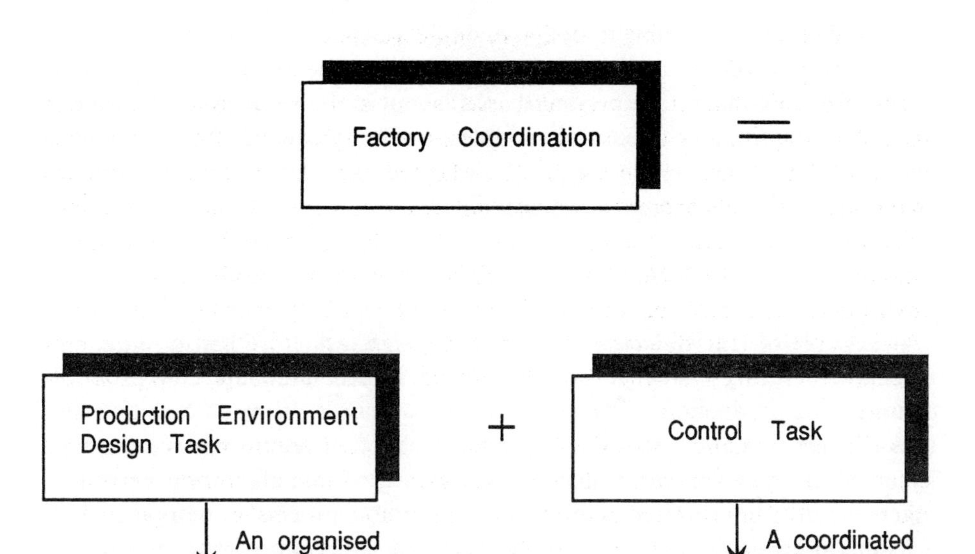

Fig. 2.2. The link between production environment design and control.

of manufacturing processes used in each cell. These two factors illustrate the importance of the design of the production environment in the integration and the control of the flow of products throughout a factory. The production environment design task influences the level of autonomy of each work group by the extent to which each work group is responsible for the manufacture of each product family and influences the variety of processes being used by each work group by determining the resources in each cell.

Traditionally the design of the production environment was considered separate from the control of product flow, but there is a close relationship between the two, as Lubben (1988) argued when he wrote that 'the physical layout significantly influences the efficiency of the production system.' Referring to the Japanese awareness of process design as Gunn (1987) recognized when he stated that 'by concentrating on simplifying the process . . . managers of Japanese firms attain greater control over the manufacturing process and the quality of the product.' Within the factory coordination architecture, we recognize a link between the tasks of production environment design and control as illustrated in Fig. 2.2.

Mitchell (1979) proposed that 'planning should divide the problem and provide the means by which the accountable people can cope with the proliferation of variety within their area of responsibility'. Within FC, we take this approach by linking the tasks of production environment design and control.

The production environment design module helps to reduce the variety of possible production related problems by organising the manufacturing system in so far as possible into a product based layout and allocating new products to product families. The control module provides guidelines and goals with which each work group can manage its activities and deal with any problems that occur within its scope of responsibility.

2.3.1 The production environment design task

A key role of the *production environment design* task (PED) is to *reorganize* the manufacturing system to simplify it and to accommodate new products coming into production. The *production environment design* task within FC uses a range of static data and the future production requirements generated by an MRP type system. Using the experience of manufacturing personnel together with different techniques, the production process is reorganized to accommodate new products and increase the efficiency of the production process, subject to various production constraints and the manufacturing goals of the organization.

The static data includes a Bill of Materials (BOM) and a Bill of Process (BOP) for each new product. A *bill of materials* defines the structure of a product by listing the names and quantities of each component of each assembly and sub-assembly in a product's structure (Orlicky, 1975). *A bill of process* describes the process steps involved in the production of a product in terms of the required resources, operation procedures, process times and set-up procedures. In short a BOM gives a description of a products' components and sub-assemblies, while a BOP gives a description of a products' process requirements.

As indicated earlier the main aim of the production environment design task is to reorganize the initial layout of the manufacturing system, rather than create the layout initially. The initial layout of the manufacturing system should be *product based.* The initial creation of a product based layout is a once-off activity, which can be accomplished using a methodology such as Group Technology. Burbidge (1989) defines *process based* layout as involving organization units which specialise in particular processes, where as *product based* layout involves units which specialize in the completion of groups of products or subassemblies (Fig. 2.3). Each cell in a product based layout is associated with the manufacture of a particular product family.

It is clear that for the majority of manufacturing plants it will not be possible to define independent product based cells. In fact the product based cells sketched in Fig. 2.3 might be considered ideal, in that each product family is completed within its associated cell. Greene and Sadowski (1984) argued that a situation may occur where different components are manufactured

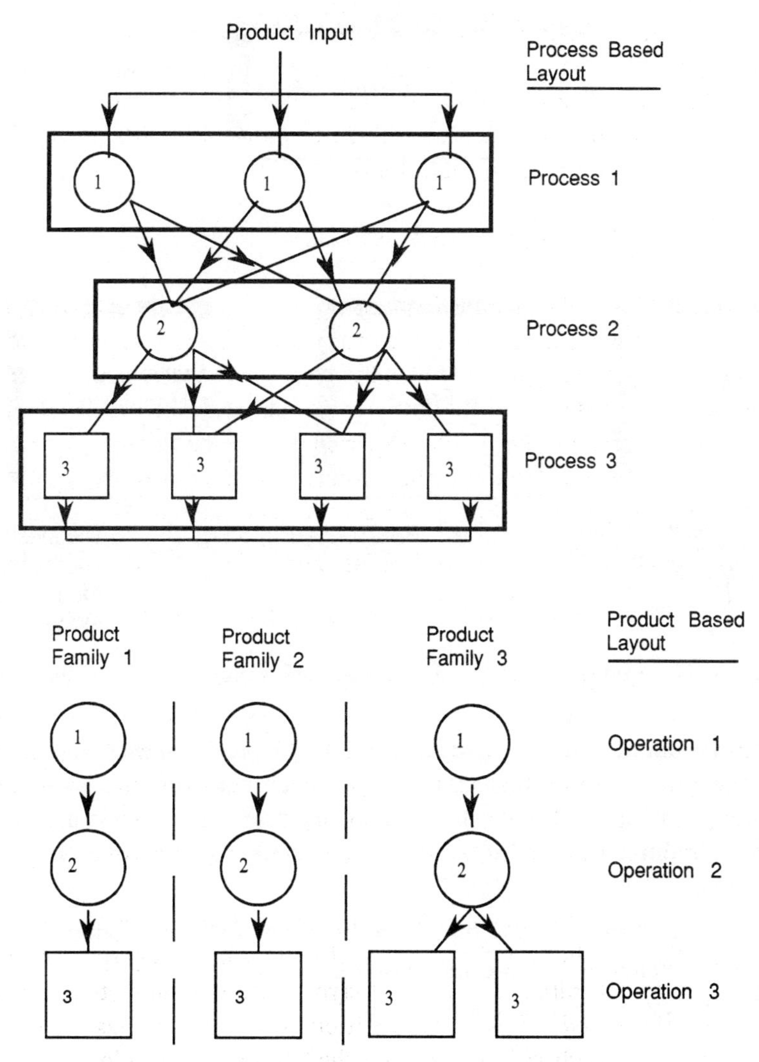

Fig. 2.3. Process based layout vs. product based layout.

in individual cells and perhaps assembled in another cell. Also it may be necessary to share expensive equipment between cells. For example, the expensive equipment used in a heat treatment process, is usually separated into one cell on a shop floor, which all products share, rather than having individual heat treatment ovens and associated equipment in each cell. It is this failure to develop completely independent cells which make the control task within FC necessary. In fact the degree of interdependence between the cells determines the complexity of the FC task.

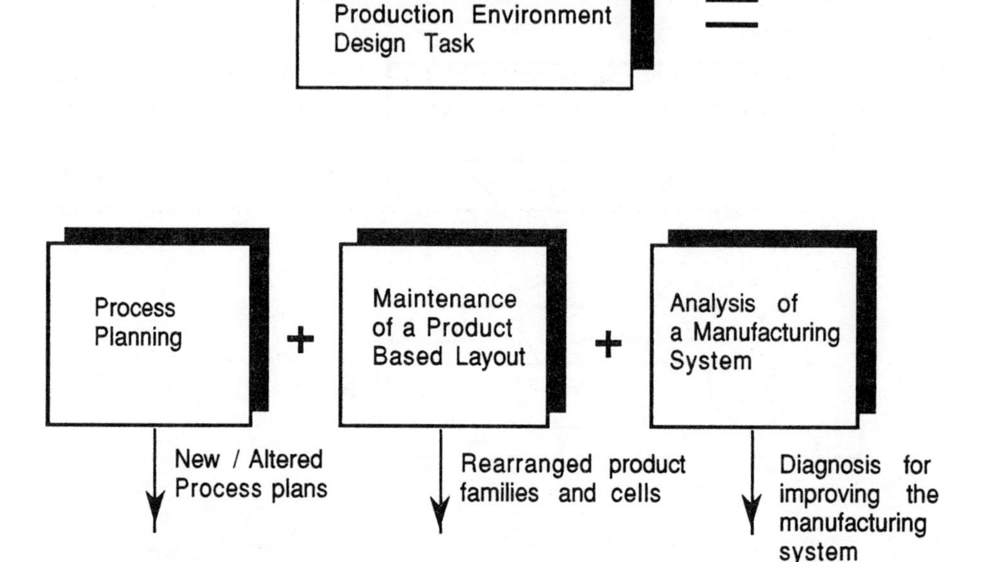

Fig. 2.4. The production environment design task within factory coordination.

Using an initial product based layout, the *production environment design* module integrates new products into the production environment with minimal disruption and reorganization to the existing product based structure. The various procedures involved in *production environment design* may include the following (Fig. 2.4):

- the generation and detailing of process plans for new products,
- the maintenance of a product based layout,
- the analysis of a production system.

Process Planning The main function of the process planning procedure is to generate a bill of process for each new product. By including the process planning function within the production environment design task, we are attempting to prevent any process plan proliferation. Nolen (1989) noted that the problem of plan proliferation can go unnoticed, because 'changes to process plans, are made routinely, remain largely undocumented and are accomplished by a variety of individuals using formal and informal systems'. One consequence of process plan proliferation is the generating of a variety of product flow paths, which can reduce the benefits of a well organized product based layout. In our opinion, the inclusion of the process planning function helps in production environment design to standardise and reduce the complexity in the planning task.

When new process plans are being developed, there may be two steps to the development process:

1 *Determine production requirements.* Manufacturing personnel decide on suitable process requirements for each product given the order constraints.

2 *Develop plan.* Based on the chosen process requirements, manufacturing personnel assign specific resources to carry out the operations. Other details such as specifying tooling and set-up requirements are also added to the process plan at this stage.

The process planning task also allows existing process plans to be altered, if such a need arises. The procedure used by the process planning system may be either *variant* or *generative.* We shall discuss generative and variant process planning in more detail later in Chapter 7.

Maintenance of the product based layout The product based layout must be maintained whenever a group of new products are introduced into production. Using a set of procedures based on Group Technology principles, the set of new products are integrated into the existing range of product families and any necessary reorganization is carried out on the product based layout. The reorganization of the product based layout occurs with minimal disruption to the daily production activities. Shambu (1989) defines Group Technology as a manufacturing philosophy that attempts to rationalize batch production by making use of design and/or manufacturing similarities among products. Families of products are established, based on the identified design or manufacturing similarities. With the formation of product families and cells, the factory layout changes from being process based to being product based, as previously illustrated in Fig. 2.3. The maintenance of a product based layout task might consist of the following steps:

1 *select analysis criteria*;
2 *develop design proposals*;
3 *analyse design proposals.*

The *select analysis criteria* involves choosing appropriate production criteria, such as throughput times, work in progress levels etc, which are used to examine each proposal in the analyse proposal step. This analysis provides a method for determining the most suitable proposal. The *develop design proposals* step involves the generation of feasible product families and corresponding cells which cater for the production of the families. This procedure may involve allocating a range of new products among existing product families or generating a series of new product families from the range of products being manufactured. The procedure acts as a decision support system to help manufacturing personnel to decide on suitable product families and cells.

During the development of proposals for developing product families and cells, it is necessary to be able to develop a number of different proposals. According to Chang and Wysk (1985) it should be possible to generate several bills of process for each product, because a bill of process is not unique. Choobineh (1984) argued that by using a selection of different bills of process, a range of different product family proposals should be developed to cater for fluctuations in demand by allowing more flexibility in the loading of cells. Each proposal contains information on product families, cells and the assignment of product families to particular cells. Each proposal is passed to the *analyse design proposals* step in order to evaluate it.

The *analyse design proposals* step seeks to determine the most suitable proposal from the range already developed in the develop proposals step. The analysis stage assesses each proposal from the perspective of each of the selected criteria. Thus the greater the variety of criteria, the more detailed the evaluation of each proposal. Since the chosen proposal determines product families and cells for the entire factory, it must satisfy the overall production goals for the factory, rather than the goals of a particular cell.

The process behind the maintenance of a product based layout should be considered as an iterative one. If a proposal is accepted, the new data is passed to the relevant production personnel and the products and cells are organized accordingly. The evaluation may indicate that another proposal should be developed, because the feedback from the analysis has indicated certain problems with the current set of proposals. Alternatively the evaluation may be weak in the analysis of certain production characteristics and the best option is to add extra criteria in order to fully assess those particular production characteristics. The analysis is a very important part of the maintenance task, because any excessive capital investment, uneven distribution of products, low utilization or excessive queuing times can be identified, analysed and effective action taken. The success of a product based layout also depends on certain characteristics of the production process such as set-up times, materials handling issues and work in progress buffer levels. For example large set-up times can increase the production lead time for a product and cause increased levels of work in progress.

The final procedure involved in the production environment design task is responsible for analysing the manufacturing system, in order to detect such potential problems.

Analysis of a manufacturing system The main function of the analysis of a manufacturing system is to present information on various characteristics of the production environment such as set-up times, throughput times and quality levels. The main purpose of this analysis is to pinpoint particular areas of the production process, where there is room for potential improvement. In each type of production environment, different categories of information are filtered and examined. For example, in a health care products production environment, product quality data is considered important and the monitor must track and maintain a record on the progress of each batch through the system. In the analysis procedure three steps may be taken:

1 *Collate information.* All the necessary information is firstly gathered by each of the PAC monitors and then it is collated by the FC monitor.
2 *Analyse alternatives.* Using various theories relating to manufacturing systems design, manufacturing personnel identify potential causes for any production problems from the collated information.
3 *Develop diagnosis.* Based on the analysis if alternatives, a proposed solution for solving a particular problem is developed.

The manufacturing systems analysis is based on the notion of continuous improvement to a production process, to simplify the control task. Schonberger (1982) places a strong emphasis on simple factory configurations by arguing that by simplifying the process, the products will flow more efficiently. The control task is relatively easier when operating in a well organized factory configuration, because there is less variability and more stability in such a structured production environment. In fact the greater the effort expanded in Production Environment Design (PED) and the degree to which the resulting cells are independent, the simpler the control task within FC becomes.

The thinking behind the manufacturing systems analysis is one of continuous pursuit of excellence. We feel that initially the analysis should be regarded as a filter and medium for presenting information on the performance of a manufacturing system, from which the manufacturing personnel can decide which particular areas merit further attention. The analysis could be developed, using the expertise and experience of the manufacturing personnel perhaps encoded in a decision support system. This type of support system not only presents information, but recommends possible courses of action for solving the problems. Manufacturing personnel can then choose to follow the recommended course of action or to use other approaches based on their own experience.

Overview of the PED task The approach within the PED task is as follows. After the initial establishment of the product based layout, new products are integrated within the existing families, a standardized process planning procedure is used to develop a selection of process plans for each new product, and the performance of the manufacturing system is continuously analysed, so as to ensure that the benefits of product based manufacturing are maintained. The integration of new products continues until there is a complete change in the product range or the families grow so large that the benefits diminish. If either of these situations occur, new families and cells have to be established.

2.3.2 The control task

The production environment design task ensures that an efficient product based manufacturing system is maintained within a manufacturing system, which involves the use of product families and autonomous work groups (Chapter 7). Burbidge (1975) argued that product based manufacturing facilitates the distribution of responsibility for the production of a family of products to each work group. With this distribution of responsibility the manufacturing system is better equipped to deal with any production problems or fluctuations in demand. Beckhard (1969) recognized that a potential problem with the distribution of responsibility is the possibility that each work group may not consistently follow the organizations' goals and he recommends the use of guidelines to coordinate the activities of each work group. Therefore, in our view the main purpose of the *control task* is to coordinate the activities of each PAC system through the provision of schedule and real time control guidelines, while recognizing that each PAC system is responsible for the activities within its own cell.

The time horizon for coordination of the flow of products by the *control task* varies depending on the manufacturing environment. However it is influenced by the the time horizon of the master production schedule (MPS), since the goal of the control task is to satisfy the production requirements and constraints set by the MPS. The *control task* involves developing schedule guidelines using a *factory level scheduler,* implementing these guidelines and providing real-time guidelines for each of the PAC systems using a *factory level dispatcher* and monitoring the progress of the schedule using a *factory level monitor.* The FC control task is in many ways, a higher level recursion of a PAC system (Fig. 2.5).

As in PAC, an important characteristic of the control problem within factory coordination is *combinatorial explosiveness.* Fox (1983) defined *combinatorial explosiveness* as the exponential growth in the number of possible schedules along each dimension such as operations, machines, tools, personnel, or orders amongst others and the presence of unexpected events which can thwart any planned solution We feel that the problem of combinatorial explosiveness may be minimized through :

- good manufacturing systems design;
- choosing the correct level of problem formulation;
- effective use of constraints and assumptions.

Good manufacturing systems design as we have seen involves the creation, where possible of a product based layout, and the introduction of programs

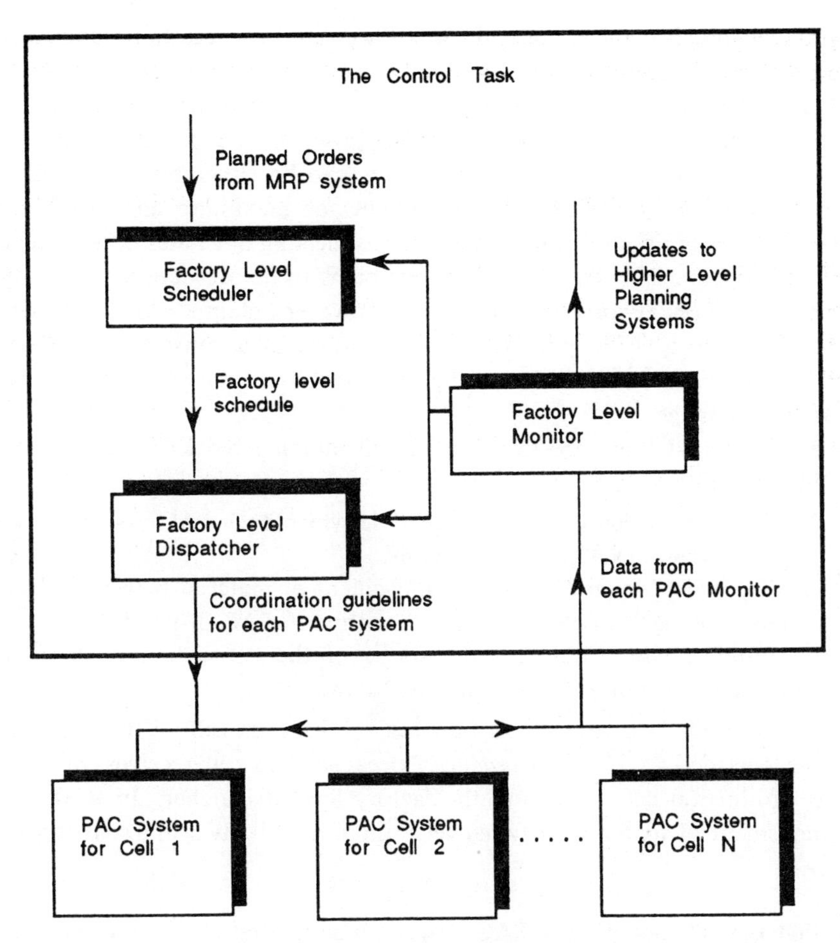

Fig. 2.5. Data exchange between the control task of factory coordination and a number of PAC systems.

for set-up reduction, good manufacturing practice and the use of minimum batch sizes within a manufacturing system. The end result of these practices is the simplification of the process and the reduction of variability in the process. With increased predictability in the production process there is more flexibility to deal with fluctuating demand. This increased predictability allows production personnel to organize the production system to cater for the variations in demand without a major overhaul such as the transfer of products to other resources, or the transfer of personnel to different tasks.

Rickel (1988) argued that by *formulating and solving a higher level problem,* the range of feasible solutions for lower level problems is reduced. Within the control task, we are using a similar approach by developing a factory level schedule which all the PAC systems on a lower level can use to organize their own production activities.

By *identifying constraints or using assumptions* to bound the space of possible solutions, the complexity of the planning problem is reduced. Smith and Fox (1984) recognized that these constraints can be either a series of casual restrictions, such as precedence requirements or maximum allowable times and/or a series of organisational goals such as meeting due dates or minimizing work in progress. The use of assumptions helps to decide between a number of feasible choices at some decision point. However assumptions are retracted if they lead the decision process to the violation of a constraint or a poor solution. With a reduction in the range of possibilities, the coordination task is easier to manage.

We believe that there are three individual building blocks associated with the FC control task :

1 A scheduler which develops a schedule which each PAC system uses as a guide when developing its schedule.
2 A dispatcher which controls the movement of material between cells on a real time basis and communicates with each PAC dispatcher.
3 A monitor which observes the status of the entire factory based on information coming from each PAC monitor.

In addition to these three basic building blocks, each cell on the shop floor may be regarded as a *virtual producer,* because each cell receives guidelines on its production activities from the factory level dispatcher. In relation to the movement of materials *between* cells, there may be two different types of *mover.*

The first type of mover The PAC mover which is used to organize materials handling *within* the cell, can also coordinate the materials movement *between* cells, provided that the material handling between cells is a simple task. Information on the next cell on a product's process routeing can be given to the mover by the particular PAC dispatcher of the cell which the batch is leaving.

The second type of mover A factory level mover, which operates on the same principles as the PAC mover, except that it is concerned with organizing the materials flow *between* the cells on shop floor. The factory level mover receives all its instructions from the factory level dispatcher. In a shop floor, where there is a variety of materials movement between cells, there may well be a requirement for a factory level mover.

In relation to the movement of materials between cells, we will use the first type of mover. We shall now describe the basic building blocks of the control task in factory coordination: scheduler, dispatcher and monitor.

The scheduler The FC scheduler is concerned with *predicative scheduling*, which Bowen *et al.* (1989) describe as 'the planning of work for an upcoming period to best optimise the system as a whole'. Ideally, the scheduler develops a plan to coordinate the flow of products from cell to cell within their final production due dates. However the plan cannot handle any unpredictable events which may occur, such as resource breakdowns or component shortages, which are the responsibility of the factory level dispatcher. The scheduler may have different strategies for developing schedules, depending on the production environment and the efficiency of its design. A production environment might be classified in terms of a job shop, batch manufacturing or a mass production environment. For example in a production environment which is distinguished by a large production volume and a standardized product range, the process is predictable with little potential for variability. In such an environment, a factory level schedule can be very detailed in its guidelines to each of the PAC systems. This is contrast to a production environment, where there is a large variety of products being manufactured with fluctuating production volumes and the potential for variability and unpredictability is greater. Here the corresponding factory level schedule provides guidelines for each of the PAC systems, and allows for the increased variability by including less detail, and placing greater emphasis on the autonomy of each cell.

There are two approaches to developing a factory level schedule ; infinite scheduling or finite scheduling. Vollmann *et al.* (1988) draw a distinction between infinite and finite scheduling techniques:

- *Infinite scheduling* produces a schedule without any consideration for each resource's capacity or the other batches to be scheduled. The assumption is that each resource has infinite capacity.
- *Finite scheduling* produces detailed schedules for each product with start and finish times. With finite scheduling, each resource has finite capacity. Therefore full consideration is given to the resource's capacity and the other batches being scheduled.

It is clear that MRP type systems use infinite scheduling techniques, while finite scheduling techniques are more appropriate for PAC systems. In so far as is feasible, we suggest that finite scheduling techniques may be used with the FC scheduler.

The planning procedure within the scheduler allocates product batches to each working period based on the overall requirements dictated by the master production schedule and detailed through the requirements planning process. The span of the working period is dictated by the particular planning horizon used in each production environment. Rickel (1988) suggested that the planning horizon used by a scheduler is dependent on the production environment and the planning horizon of the master production schedule.

We envisage that the scheduling procedure for FC would observe the following steps:

- *select appropriate criteria*;
- *allocate the requirements*;
- *develop a schedule*;
- *analyse the schedule.*

The *select appropriate criteria* task is similar to the selection of analysis criteria in the procedure for maintaining a product based layout. Criteria such as number of daily/weekly finished batches, average throughput times, average work in progress levels and average resource utilization, are selected to be used in the later *analyse the schedule* task. Chryssolouris (1987) recognized the importance of using a multiple criteria analysis, when he argued that a methodology involving only one criterion 'assumes that the manufacturing environment is static when in fact, priorities, requirements and conditions are constantly changing'.

The *allocate the requirements* task involves the transformation of the requirements, stated in the master production schedule and detailed through the requirements planning procedure, into well-defined batch orders with due dates spread throughout the time horizon. For example using the production smoothing approach to planning in Just in Time systems, the batch orders are averaged evenly throughout the time horizon. In other types of production systems, the spread of orders may not be so even, with particular product requirements being fulfilled, before production of other products commence.

The *develop a schedule* stage involves developing a schedule proposal to suit the flow of products in a particular environment, which is analysed for its feasibility in the *analyse the schedule* step. The main data inputs to the *develop a schedule* step include data on all resources and products within the factory. This step may also receive feedback from the factory level monitor, regarding the development of a new schedule, because of some problems on the shop floor. The schedule proposal takes account of the production due dates of the required batches, in determining coordination guidelines to enable the product flow to meet these due dates.

The *analyse the schedule* task examines each proposed schedule in relation to its suitability and efficiency, with respect to a series of production criteria. The use of a range of criteria helps to ensure that the schedule proposals are analysed from the perspective of the overall production system. The schedule analysis can be completed using simulation techniques, operations research techniques, or artificial intelligence techniques. From the analysis, feedback can be given to the *develop proposal* step, which can be used to generate better schedule proposals.

Once a schedule proposal is accepted, it is regarded as a plan which can be used to coordinate each cell with respect to overall production goals. However deviations from the plan may occur because of the occurrence of unexpected events and of tradeoffs to decide between conflicting requirements and goals, when the proposal was being developed. As in PAC, it is the function of the dispatcher to implement the schedule and to manage any of the unexpected events through real time control at a factory level.

The dispatcher The factory level dispatcher implements the factory level schedule by passing the relevant guidelines relating to each cell to the appropriate PAC system. In relation to the provision of real-time guidelines, the factory level dispatcher only provides such guidelines when it can make an effective contribution to a problem arising in a manufacturing environment. Such guidelines include changes in product priorities and/or process data because of fluctuating production requirements and the regulation of the flow of raw material and work in progress stocks between cells.

In order to have a detailed picture of the production environment, the dispatcher uses dynamic and static data. Bills of process for each product, and the factory level schedule are examples of static data used by the dispatcher. Examples of dynamic data include the location of each work in progress batch, the status of the inter cell transportation system and the status of the input buffer of each cell. The dispatching task can be carried out:

1 by a production manager using his/her own expertise;
2 by a production manager, who uses distributed software systems containing rules and other forms of intelligence in a decision support system;
3 by a distributed software system, which automatically controls the factory, using various forms of intelligence such as dispatching rules.

The dispatcher task might be summarized in the following general procedure:

1 *receive information;*
2 *analyse the alternatives;*
3 *relay decisions.*

During the *receive information* step, the dispatcher gathers all the necessary dynamic data from the factory level monitor and static data from the manufacturing database and the factory level scheduler. Once the information is collated, the dispatcher has a detailed picture of the shop floor. With this knowledge, the dispatcher can develop feasible solutions to problems that may occur in the *analyse the alternatives* step. After selecting the most suitable option, the final step *relay decisions* is carried out, which passes the relevant information for a solving a particular problem to the relevant PAC systems.

The factory level dispatcher has control over the entire manufacturing system in a factory. Thus the real time scheduling task is open to a wide range of possible problems and solutions. This wide variety of problems and solutions makes the dispatching task potentially quite complex. *In order to reduce the complexity of the dispatching task, factory coordination places a strong emphasis on production environment design and factory level scheduling and only deals with problems which are outside the scope of the individual PAC systems.* This ensures that each PAC system has full responsibility for all problems within the individual cells.

The dispatcher through passing on the schedule guidelines to each PAC system and carrying out any necessary real time scheduling, ensures that all cells are coordinated to satisfy the overall goals for the production system. For example if a resource breaks down within a particular cell, that cell may become overloaded. Therefore the cell's PAC system can use the FC dispatcher to select an alternative cell to handle some of its excess production or to alter the priorities of some associated batches.

Another task of the factory level dispatcher is the regulation of the flow of work in progress and raw materials moving into and out of each cell. As batches move between cells through the factory, the dispatcher directs the movement using the intercell transportation system. The levels of raw materials in the central storage point of each cell are monitored by the FC monitor and if they go below a certain threshold level, the factory level dispatcher will notify the raw materials storage points in the warehouse to replenish the cell's central storage point. The movement of raw materials *within* the cell is the responsibility of the PAC dispatcher and mover.

In summary, the factory level dispatcher task involves :

- issuing schedule guidelines to each of the PAC systems from the factory level schedule;
- completing any real time scheduling concerning changing due dates or alternatives routes in other cells as the batches move through the process;
- ensuring that the batches follow the correct production sequence and that sufficient quantities of raw materials are available at each cell.

The monitor We envisage the factory coordination monitor as having two tasks: the provision of accurate reports to management and the delivery of timely and accurate data to both the scheduling and dispatching functions and higher level planning systems. With access to accurate data from the monitor, the scheduling and dispatching activities may function more efficiently. One of the data exchanges with the scheduler may involve a request for a new schedule to be developed, because of some problem on the shop floor. From the data provided by the monitor, the dispatcher is aware of the current status of the shop floor (e.g. work in progress buffer levels, raw material stock levels etc). The reports to management summarise the production situation throughout the factory by filtering relevant data from each PAC monitor and compiling an overall picture of the performance of the manufacturing system. More detailed data concerning measurements inside each cell can be found in the reports provided by the particular PAC monitor.

The efficiency of a factory level monitor can be assessed by the convenience and simplicity of its operations for manufacturing personnel, and the promptness and accuracy of the information that the factory level monitor generates. As with the PAC monitor, the factory level monitor procedure might involve the following four steps:

1 *capture the PAC data*;
2 *analyse the PAC data*;
3 *provide decision support*;
4 *provide historical reports.*

The *capture the PAC data* step involves filtering and condensing data from each of the PAC monitors and gathering it into an organized form. According to previously identified monitoring reference goals, the data is then analysed in the *analyse the PAC data* step and any necessary reports are prepared for management. The monitoring reference guidelines help identify different types of data, which should be analysed and presented by the monitor. For example in a production environment producing high quality products, the monitor could be instructed to provide all data on the quality levels of each batch.

In the *provide decision support* step, higher level planning systems and the factory level scheduler and dispatcher are kept fully informed of all activities throughout the factory through data generated by the monitor. This data feedback helps the dispatcher to carry out its real-time scheduling task. The data feedback to to the scheduler may include a request for a new factory level schedule. The factory level monitor also provides relevant data for higher level planning systems. For example, the factory level monitor can provide regular updates on lead times to an MRP type system. The regular update on lead times enables the MRP system to produce more accurate plans for materials requirements. The *provide historical reports* step involves providing reports to management, describing the performance of the factory in relation to different criteria such as fulfilment of product orders or product quality.

The filtering and condensing of data by the factory level monitor is important in order to avoid an overabundance of irrelevant information. Ackoff (1977) maintained that an explanatory model of the decision process within a production environment can assist in determining which information to provide to each department. This decision process model articulates the different data requirements for each department within an organization and how they influence the activities in other departments. For example when the sales department gives a commitment to a customer for an order at a certain date, it is influencing the production planning task within manufacturing, who have to use the due date set by the sales department, when scheduling the production of that particular order. Problems such as capacity overload can occur, when planning the production of a customer's order using the particular due date set by the sales department. Information drawn from the factory level monitor can assist the sales department in determining a reasonable due date to a customer.

In summary, the factory level monitor should be regarded as a filter for all data coming from each PAC system. Once the data is filtered and analysed, it can be used by:

- the scheduler and the dispatcher in their decision making process;
- higher level planning systems such as MRP systems, through for example more accurate estimates of manufacturing lead times;
- management in assessing the overall manufacturing performance of a factory.

Overview of FC We shall conclude this section on FC by summarizing the FC architecture (Fig. 2.6);

- FC involves a *production environment design* task and a *control* task. By combining these two tasks in a single architecture, we are arguing that by improving the efficiency of the manufacturing environment through the production environment design task, the control task is greatly simplified and the control system can coordinate the flow of products more effectively.
- The *production environment design* task is concerned with reorganizing an product based layout to maintain its efficiency. It uses a selection of techniques involving process planning, maintenance of a product based layout and manufacturing systems analysis.
- The *control* task coordinates the flow of products between cells to ensure that their production due dates are fulfilled with the highest quality standards and lowest costs. This is achieved by passing appropriate guidelines to the individual PAC systems.

2.4 Conclusions

The shop floor control task is an important and complex task in modern manufacturing. In this chapter we have described one approach to managing this control task efficiently which involves FC and PAC systems coordinating the flow of work through a factory. In describing this approach, we have concentrated on what tasks need to be included, rather than how these particular tasks should be carried out.

FC consists of two tasks; a production environment design task and a

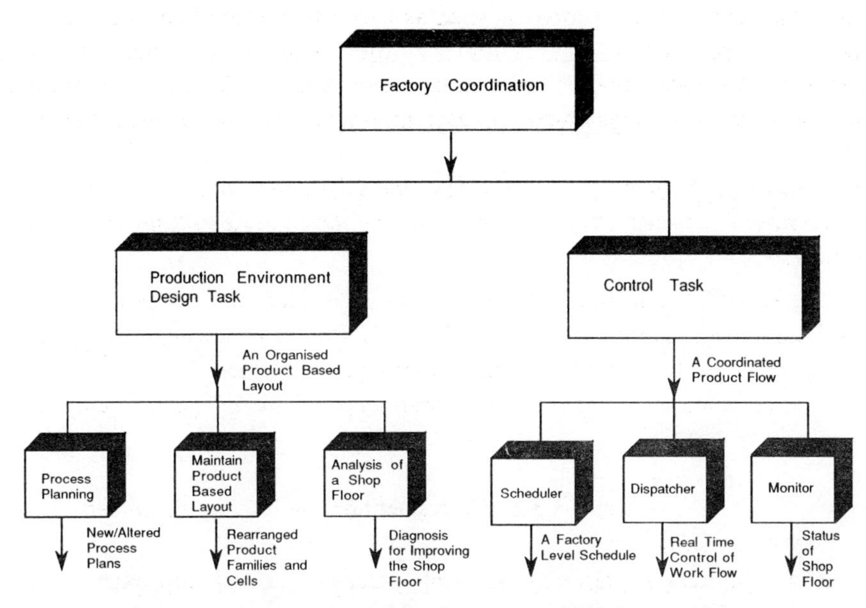

Fig. 2.6. An overall picture of the factory coordination architecture.

control task. The production environment design task is concerned with the reorganization of a manufacturing environment to support product based manufacturing and to ensure the continuous improvement of the manufacturing environment. The FC control task involves using a scheduler, dispatcher and monitor. The scheduler uses different rules to develop a suitable factory level schedule. The dispatcher implements the schedule and is charged with real time control of the workflow within a factory. The monitor provides data on the status of the factory and the progress of the schedule. With these three factory level systems all cells are coordinated to balance the flow of products through a factory. By establishing a relationship between the control and the production environment design tasks, we are trying to organize an efficient manufacturing environment so as to reduce the complexity of controlling the flow of products within the environment.

With the layout of the shop floor being as close as possible to a product based layout, each PAC system has a definite area of responsibility. Using the guidelines from the FC control task, each PAC system controls the workflow *within* each cell. To complete this control task, a PAC system uses a scheduler to develop a schedule based on the guidelines in the factory level schedule, a dispatcher to control the workflow on a real time basis and a monitor which gives progress reports on the schedule. Thus the FC control task and the PAC task are similar, although operating at different levels.

3

A structured functional model for shop floor control

3.1 Introduction

In the last chapter, we described an architecture for a new approach to shop floor control. The architecture consisted of two levels; a Factory Coordination (FC) task operating at factory level and a number of production activity control (PAC) systems each operating at cell level. In this description we adopted a bottom up approach to describing the architecture. This approach started with a description of the PAC task within each cell of the factory. Each of the five basic building blocks of PAC, and their relationship with each other was outlined. Then we went on to describe the next level in the architecture, that of FC. In our description of FC, we recognized the link between the production environment design element and the control element. Then taking each of the two elements in turn we described the individual tasks of each one.

The purpose of this chapter is to provide an alternative perspective of the same architecture using a modelling technique for describing architectures, known as Structured Analysis and Design Technique (SADT). This perspective contrasts with our previous bottom up approach by adopting a top down analysis of the activities of FC and PAC. In our top down analysis, we will start with an overall description of the context of FC and PAC and gradually decompose the context to explain the individual tasks of FC and PAC. The overall structure of this chapter is as follows:

1 We describe SADT, a formal top-down descriptive methodology used to describe a functional model of FC and PAC.
2 We outline an overview of the SADT model of FC and PAC. This discussion highlights the interactions with other production related activities of the manufacturing organization. One of the tasks within this context, we broadly describe as a coordinate factory task. The coordinate factory

task consists of a production environment design element and a control element.

3 The production environment design aspect of the FC task is described.

4 Finally, the coordination tasks, comprising the control tasks within FC and PAC, are described.

It is important to reiterate that the emphasizes of this part of the book is to describe what tasks should be realized within FC and PAC, rather than how these tasks should be realized.

3.2 Structured Analysis and Design Technique

In order to illustrate the functions and concepts of both FC and PAC, we have used a formalized methodology for documenting the architecture of large and complex systems. This methodology is known as SADT*, Structured Analysis and Design Technique, 'which can be used to cope with complexity through a team oriented, organized discipline of thought and action, accompanied by concise, complete, and readable word and picture documentation' (Ross, 1985). As we only give a brief overview of the technique, we refer those readers who are interested in a greater understanding of SADT to the work of Ross (1985) and Bravco (1985).

SADT is a methodology which provides a coherent, integrated set of methods and rules that together form a disciplined approach to analysis and design. The basis of the SADT approach is founded on seven closely interrelated-concepts (Ross, 1985), which are now described.

1. Understanding via Model Building The advantage of building a graphical model of a system is that it can express ideas and relationships with clarity and effectiveness. SADT is applied to a problem by building on paper a model which attempts to express the structure of that which is being modelled.

2. Top Down Decomposition SADT is a structured decomposition, which facilitates the orderly breaking down of a complex subject into its constituent parts. The analysis performed on any problem is top-down, modular, hierarchical and structured. Application of SADT starts with the most general description of the system under study. This description is represented by a box, and it is then decomposed into a number of more detailed boxes, each of which represents a major function of that single box. In turn each of these boxes is decomposed, gradually exposing more detail and information along the way. This structured approach to decomposition is illustrated in Fig. 3.1.

* SADT is a trademark of SofTech Inc.

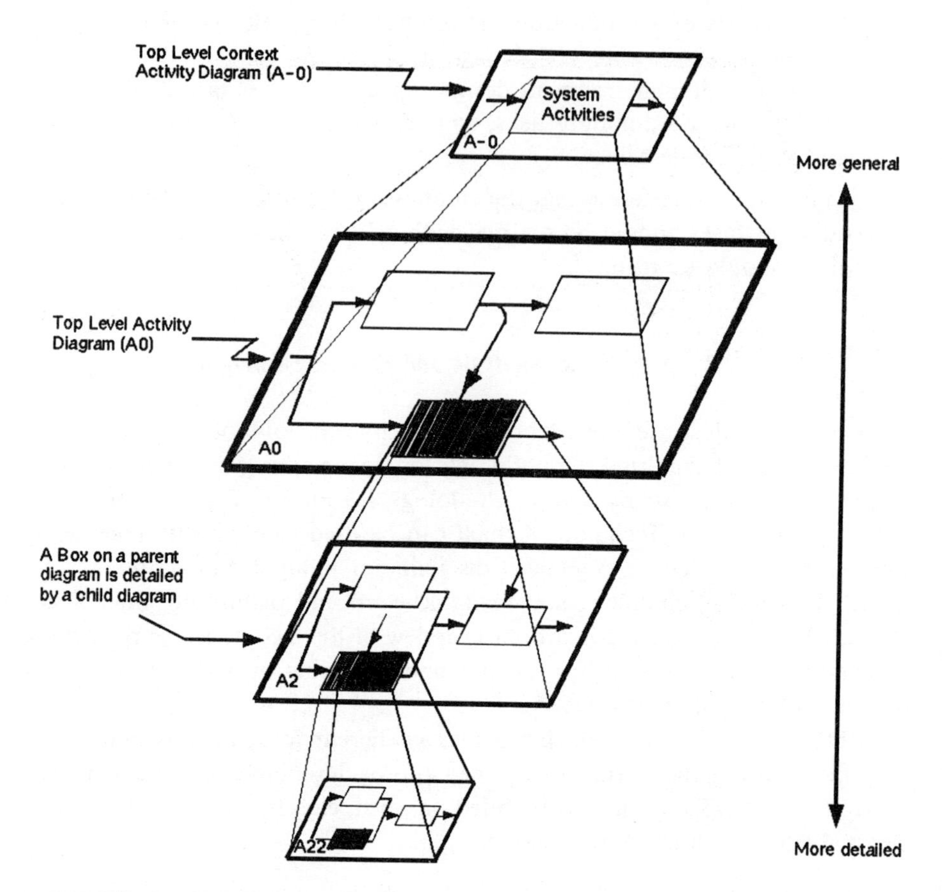

Fig. 3.1. SADT model showing structured decomposition (Ross, 1985).

This top down approach reduces the level of detail revealed at any stage, and the number of boxes that may be used in any one diagram is limited to six. This ensures a uniform representation of successive levels of detail, and reduces the complexity of each individual diagram. Each box in an SADT model is illustrated in precise relationships to other boxes by means of interconnecting arrows. When a box is decomposed into sub-boxes, the interfaces between the sub-boxes are shown as arrows. The title of each sub-box and its interfaces define a bounded context for the detailing of that sub-box. The terms parent and child are used to denote this relationship between a box (parent box), and the diagram (child diagram) presenting the sub-boxes into which the parent is decomposed (Fig. 3.2). In other words, the child sub-box is restricted to contain only those elements that lie within the scope of its parent box. Further, the child box cannot omit elements which are part of the contents of its parent box.

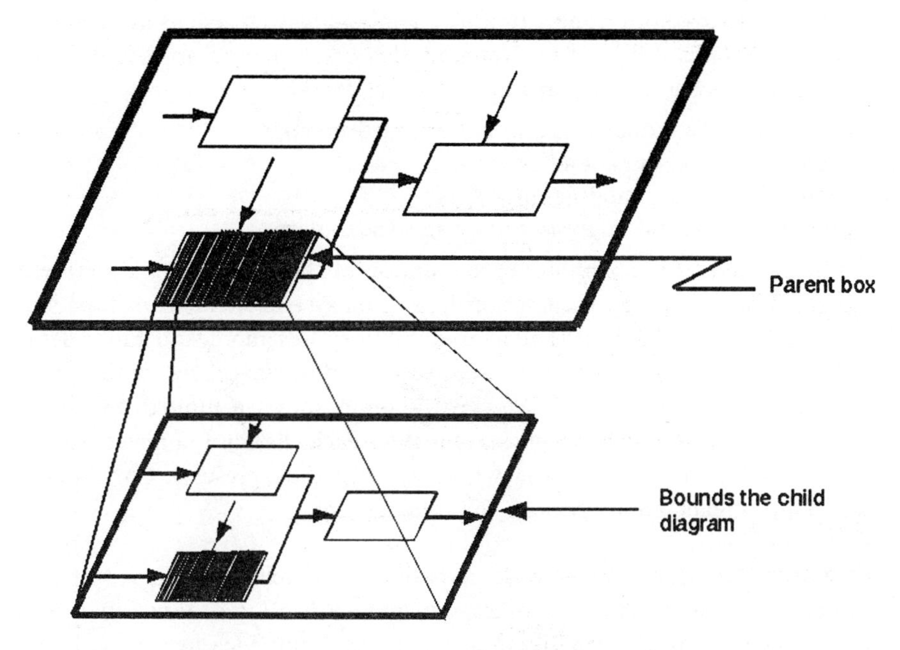

Fig. 3.2. Parent and child relationship (Ross, 1985).

3. Functional Modelling versus Implementation Modelling A functional model is always the starting point of an SADT study. This is a representation of what the problem is, rather than how the problem will be solved or the solution implemented. The purpose of this procedure is to ensure that a problem is fully and clearly understood before the details of the solution are decided. When the functional model has been developed, it is then used to start development of an implementation model.

4. Dual Aspects of a System SADT attempts to understand and describe a problem in terms of two major aspects; activities and data. These aspects are usually examined together, but at any one time the emphasis is on one of them. Thus, an SADT model may have two parts, an activity decomposition and a data decomposition. The activity decomposition details the happenings, such as activities performed by personnel, machines or software, (as activity boxes), while illustrating the things that interrelate them (as data arrows). The data decompositions details the 'things' of the system, such as documents or data, (as data boxes), and shows the activities of happenings that interrelate them (as activity arrows). Within our model of PAC and FC, we only use activity modelling.

5. Graphic Format of Model Representation Analysis and design results are difficult to express both concisely and unambiguously when using a natural language. SADT is designed to overcome this obstacle, and an SADT model is represented in a graphical language in order to:

(a) Expose detail gradually and in a controlled manner;
(b) encourage conciseness and accuracy;
(c) focus attention on module interfaces;
(d) provide a powerful analysis and design vocabulary.

An SADT model is a graphic representation of the hierarchical structure of a system, clearly revealing the relationships of all system elements or functions. This model is structured so that it gradually exposes more and more detail. An SADT model is an organized sequence of diagrams, each with concise supporting text. A high level overview diagram shows a limited amount of detail about a well bounded subject. Further, each diagram connects exactly into the model to represent the whole system, thus preserving the logical relationship of each component to the total system.

6. Support of Disciplined Teamwork Analysis of complex systems requires the coordination of the creation, modification, and verification of functional specifications. These steps require disciplined and coordinated teamwork. The ideas of project personnel must be communicated at every step and analysis level to ensure that the SADT models produced reflect the best thinking of the team. Therefore using SADT provides a team of developers with a standard methodology which can be understood by all, thereby reducing misunderstanding.

7. All Decisions and Comments in Written Form SADT methods require that all analysis, design decisions and comments are in written form, and it includes procedures which retain written records of all decisions and alternative approaches as they unfold during the project.

SADT activity diagrams are composed of boxes, and arrows connecting the boxes. A rectangle is the only shape of box used, and the name of the activity is written inside the box as its title. On activity diagrams (which are used exclusively in this chapter), the title must be a verb. Arrows are used to connect boxes on a particular diagram, and the four sides of a box are assigned a specific meaning which defines the role of the arrow which may enter or leave that side of the box. Fig. 3.3 shows the four types of arrows. These are input, control, output and mechanism. The first three arrows are called interface arrows, while the mechanism arrow is known as a support arrow. These conventions are strictly obeyed on all SADT diagrams, e.g. an arrow never leaves the left side of a box, and the arrow entering the left side of a box always represents an input etc.

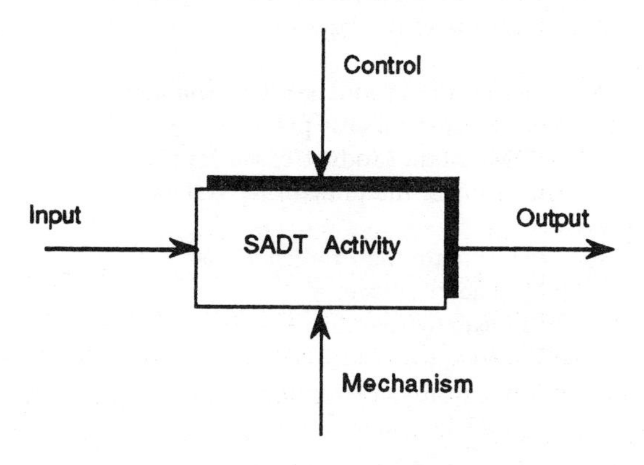

Fig. 3.3. Input, output, control and mechanism of an SADT box (Ross, 1985).

In the case of activity diagrams, the interfaces between activity boxes are normally data items, since data is commonly passed back and forth between activities. The mechanism arrow is an exception, since it represents the mechanism necessary to realise the box (i.e. it could be a person who carries out a specific activity). There is a basic difference between an input, which (in the activity view) is converted by the activity into the output, and a control, which rather constrains how the activity functions in converting the input into the output. The control thus serves a completely different role than the input, and this distinction is important to the basis understanding of the system's operation.

SADT has a referencing system which is based on the hierarchy represented by the diagrams (as illustrated in Table 3.1). This hierarchy is expressed by the assignment of a *node number* to each activity diagram. By convention, the top-level context diagram (only one box) has node number A-0, and its child (the top-level diagram) has node number A0, and its children are numbered A1, A2, etc. All remaining boxes in a model are assigned node numbers as follows. Each box on a diagram contains a *box number* written in its lower right hand corner. Each diagram has a node number in its lower left hand corner derived from its parent diagram and the box which it details. The diagram node number consists of the parent diagram's node number followed by the box number.

After having described some of the principles of SADT, we shall now describe the SADT model of the FC and PAC in the section following.

A-1 Context for Coordinate Factory
A-0 Coordinate Factory (Context Diagram)
A0 Coordinate the Factory

A1 Design the Production Environment
  A11 Develop process plans
  A12 Maintain product based layout
  A13 Analyse manufacturing system

A2 Coordinate the Product Flow
  A21 Schedule Factory
  A22 Dispatch Factory
  A23 Move between Cells
  A24 Control Cells
    A241 Schedule Cell
    A242 Dispatch Cell
    A243 Move
    A244 Produce
    A245 Monitor Cell
  A25 Monitor Factory

Table 3.1. *Node index table for FC and PAC*

3.3 Overview of SADT model for FC and PAC

In our overview of the SADT model, we shall initially look at the overall context of a shop floor control system (Fig. 3.4). In the sections following, we shall gradually decompose a section of the context diagram to describe the FC and PAC tasks. This context consists of three elements:

Coordinate Factory This task involves the maintenance of a shop floor to ensure that it operates efficiently and the organization of product flow on a shop floor to ensure that all production requirements are fulfilled.

Reconfigure Factory The purpose of this task is to rearrange the shop floor layout. This may occur if a shop floor is being changed to support product based manufacturing or if a range of new equipment or products are being introduced.

Produce Goods This task involves the actual production of materials into final products for fulfilling customers orders.

The coordinate factory task provides instructions to the produce goods task which ensure that all products are manufactured to a high quality standard

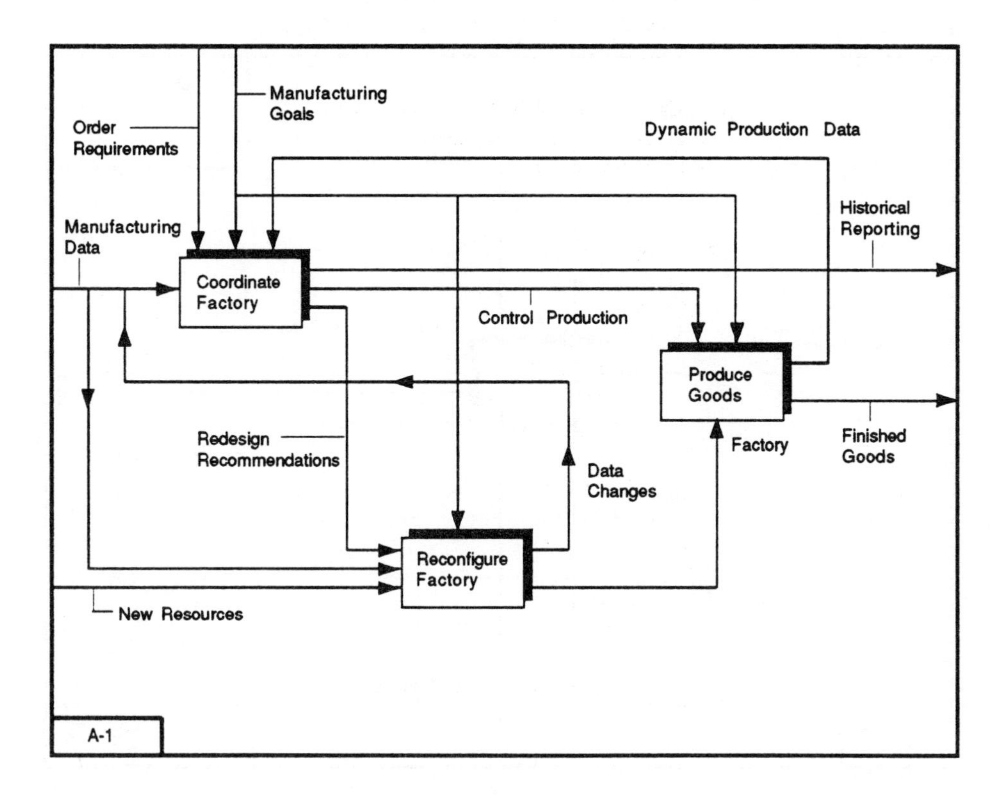

Fig. 3.4. A-1: Context diagram for coordinate factory.

and in time to meet their due dates. The reconfigure factory task provides the resources which enable the produce goods task to operate. By organizing the product flow on a shop floor, the coordinate factory task may provide recommendations to the reconfigure factory task, in order to ensure that the shop floor is configured efficiently. We shall now focus on the coordinate factory task and break it down into its constituent parts to explain FC and PAC.

Coordinate Factory The SADT representation of the coordinate factory activities is displayed in Fig. 3.5, and here the boundaries at the operational level of the PMS hierarchy are highlighted, along with the inputs, outputs, controls and mechanisms. We will now discuss each briefly.

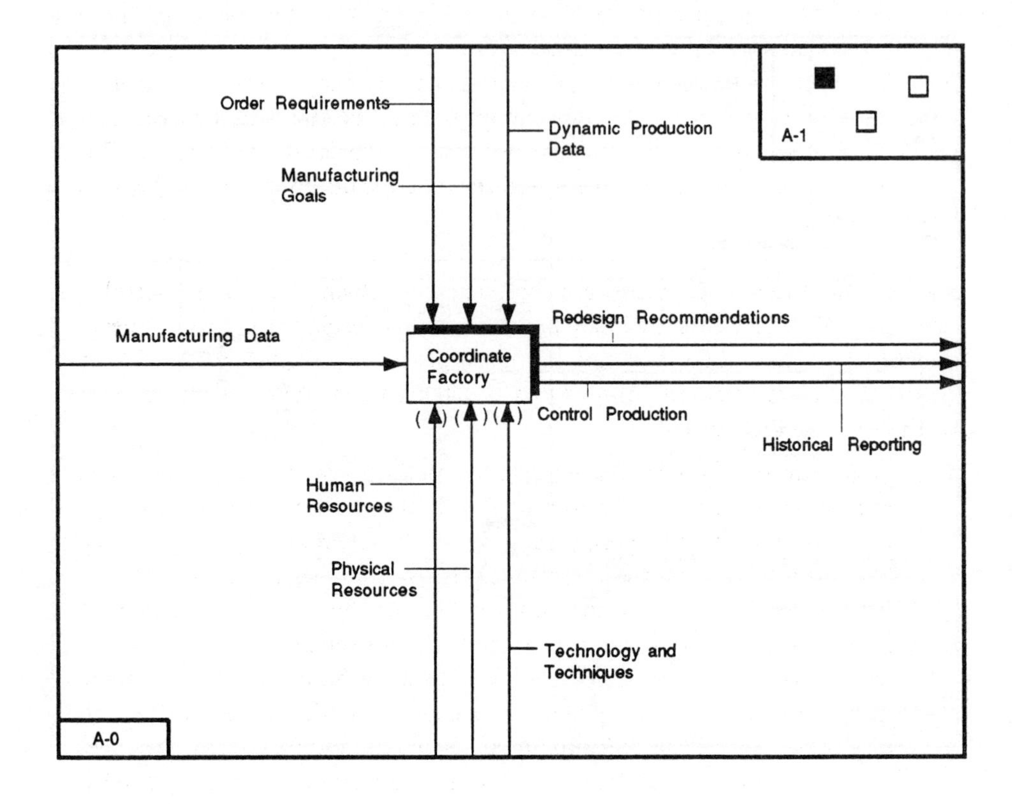

Fig. 3.5. A-0: Coordinate factory.

Inputs

1 Manufacturing data, which is composed of static data and dynamic data. The static data basically describes how to manufacture and assemble the products, and includes data on process times, product structures, product routings etc. The dynamic data describes the current state of the manufacturing environment, with information on work in progress levels, quality and performance measures etc.

Control elements

1 Order requirements from a higher planning system of the PMS (typically an MRP type system);

2 Manufacturing goals, which are ideals set by forward thinking management on measures the company should adopt in order to achieve

manufacturing excellence. These might include high operator satisfaction, low work in progress, flexible and responsive manufacturing, total quality control, and timely production.

3 Dynamic production data, received from the shop floor and includes important quasi real-time information such as the status of workstations, work in progress levels and the progress of the production plans. This data is essential for intelligent and informed shop floor control decision making.

Mechanisms The mechanisms are the means by which shop floor control is achieved. Note that the enclosed brackets around each of the arrow mechanisms imply that these will not be decomposed any further in the SADT model. There are essentially three types which exist to varying degrees in each of the FC and PAC tasks:

1 Human resources, which Deming (1986) describes as 'the company's most important resource'. Referring to the need for a new type of management, Skinner (1985) poses the question 'what kind of managers and managements are needed to become positive rather than negative factors in bringing the factory of the future to bear on our industrial problems ?' He then suggests prescriptive characteristics for the men and women who will be needed, which include having a strong technological base, a broad view of the different business functions, good interdisciplinary skills, and a positive view of the inevitability of change in manufacturing.

2 Physical resources, which are the facilities, equipment and tooling needed to carry out the manufacturing tasks.

3 Technology and techniques, which can be used to aid the activities at the operational level. The technology includes: expert systems, relational databases, distributed networking facilities and software to hardware integration. The techniques used include manual and computer based methods such as Gantt Charts, SADT, heuristics, algorithms, simulation tools, etc.

Outputs The three outputs in the diagram show the types of information which emanate from the FC system, namely:

1 Historical reporting, which is an essential facility that enables management to obtain a clear picture of the overall performance of the production system. This type of information can be used to pinpoint production problems, and assess overall production trends.

2 Control Production is ultimately the instructions issued to the production facility detailing where and when to commence operations on particular jobs.

3 Redesign recommendations, provided by the production environment design task of FC, which may suggest possible beneficial reorganizations of product flow for manufacturing cells within the production facility.

Thus far, an overall attempt has been made to described the operational issues of shop floor control at the highest level (i.e. A-0 coordinate factory), and with the aid of SADT, a model showing the relationships between the different factors which influence shop floor control has been outlined. As we continue the discussion of the FC and PAC tasks, we shall be looking in more detail at each task. Table 3.1 summarizes the main elements for the SADT model of FC and PAC and how they are interrelated.

3.4 A0: Coordinate the factory

This task is concerned with the organization of the layout of the shop floor to support product based manufacturing and the coordination of the flow of products throughout a factory so as to fulfil the production requirements as stated in the master production schedule and detailed through an MRP type system. The data requirements to complete this task include:

1 Production requirements which includes details on due dates and order quantities as stated by a requirements planning system;
2 product data which includes the list of new and current products being manufactured and their respective bill of materials;
3 process data which includes the bill of process for each product and details on all resources contained within a factory.

Looking at Fig. 3.6, we see that the coordinate factory task consists of two main tasks;

1 design the production environment, which is concerned with maintaining an efficiently organized shop floor;
2 coordinate the product flow, which organizes the flow of products throughout a shop floor. This control task is organized on two levels ; the factory level and the cell level.

Relating these two tasks to FC and PAC, we shall see that FC is concerned with the production environment design task and the factory level section of the control task which organizes product flow between cells. PAC is concerned with the cell level of the control task, which involves organizing product flow within a cell. The complexity of the control task is reduced, by operating within a well organized production environment. By organizing the flow of products between cells, the control task ensures that all production activities within each cell are coordinated with the due dates of the products. We shall now describe the design and control tasks in greater detail.

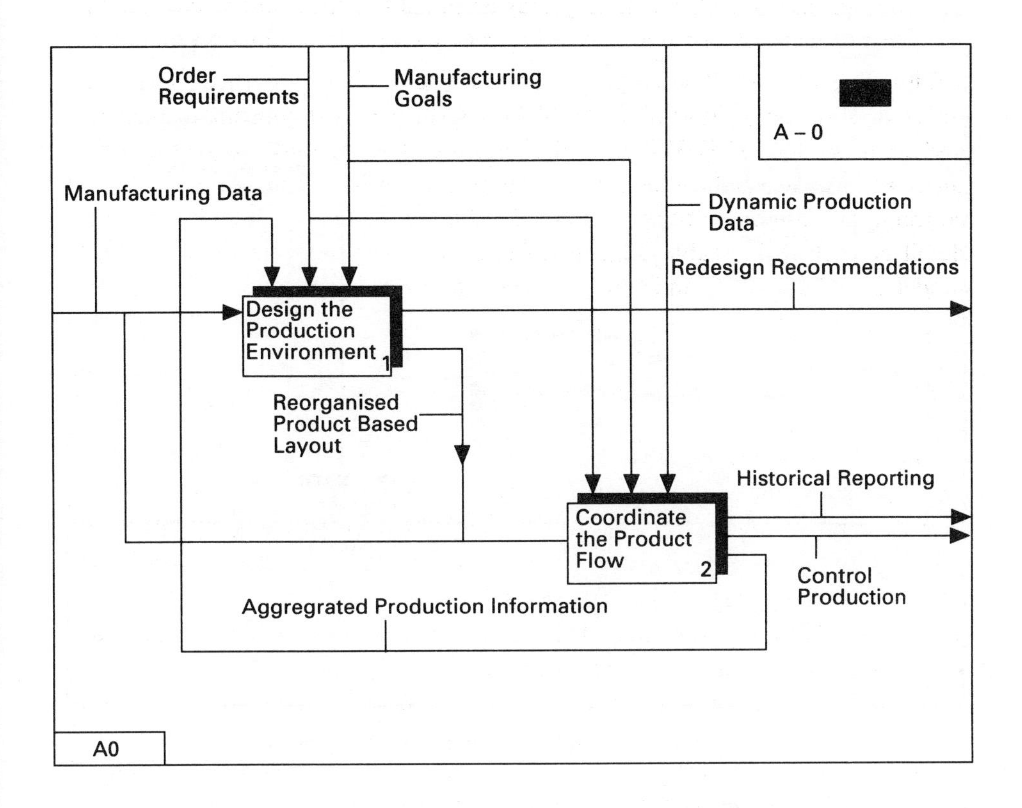

Fig. 3.6. A0: Coordinate the factory.

3.5 A1: Design the production environment

The production environment design task allocates new products to existing product families and completes any necessary rearrangement of a layout to ensure that that a shop floor performs efficiently and supports product based manufacturing as much as possible. We expect that the production environment design task may have to compromise with some manufacturing environments and maintain a less than ideal product based layout. There is an important distinction between the production environment design task and the reconfigure factory task. Any task which is concerned with the initial design and introduction of product based manufacturing to a shop floor is dealt with by the reconfigure factory task, as illustrated in Fig. 3.4. Subsequent reorganizing of the product based layout after it have been introduced, is part

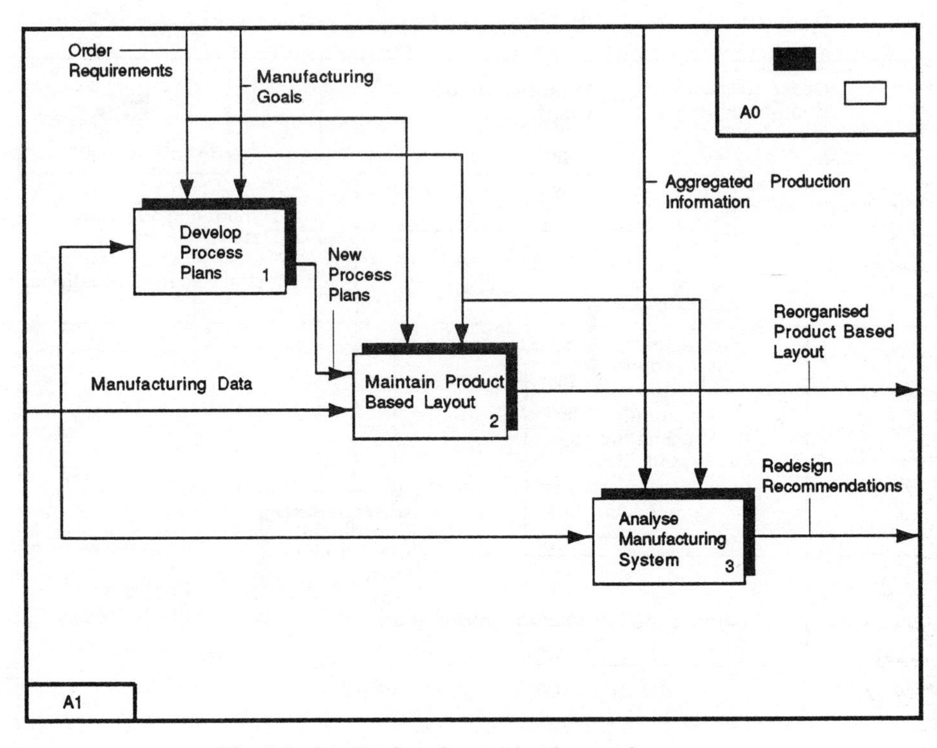

Fig. 3.7. A1: Design the production environment.

of the production environment design task. When reorganizing the production environment, production personnel may use three procedures (Fig. 3.7):

1 A11: develop process plans;
2 A12: maintain product based layout;
3 A13: analyse manufacturing system.

In the develop process plans step, production personnel generate process plans for new products and alter existing process plans. The new process plans are based on the general operational requirements for each product. For example a general operational requirement for a product may be a turning operation. The process plan identifies which particular resource should carry out the turning operation.

The objective of the maintain product based layout step is to integrate new products into existing product families or if necessary, create new product families. The methodology used in this step is Group Technology. The procedures in this step allocate new products with minimal disruption to the ongoing manufacture of products.

In the analyse manufacturing system step, manufacturing personnel identify production problems by analysing data which have been collated from the factory level monitor and then develop solutions to solve the problems. The problems may be caused by process characteristics such as large set-up times, large work in progress levels or high scrap rates. Through this analysis, manufacturing personnel are seeking to improve the efficiency of the manufacturing process.

Production personnel have to complete the design of the production environment within various production constraints while seeking to meet various manufacturing goals. These constraints and goals can limit the range of feasible options in relation to incorporating new products into the production environment. The production constraints can include limitations on expenditure of equipment, a ban on operator overtime or restricted use of certain resources. Increased emphasis on product quality and a reduction of production lead times are examples of manufacturing goals.

A11 : Develop process plans The process planning function is a procedure where new process plans are developed based on the operational requirements for each new product and existing plans are altered to take account of various production changes on the shop floor (Fig. 3.8).

When a new process plan is being developed, one may use a two stage procedure involving:

1 A111 : determine production requirements;
2 A112 : develop plan.

The determine production requirements step involves establishing the process requirements for each process step in the production of a new product. For example, a product may require one side to be milled and a hole inserted on another side. Therefore the operational requirements for the product include a milling operation and a drilling operation. When determining the operational requirements, there is no reference to specific resources within a factory. One of the key constraints in choosing the operational requirements for a product is its design requirements.

This list of operational requirements is used in the develop plan step to generate a process plan for each new product. In the develop plan step, each operational requirement is taken in sequence, and a suitable range of resources which are capable of completing the operational requirement is found. The user may select whichever resource he/she thinks is suitable. Other resources may than be regarded as alternatives, which can be used to alleviate any capacity overload problems that may occur in the future. The procedure for selecting a suitable resource for each operation step can be completed using a decision support system, but is subject to production constraints and manufacturing

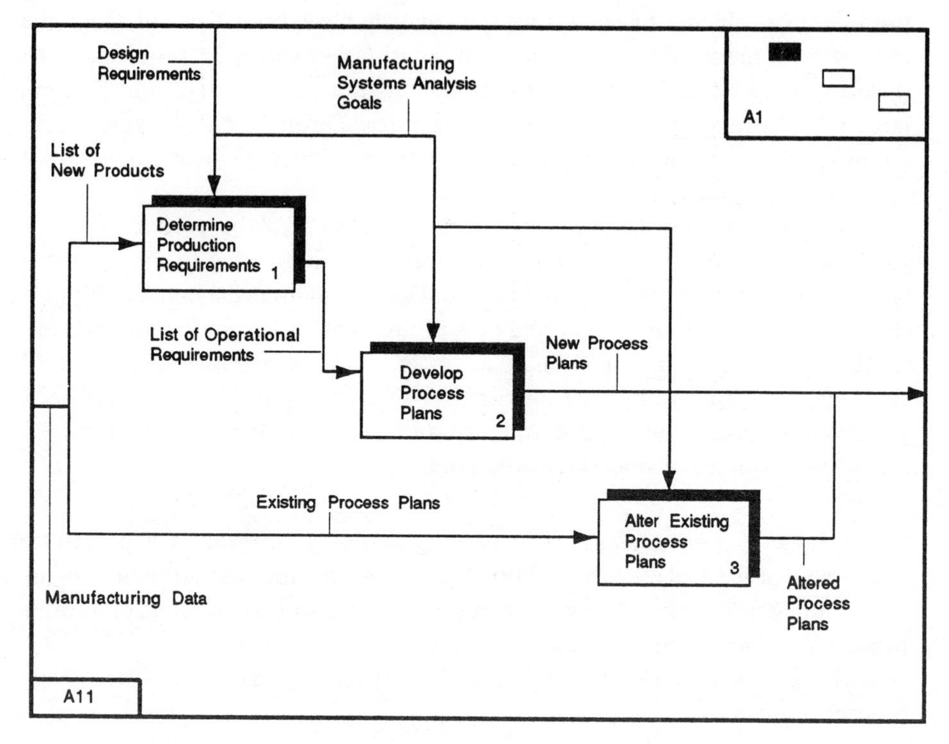

Fig. 3.8. A11: Develop process plan.

goals. In addition to choosing a particular resource for completing each operational requirement, the user may add some details on set-up procedures, quality standards, fixture and tooling requirements.

As the specification for each operational requirement is being completed, the user should be aware of the need to ensure that all operational details increase the efficiency of the operation. For example, several different set-up procedures may be technically suitable for a particular operation, but some of them may involve cumbersome steps which may lead to a large set-up time. Therefore as a process plan is being developed, the user should take note of any feedback from the manufacturing systems analysis task.

The procedure for changing existing process plans consists of one step:

A113 : alter process plans.

The facility to alter existing process plans allows a user to recall an existing process plan and make changes to any of the operation steps in the plan. The alterations may be necessary because of various problems on the shop floor; problems such as a particular resource becoming obsolete and being

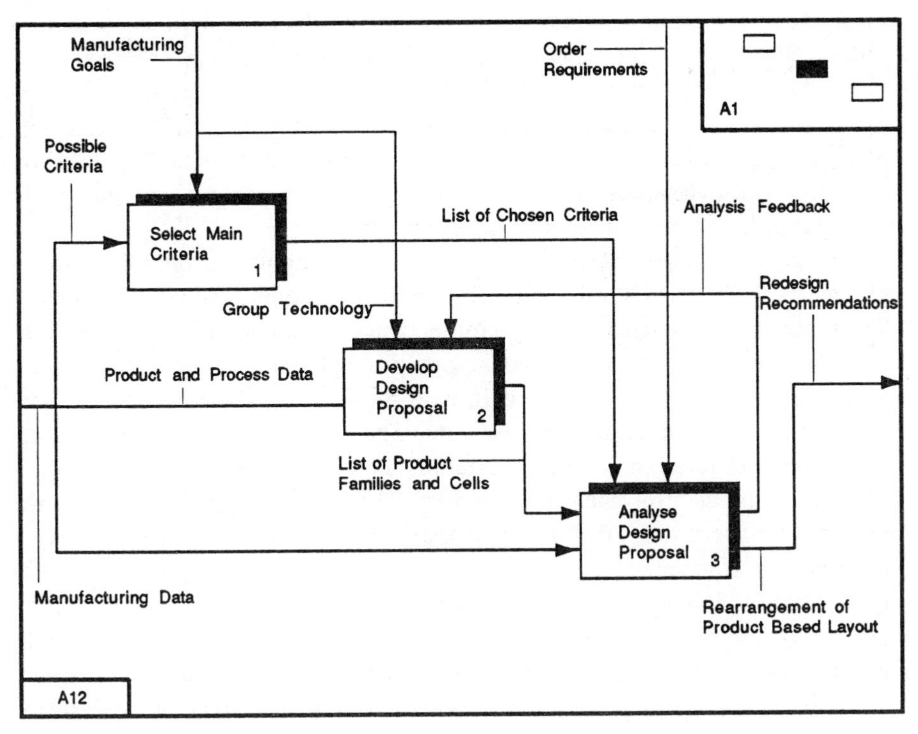

Fig. 3.9. A12 : Maintain product based layout.

replaced with a new resource or a particular resource which is constantly overloaded and some products may need to be diverted from it to smooth out the production load.

The process information in the new process plans may be used in the allocation of new products into product families. This allocation process is discussed in the section following under the task – A12, maintain a product based layout.

A12: Maintain a product based layout The purpose of this module is to reorganize a product based layout to accommodate any new products being introduced. The reorganization involves allocating the new products into existing product families or creating new product families. The maintenance procedure involves three stages (Fig. 3.9):

1 A121 : select analysis criteria;
2 A122 : develop design proposals;
3 A123 : analyse design proposals.

In the select analysis criteria step, the user chooses various criteria such as production lead times or work in progress levels, which are used as evaluation criteria when examining each proposal in the analyse design proposals step later. By choosing a wide range of criteria each proposal receives a detailed examination. The selection of the evaluation criteria may be based on the manufacturing goals of the organization. For example, if an organization wish to reduce the time taken to fulfil a customer's order, then changes in production lead time is a suitable criterion.

The develop design proposals step involves drawing up a proposal involving the rearranging of existing product families to accommodate new products. This procedure is carried out by manufacturing personnel using Group Technology principles. The Group Technology principles can be incorporated into a decision support system. The main data inputs for the develop design proposals step are product and process data. The product data includes a listing of the new products and a bill of materials for each. The process data consists of a bill of process for each new product. These two data inputs provide the material for the Group Technology principles to allocate new products based on similar components and/or similar manufacturing processes. Each proposal is passed to the analyse design proposals step.

As with the previous step – select main criteria – the manufacturing goals of an organization has a key role in the development of product families. The rearrangement of product families may also involve some changes to the existing combination of resources into cells. These changes to cell organization may only involve altering the data on the resource to show its new cell without actually moving the resource. In order to enhance the feasibility of future design proposals, the step – develop design proposal – receives feedback on any previous proposals from the later step – analyse design proposals.

The analyse design proposals step examines each proposal in relation to the particular characteristics of a production environment. The examination is undertaken using either operations research, simulation or artificial intelligence techniques, as well as the experience of the manufacturing personnel, to provide feedback on a proposal. Considerations in the analyse design proposals step include the evaluation criteria chosen in the select analysis criteria step and the estimated order requirements of the new products. The order requirements are regarded as constraints because particular demand levels may invalidate a proposal due to production underloading or overloading of a particular cell.

The inputs to this step include manufacturing data, the proposed new product groups and cell arrangements. Manufacturing data includes the process plans for the new products. This information may be required to carry out a detailed analysis of the new proposals. The end result of this maintenance procedure is a proposal for rearranging a product based layout to integrate

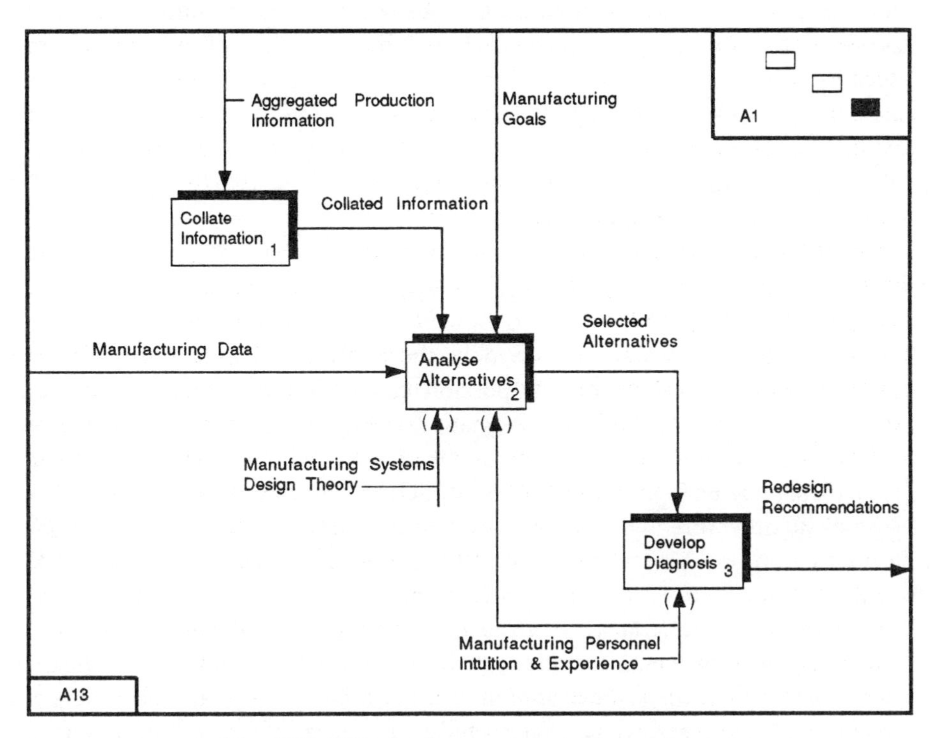

Fig. 3.10. A13: Analyse manufacturing system.

a range of new products, either through the extension of existing families or through the creation of new ones.

3.5.1 A13: Analyse manufacturing system

The purpose of the manufacturing systems analysis is to provide a diagnosis of potential production problems within a production environment. These production problems can include large production lead times, large set-up times or low quality levels. This diagnosis can be used to improve the efficiency of the production system. The analysis procedure might consist of three steps (Fig. 3.10):

1 A131 : collate information;
2 A132 : analyse alternatives;
3 A133 : develop diagnosis.

In the collate information step, aggregated production information is received from the FC monitor. This information from the FC monitor is filtered to give a detailed picture of the various problems areas throughout the manufacturing system. This information may be categorized according to different production characteristics, which affect the efficiency of the production process, such as:

1 Production lead times;
2 set-up times;
3 process times;
4 materials handling times;
5 queue times;
6 quality levels;
7 work in progress levels.

This collated information is used in the analyse alternatives step to help manufacturing personnel identify possible causes and solutions for these production problems. In this analysis, manufacturing personnel may use various ideas from manufacturing systems design theory. Manufacturing systems design theory includes ideas for set-up reduction, product design principles for ease of manufacture and cellular manufacturing amongst others. The cause of a production problem may be related to an inefficiency in some section of the production process. For example, a large production lead time may be caused by a large set-up time at a particular operation along a process route.

An organization's manufacturing goals may place a priority on solving some problems before others. For example if a manufacturing goal is to produce a high quality product and the analysis indicates a large scrap and rework rate at a particular section of the process, then this problem will receive high priority. The output from the analyse alternatives step is a series of theoretically possible alternatives for solving a particular problem.

In the develop diagnosis step, the proposed solutions for each problem are assessed in relation to their effectiveness for the particular manufacturing environment in which the problem has occurred. The resultant diagnosis consists of a number of alternatives which may provide a solution to the problem and is passed onto the relevant manufacturing personnel as a list of recommendations.

Overview of the design the production environment task One of the objectives of this task is to reorganize a shop floor to accommodate new products and to ensure that the shop floor is performing efficiently. In order to achieve this, three tasks are required: process planning for developing new process plans and altering existing ones, a group technology based maintenance procedure for allocating new products into product families, and an analysis procedure for diagnosing any problems on the shop floor. We argue that by organizing a shop floor efficiently, the planning and control of production activities is easier. Thus the link between the the design of the production environment

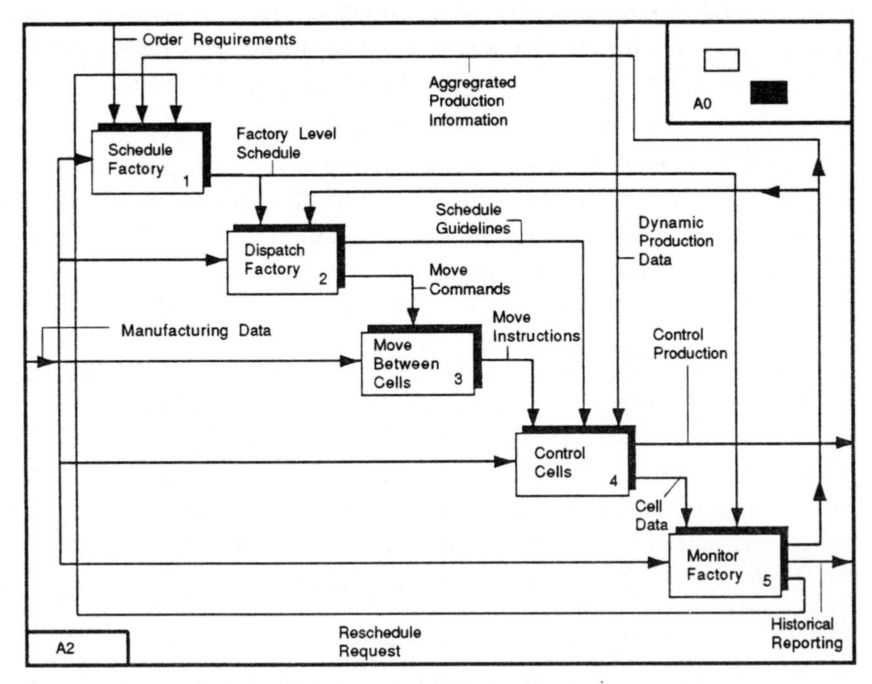

Fig. 3.11. A2: Coordinate product flow.

and the planning and control activities. We term the various planning and control activities under the heading of coordinate the product flow, which we shall discuss in the next section.

3.6 A2: Coordinate the product flow

This control task within FC organizes the flow of products between all cells within a factory (Fig. 3.11). The control task can be complex because of the various production constraints and manufacturing goals which relate to the entire manufacturing systems. Issues such as delivery dates, work in progress levels and utilization on capital intensive equipment, combine with manufacturing goals such as the maintenance of high product quality and decreased production lead times to present a series of conflicting objectives which require trade-offs. This task is the responsibility of production personnel in charge of factory level planning and control.

The main manufacturing data requirements for the control task include:

1 Process and product data, which gives the current list of products being manufactured and their process requirements;

2 the production requirements of each product as stated in the master production schedule, which is developed at the higher level master planning stage.

The control task consists of the following activities:

1 A21: schedule factory involves the development of a factory level schedule, which is used to coordinate the production activities of all cells;
2 A22: dispatch factory involves the coordination the flow of products and materials between cells on a real-time basis;
3 A23: move between cells involves the transportation of batches between cells;
4 A24 : control cells involves the control of production activities within each cell according to the guidelines given by the dispatch factory task. This is the main function of a PAC system.
5 A25 : monitor factory involves the collection of all relevant data on how the production environment is performing from the PAC monitors. This data is passed as real-time feedback to the factory level planning and control systems and as reports to management and higher level planning systems.

A21: Schedule factory The FC scheduler is used by production personnel to develop a factory level schedule to coordinate the flow of products between cells throughout a factory. It is an offline procedure which might consist of the following four steps (Fig. 3.12):

1 A211 : choose criteria;
2 A212 : allocate requirements;
3 A213 : develop schedule;
4 A214 : analyse schedule.

The choose criteria task involves selecting relevant production criteria such as reduced production lead times or fulfilment of due dates to be used to evaluate a factory level schedule in the analyse schedule step.

The allocate requirements task involves examining all the production orders contained in the master production schedule and downloaded through an MRP type system. Following this examination, the due dates are distributed throughout the time period of the master production schedule. This distribution of production orders depends on production constraints such as the priority attached to each order or the availability of materials. Aggregated production information relating to various manufacturing goals such as the ability to fulfil a production due date, also play a role in the distribution of order due dates. Once the production orders have been distributed, the allocate requirements step is used to cater for changes in production due dates and reassigning due dates to rushed orders that are urgently required by a customer.

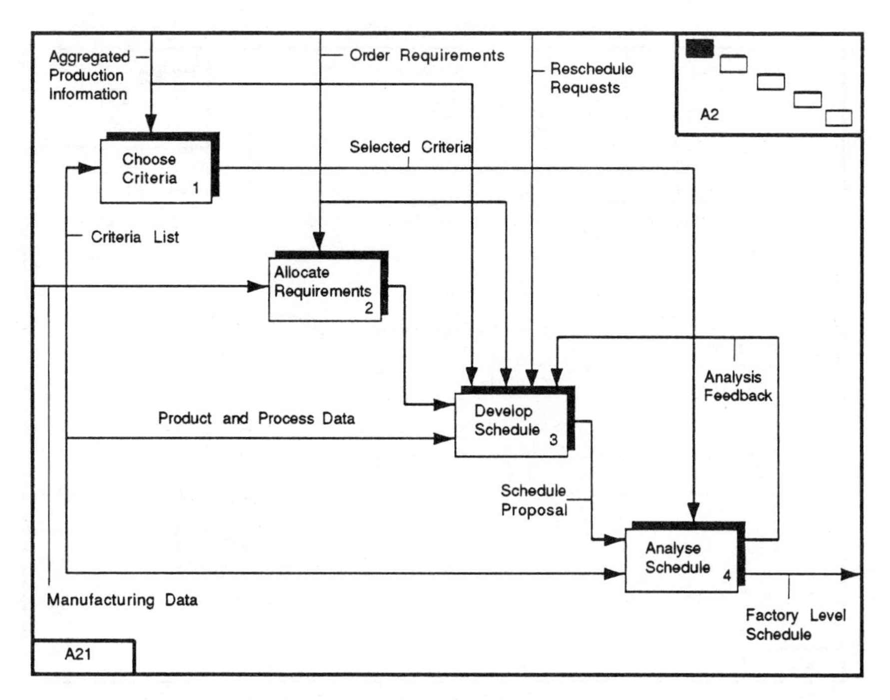

Fig. 3.12. A21: Schedule factory.

The develop schedule step involves using the distribution of due dates as decided by the allocate requirements step together with product and process data to develop a schedule proposal for a particular period of time. The time horizon of the schedule is clearly much shorter than the time period spanned by the master production schedule. As with the other steps, production constraints and manufacturing goals, in the form of aggregated production information, influence the development of a schedule proposal. Production personnel develop a schedule proposal using their expertise and may also use a decision support system. The decision support system may incorporate scheduling algorithms, operations research techniques and artificial intelligence techniques, which together provide a library of rules with which to develop different types of schedules. Once the proposal is developed it is passed for analysis to the analyse schedule step. Occasionally a new schedule may have to be developed because of a reschedule request from the FC monitor. The need for a new schedule may arise because of circumstances on the shop floor, which are hindering the progress of the original schedule.

The analyse schedule step evaluates each schedule proposal according to the list of criteria chosen in the select criteria step. The evaluation provides feedback to the production personnel on the feasibility of the proposal in relation to the current production circumstances. The experience of the pro-

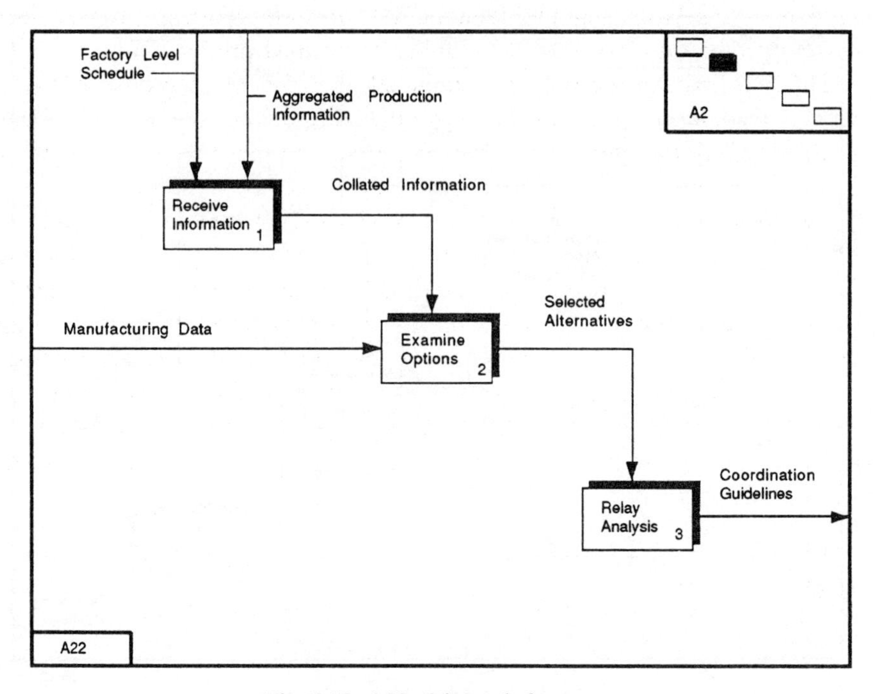

Fig. 3.13. A22: Dispatch factory.

duction personnel who are familiar with the particular characteristics of the manufacturing system may well be invaluable. The outcome of the schedule analysis is a schedule, which is either deemed to be suitable for the shop floor or else requires further development before being passed onto the shop floor. In either case, feedback on developing future schedules will be passed back to the develop schedule step. Each proposal which has been evaluated as being suitable is then passed to the factory level dispatcher.

A22: Dispatch factory The factory level dispatcher coordinates the flow of products between cells on a real time basis. This coordination task may involve the following three steps (Fig. 3.13):

A. A221 : receive information;
B. A222 : examine options;
C. A223 : relay analysis.

The receive information step involves taking static data from a manufacturing database and dynamic data from the FC monitor. The static data is data which is not changing on a real time basis and includes such information as product's process routes and production requirements. The dynamic data includes data which is changing constantly and enables the dispatcher to have a detailed knowledge of how efficient the production system is functioning on a real-time basis. Sample dynamic data includes work in progress levels, current product quality levels or number of finished products.

This information needs to be examined and filtered according to the dispatching goals being used. For example, dispatching might give a higher priority to certain batches because they are behind schedule or restrict the use of a certain raw material because it is below its safety stock levels.

The examine options step uses this condensed information and finds suitable solutions to the particular problems. The range of possible solutions is constrained by the schedule guidelines developed by the FC scheduler and the dispatching goals. This analysis is carried out by production personnel, who may use techniques such as expert systems, simulation, algorithms, or heuristics, as a decision support aid. The FC dispatcher is concerned with ensuring that all products are following the correct sequence of operations and are transported between the correct cells. The dispatcher is also concerned with solving any particular problems that may cause a deviation from the planned factory level schedule. In relation to the movement of :

Raw materials The dispatcher informs the raw materials storage facility about a requirement for additional raw materials in any cell.

Work in progress The dispatcher may use either a factory level mover or a PAC mover to transport work in progress stocks between cells.

The final step – relay analysis – transmits the operational or routing related decision to all relevant PAC dispatchers either manually or by using distributed software systems.

A23: Move between cells One of the responsibilities of the dispatch factory task, involves issuing commands relating to the movement of materials between cells. The move between cells task is charged with organizing this movement of materials either manually or by some automated means (e.g. automated guidance vehicle). The move between cells task may be carried out by either a factory level mover or a PAC mover. In our opinion, the greater the complexity and variety involved in the movement of materials, the more likely it is a factory level mover that is required for the task.

A24: Control cells Fig. 3.14 illustrates the SADT representation of the PAC task. The overall goal of the PAC task is to develop and implement a schedule based on the guidelines received from FC. By using these guidelines, each PAC system is aware of other PAC systems requirements from it. In the diagram, the different inputs, outputs and controls for each of the modules are clearly indicated, and these are now summarized.

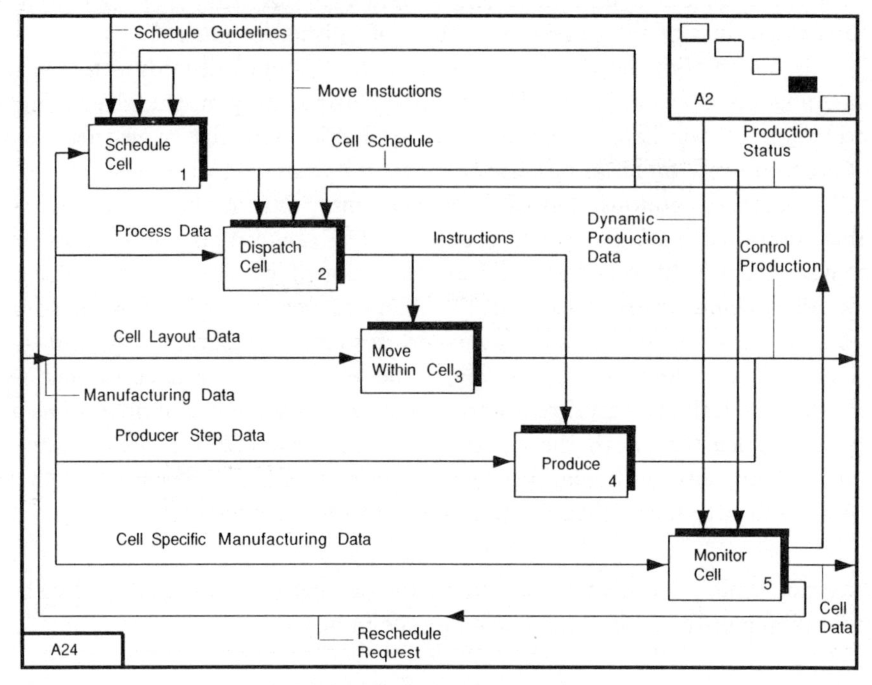

Fig. 3.14. A24: Control cells.

A241: Schedule cell The main input to the schedule activity is the manufacturing data which describes how each of the products are processed within the cell. The Schedule activity is then subject to production constraints and manufacturing goals, before releasing the cell schedule to the dispatch activity.

A242: Dispatch cell The dispatch activity takes the cell schedule from the schedule activity and attempts to implement this based on the state of the cell. This dynamic status of the cell is the constraint illustrated as an arrow coming from the monitor activity. The dispatch activity must have process information to enable it to make real time decisions such as routeing a job to an alternative workstation. The dispatch activity issues commands to the move and produce activities and it also can request a new schedule from the schedule activity.

A243: Move The move activity responds to commands issued by the dispatch activity. It then translates these commands into specific device instructions for the moving devices on the shop floor. The move activity also sends dynamic information to the monitor activity.

A244: Produce The produce activity is similar to the move activity in that it also responds to commands issued by the dispatch activity. It then translates

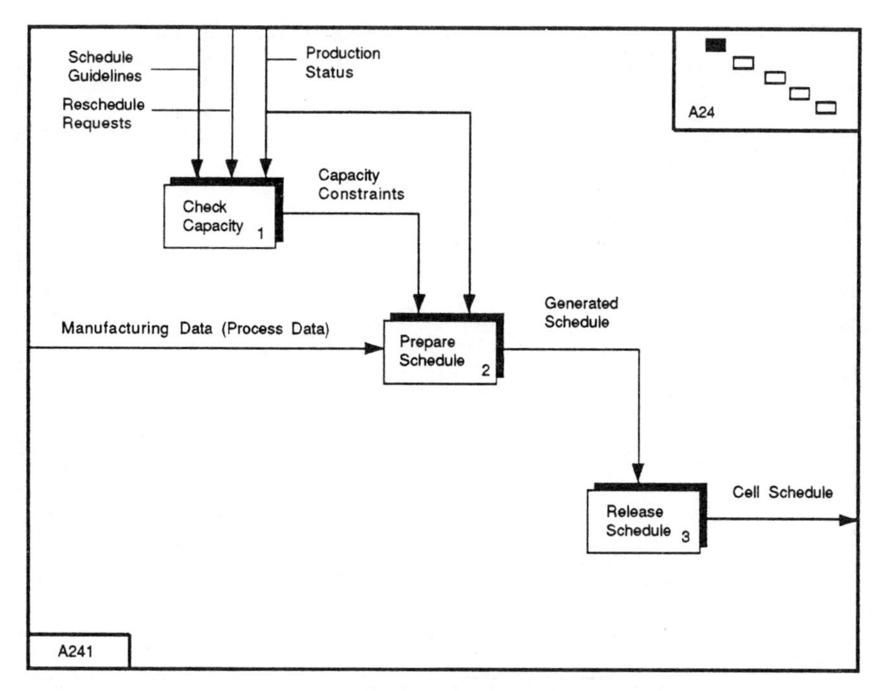

Fig. 3.15. A241: Schedule cell.

these commands into specific instructions for the producing devices within the cell. This activity also sends device status to the monitor activity.

A245: Monitor cell Finally, the monitor activity receives the dynamic data from the move and produce activities, and relays this back to the schedule and dispatch activities.

We now go on to describe the three main activities of the PAC task, namely schedule cell, dispatch cell and monitor cell.

A241: Schedule cell Fig. 3.15 illustrates a further breakdown of the schedule cell activity (A241). We have sub-divided this activity in three distinct activities, and these are now described.

A2411 Check capacity The first activity is to check the capacity of the cell to ensure that the schedule guidelines received from FC can be fulfilled. Static data on processing times at workstation and dynamic data on current workstation availability are some of the data requirements for calculating the cell capacity. The means of carrying out the capacity check will vary depending on the manufacturing environment, and these would range from manual calculations to the use of computer tools such as simulation. The output of the check

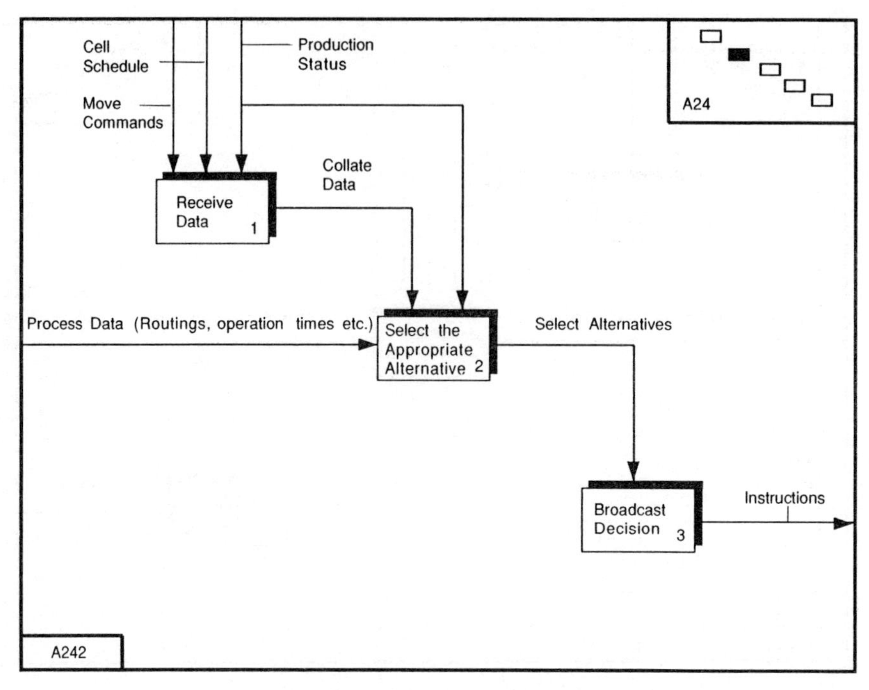

Fig. 3.16. A242: Dispatch cell.

capacity activity is then used by the prepare schedule activity. If the capacity available falls behind the capacity required, the FC system is notified, and new guidelines will then be issued.

A2412 Prepare schedule The second scheduling activity is to prepare a feasible schedule. It is more than likely that this activity will involve some computational method such as simulation. The type of activity to be used will be specified by the scheduling goals.

A2413 Release schedule The final activity releases the schedule to the dispatch and monitor activities with in cell. In an information technology environment, this usually would be performed by a distributed software system.

A242: Dispatch cell The dispatch cell activity may be sub-divided into three activities which combine together to ensure that the schedule received from the schedule activity is adhered to. These three activities are now outlined.

A2421 Receive data Essentially, the dispatcher works by reacting to information it receives from the other PAC modules. Therefore, it receives information from the schedule activity and from the monitor, and this information must

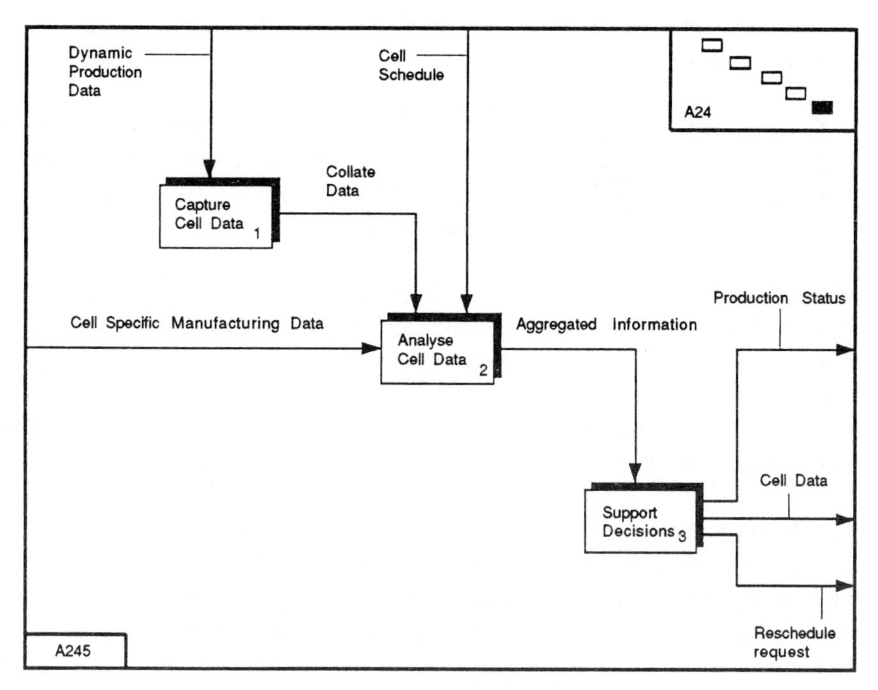

Fig. 3.17. A245: Monitor cell.

then be collated and organized so that informed and accurate decision making can be made.

A2422 Select the appropriate alternative Once the relevant information has been collated, the dispatch cell activity must activate a selection process to analyse whatever alternatives are available. This is the decision making activity within the dispatcher, and an example of multiple alternatives is when a job may be processed on any one of a number of machines. The dispatcher can use various techniques to select the most appropriate alternative, such as expert systems or simulation.

A2423 Broadcast decision The final activity is to broadcast the course of action decided by activity A222 on to the relevant location.

A245: Monitor Finally, the monitor activity can also be sub-divided into three separate activities, and these are now described.

A2451 Capture cell data The first activity of the monitor is to collect manufacturing specific data from the shop floor, and this is achieved using data collection systems. This data must then be collated and passed on the data analysis activity of the monitor.

A2452 Analyse cell data The data collated is then analysed with a view to highlighting the important manufacturing characteristics. This data may be analysed according to previously defined reference goals, and also static measures for the purposes of comparison.

A2453 Support decisions Finally, this data may be used to provide decision support to the PAC activities of schedule and dispatch. A common example of decision support is when the monitor informs the dispatcher when raw material levels have been depleted below specified reorder points. The monitor also must evaluate whether or not there is a need for a new schedule, and thus may request a new schedule from the cell scheduler.

A25: Monitor factory The FC monitor provides a detailed assessment of how efficiently the manufacturing system is performing. This feedback is given in the form of reports to management, updates to higher level planning systems, or on a real time basis to the FC scheduler and FC dispatcher. With this timely and accurate flow of information, the scheduler and dispatcher have a detailed knowledge of all events on the shop floor, on which to base their decisions. The monitor uses three steps to compile and pass on relevant information to management and the other FC control procedures (Fig. 3.18):

1. A251 : capture the PAC data;
2. A252 : analyse the PAC data;
3. A253 : provide decision support.

The capture the PAC data step involves gathering large amounts of data from all PAC monitors within a factory. Using monitoring reference goals and data collection systems, production personnel filter and condense a large amount of data. The monitoring guidelines may help the analysis procedure in indicating which data should be analysed and the objectives that the analysis should be aiming to satisfy. For example the monitoring guidelines can emphasis product quality data and the importance of striving for zero defects. Therefore the analysis should develop solutions which are compatible with the emphasis on product quality.

This collated data is then passed to the analyse the PAC data step, where the data is evaluated according to the monitoring reference goals to determine which information to pass on, to assist the FC scheduler, FC dispatcher and management in their tasks. The analysis involves a comparison of actual events on the shop floor with the sequence of events as planned in the factory level schedule. This comparison should indicate possible causes for the deviations from the schedule. and may form the basis for a request for a new schedule.

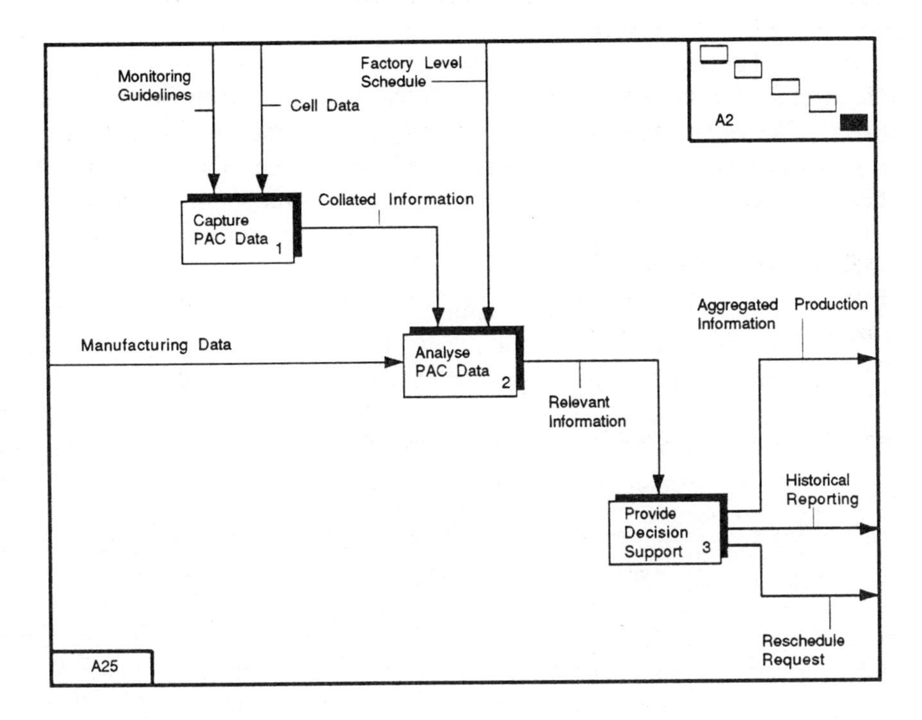

Fig. 3.18. A25: Monitor factory.

The provide decision support step involves passing any proposals and information arising from the analysis in the form of aggregated production information, to the relevant higher level planning systems and the FC scheduling and dispatching functions using distributed software systems. Data relating to the causes of the deviations may also be used in the manufacturing systems analysis task. This step also involves providing historical reports to management, detailing the performance of the factory. Based on the analysis completed in the – analyse the PAC data – step, the FC monitor may issue a request for a new schedule to the FC scheduler.

Overview of the Coordinate the Product Flow Task This task is concerned with organizing the product flow throughout a factory. This is achieved by means of a FC control task which coordinates product flow between cells, and a PAC system organizing all production activities within a cell. The FC control task has three building blocks: scheduler, dispatcher and monitor, while the PAC task uses five building blocks; scheduler, dispatcher, monitor, producer, and mover.

3.7 Conclusions

In this chapter, we have presented a formal description of our proposed architecture for shop floor control using the top down analysis methodology SADT. The architecture consists of a FC system coordinating various cells on a shop floor. Each cell is controlled by a PAC system. The cells may either be completely product based or a combination of product and process based. This description illustrates our idea that the two major tasks of shop floor control systems are:

1. A production environment design task which enables the manufacturing environment to adapt to the inevitable changes brought about by new product introduction and process changes. This is considered part of the FC task.
2. A shop floor planning and control system which ensures balanced and organized product flow through each of the cells in the factory. There are two tiers to this planning and control system; an FC control task coordinating flow between cells and a PAC system for each cell which organizes the flow within the cell.

In describing these tasks, our main concern has been to show what tasks are required. Later, in Chapters 6 and 7, we shall describe our ideas on how these tasks relating to process planning, manufacturing systems analysis, maintenance of a product based layout and scheduling, may be carried out. In the remainder of this book we shall examine suitable methods of achieving a realistic implementation of our architecture. It is our firm belief that the best possible implementation of our work is in an information technology environment, which combines the processing power of computers with the creative talents of people. In the following chapter we outline a blueprint for an information technology implementation of FC and PAC.

Part Three
AN INFORMATION TECHNOLOGY ARCHITECTURE FOR SHOP FLOOR CONTROL

Overview

In Part Two we outlined a functional architecture for FC and PAC. We now present an information technology (IT) architecture which meets the demands of the functional architecture presented earlier and outline some of the tools necessary to realise this architecture in practice.

In Chapter 4 we present the proposed IT architecture. It is based on a layered approach which recognizes three layers, namely the factory layer, the cell layer and the device layer. We define the entities within each layer and the services provided by the layer. We try to identify the communications needs of the architecture and through an example we outline the type of protocols necessary to to support communications within and between the layers.

To illustrate the communications protocols we use the Petri net formalism. Specifically we illustrate the communications between a PAC dispatcher and producer using a Petri net model. For readers not familiar with the Petri net formalism we present a short outline of Petri net modelling.

In Chapter 5 we discuss the software tools and techniques available to implement the IT architecture presented in Chapter 4. This overview of IT tools includes a statement of requirements in terms of flexibility, inter operability and portability. Further the tools are presented in terms of a computing model which highlights data communications, data management, data processing and user interfaces.

4

An information technology architecture for shop floor control

4.1 Introduction

In this chapter, we examine our proposed architecture for shop floor control from an information technology (IT) viewpoint. The aim of this IT reference architecture is to compliment the overall understanding of our architecture for shop floor control, and also to provide the basis for future software developments in this domain.

In Part Two we described the shop floor control architecture without explicit reference to the technology needed to implement this architecture. In fact, the concepts of FC and PAC can be implemented in a fully computerized or a completely manual environment. However, it is our firm belief that the most appropriate implementation path for our ideas is within an information technology environment which takes full cognisance of the role of people. This implementation method is termed the sociotechnical approach, and essentially it attempts to blend the creative talents of humans with the great processing power of computers, with a view to enhancing organizational performance. The sociotechnical approach is considered in depth in Chapter 8.

In this chapter our primary concern is to outline the basic requirements of a responsive and flexible software system which performs the tasks of FC and PAC. We believe that this chapter is the first small step along the road towards the ultimate software standardization of the different aspects of FC and PAC. Standardization is a prime goal of any proposed software system, and our case in point is no exception. Our motivations for pursuing this standardization of an information technology reference model for shop floor control are:

1 To ensure portability of the different software modules at the application level. This would mean that an application such as a PAC scheduler could

be easily transferred from one type of operating system, say VAX/VMS, to another, such as ULTRIX^{TM}*.

2 To allow for the possibility of integrating different computer applications (i.e. enabling a PAC dispatcher from Digital Equipment Corporation to work with a scheduler from another software vendor). Thus both FC and PAC would be able to operate in a heterogeneous computing environment.

3 To provide an initial framework for the implementation of FC and PAC in an information technology environment.

Essentially, our task is to develop a software model which is conceptually similar to the Open Systems Interconnection model developed by the International Standardization Organization. We employ the same technique used by OSI to develop their seven layered architecture for opens systems interconnection. This technique is known as layering and the basic principles (described in the following section) recommend that a software system should be carefully designed into individual layers, thereby encouraging a solution which is both flexible and modular. These layers then offer standard services and protocols to enable efficient and effective integration, and ultimately standardization.

The overall structure of this chapter is as follows:

1 We describe the layered approach used to develop our reference architecture, and map the functional architecture described in part two of this book onto a layered reference architecture.

2 We describe the essential components of the layered referenced architecture for FC and PAC.

3 We describe a formal diagramming technique known as Petri nets, which can be used to model the characteristics of interacting system components.

4 Finally, using Petri nets, we outline a sample protocol between two layers within our reference architecture for FC and PAC.

It is important to realize that the IT architecture described in this chapter is not a precise implementation specification, rather it provides a conceptual and functional framework which provides a useful starting point for the standardization process of shop floor control systems.

4.2 The concepts of a layered architecture

The main purpose of this chapter is to define an information technology reference architecture for shop floor control. This requires an abstract model

* ULTRIX is a trademark of Digital Equipment Corporation

which describes the different software entities, and for this purpose we use a technique known as the layered approach. We believe that this is an appropriate abstraction method to outline some of the basic requirements for an IT model of FC and PAC.

The key terms which are used to describe the layered approach in this chapter are:

1 An entity, which is an active element within a layer. An example of an entity within the context of our architecture is a PAC scheduler.
2 A service, which is a capability provided by one entity/layer to other adjacent entities/layers. An example of a service a scheduler offers to other another entity, known as a dispatcher, is a rescheduling service.
3 A function, which is part of the activity of an entity. Taking once more the example of the scheduler, a typical function is that of schedule development.
4 A protocol, which is a set of rules and formats (both semantic and syntatic) which determines the communication behaviour of entities in the performance of functions. Within PAC and FC, there may be many different protocols, depending on the entities or layers which are communicating with each other.
5 A layer is a group of similar entities.

Thus, these are the basic components required to describe an architecture using the layered approach. For the reference architecture of FC and PAC we need to define clearly:

1 The number of necessary layers;
2 the entities within each layer;
3 the services provided by each layer;
4 the protocols used to communicate between entities. Only one sample protocol will be described in this chapter, as we believe that a complete protocol definition for FC and PAC is outside the scope of this book.

ISO laid down some useful principles which should be considered when selecting layers for a reference model (ISO, 1984). It was pointed out in this document that 'it may be difficult to prove that any particular layering selected is the best possible solution. However, there are general principles which can be applied to the question of where a boundary should be placed and how many boundaries should be placed'. These 'general principles' included the following:

1 Do not create so many layers as to make the systems engineering task of describing and integrating the layers more difficult than necessary;
2 create separate layers to handle functions that are manifestly different in the process performed or the technology involved;

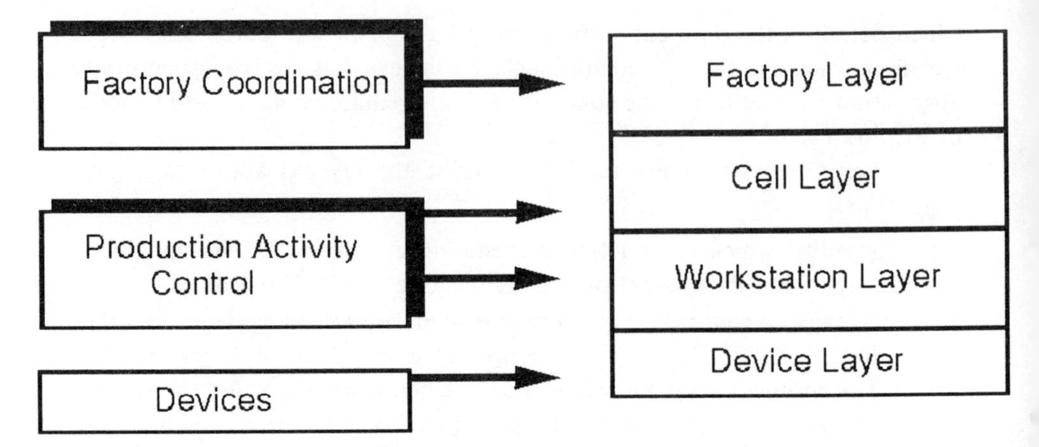

Fig. 4.1. Architectural mapping from functional to layered.

3 create a layer of easily localized functions so that the layer could be totally redesigned and its protocols changed in a major way to take advantage of new advances in architectural, hardware or software technology without changing the services expected from and provided to the adjacent layers;

4 create a boundary between layers at a point where the number of interactions across the boundary are minimized.

Thus, we now must take the existing functional architecture for FC and PAC described in Part Two of this book, and identify clearly the layers, services, entities and protocols. Our proposed mapping of the function architecture onto a layered architecture is shown in Fig. 4.1. The illustration shows that there is almost a one-to-one correspondencc bctween the levels of the functional architecture and the layers of the layered IT reference architecture.

The major difference is that two clear distinct layers have been identified for the Production Activity Control task. These are (1) the cell layer, and (2) the workstation layer. The choice of the different layers reflects the control hierarchy of our shop floor control architecture. When examined closely, there are three distinct levels of control, and each level has some degree of control over the level below. Each level also relies on information from the level below to make decisions. Thus, the relationships between each adjacent layer may be described as one between the controller and the controlled.

Firstly, let us consider the controller/controlled relationship between the factory layer and the cell layer. In our architecture, the factory layer contains the four entities of FC (production environment design, FC scheduler, FC dispatcher and FC monitor), and these combine to guide the entities in the lower cell layer. This guidance is in the form of offline planning and real-time coordination recommendations on start and finish times of jobs within each

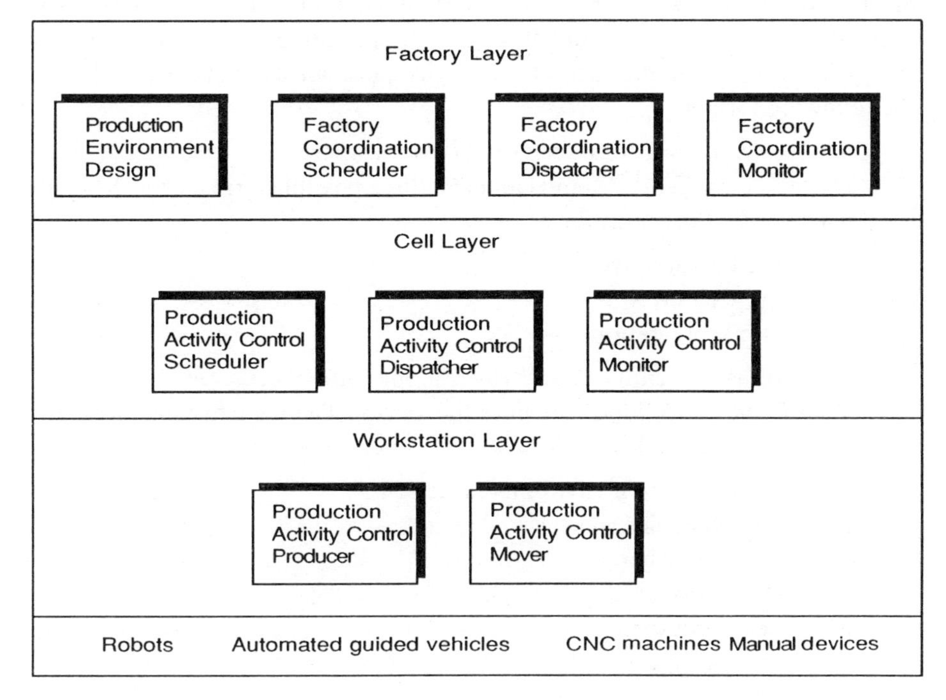

Fig. 4.2. Layered architecture for FC and PAC.

cell. As we have illustrated in Chapter 2, in addition to managing the flow of products between cells, FC also specifies the production environment design requirements for each cell.

The cell layer then directly controls the workstation layer, by issuing precise instructions on where and when operation tasks on individual jobs should be performed. These instructions are based on feedback obtained from the workstation layer. Finally, the workstation layer controls the actual devices on the shop floor, by translating the precise instructions from the cell layer into step-by-step commands for the individual devices.

The layered reference model is shown in more detail in Fig. 4.2, illustrating each of the entities and layers. In the following section, the basic services provided by each of the entities in the top three layers of our reference model are outlined.

4.3 The entities and core services of the reference architecture

Here we briefly summarize each of the functional entities and core services of each layer of the reference architecture for FC and PAC. There are nine distinct functional entities in all, and each of these provide unique services in order that the tasks of FC and PAC may be performed. We have derived a specific syntax to describe each of the services, and all services are based upon the following format:

- Service syntax = LCODE_ENTCODE$ACTION, where:
- LCODE is the layer code, and can have three possible formats which signify which layer the service belongs to:

  1 FAC: Factory layer.
  2 CEL: Cell layer.
  3 WST: Workstation layer.

- ENTCODE is the entity code, which along with the LCODE will uniquely identify the entity which provides the service. There are five distinct entity codes:

  1 DES: Production environment design.
  2 SCH: Scheduler.
  3 DIS: Dispatcher.
  4 MON: Monitor.
  5 PRO: Producer.
  6 MOV: Mover.

- Finally, the ACTION identifies the type of service provided by an entity. An example of a complete service call is FAC_DES$ANALYSE_SYSTEM, which describes one of the services provided by the production environment design module of FC.

We commence our review of entities and services by analysing the factory layer of our reference architecture. The factory layer is composed of the FC activities of shop floor control, and this layer has two overall goals. Firstly, it must organise the flow of products throughout the factory to ensure that production within each of the factory cells is well managed and evenly distributed. Secondly, it must allow for periodic reorganization of the manufacturing system to accommodate the inevitable introduction of new products or different processes and procedures.

Within the FC layer, there are four distinct entities, namely:

- Production Environment Design accommodates the introduction of new products into the manufacturing system. Depending on the particular attributes of the new products, recommendations will be made on how best to blend in the new process requirements with the existing manufacturing capabilities. This design task also performs an analysis of the production system based on information from the factory monitor, and this type of feedback can then be used to assess the effectiveness of the production environment. The services provided by this design task include:

1 FAC_DES$GENERATE_PROCESS_PLANS: When new products are being introduced, this service generates a process plan. It bases its analysis on the component and process requirements of the new product, and then attempts to match this to the most appropriate step of the process.
2 FAC_DES$MAINTAIN_PRODUCT_BASED_LAYOUT: As new products are introduced, an occasional overall check is needed to ensure that the manufacturing system maintains its product based integrity. This service is designed to cater for this need by using Group Technology principles (see Chapter 7).
3 FAC_DES$ANALYSE_SYSTEM: Based on information received from the factory monitor, this service performs an analysis to pinpoint potential weaknesses in the process layout. Important manufacturing parameters such as set-up times, high utilization and overall cell performance are singled out in the attempt to evaluate the smoothness and efficiency of the manufacturing system. This is a particularly useful tool to assess the progression of a factory towards the ideal of manufacturing excellence.

- The FC scheduler generates a realistic and achievable plan for a specified time period. This plan is then a basis for coordinating the flow of products from cell to cell. The services provided by the factory level scheduler include:

  1 FAC_SCH$DEVELOP_SCHEDULE: The schedule is developed using a previously selected strategy, such as backward scheduling or priority based scheduling. The tightness of the schedule, that is, how detailed the output is with respect to the start and finish times of operations, is also an important factor, which is partially influenced by the design and efficiency of the manufacturing system. The output of the FC scheduler are the general guidelines, which are then interpreted by the scheduling task of each cell.
  2 FAC_SCH$RELEASE_SCHEDULE: This is the final step of the scheduling process, and the newly generated schedule is distributed to the factory dispatcher and the factory monitor.
  3 FAC_SCH$PERFORM_RESCHEDULE: It the schedule is not going to plan because of some unforeseen event such as process problems or raw material unavailability, a request is made to the FC scheduler to generate a new schedule taking the latest manufacturing situation (as viewed by the factory monitor) into account.

- The FC dispatcher disseminates the newly developed schedule to each of cells. It may also perform some real-time scheduling once the development of a feasible solution is within its capability. If this is not the case, the FC dispatcher notifies the FC scheduler to request a reschedule. The services provided by the factory level dispatcher include:

1 FAC_DIS$DISTRIBUTE_SCHEDULE: This service will take the schedule guidelines passed down from the FC scheduler and ensure that they are then distributed to the cell layer.
2 FAC_DIS$MOVE_COMMAND: This service instructs a particular moving device to transport jobs between individual cells.

- The FC monitor provides an accurate picture of what is happening at each of the production cells. It aggregates the information received from each of the cell monitors, and presented this in a clear and concise format which enables informed and timely decision making within the FC layer. Relevant information on issues such as production lead times is passed back to higher level planning systems, such as MRP systems. The factory monitor must also generate historical reports for the purposes of reflective analysis at a later stage. The services provided by the factory level monitor include:

1 FAC_MON$ANALYSE_PAC_DATA: On receiving the PAC data from each of the cell monitors, a process of organizing the information in a logical and coherent manner must take place. This is the primary function of this service.
2 FAC_MON$ADVISE_MODULES: Depending on the trends of certain manufacturing parameters, the factory monitor sends advice to other entities. For example, if the schedule is obviously going out of control, it can inform the factory dispatcher to take the appropriate corrective action. The factory monitor also gives relevant information on the manufacturing environment to higher level planning systems, so as to improve the accuracy of their plans.
3 FAC_MON$GENERATE_REPORTS: Report generation is an important service provided by the factory monitor. It enables manufacturing personnel to analyse significant trends with respect to predefined production parameters, such as, set-up time variances, schedule success rate and overall production quality.
4 FAC_MON$RECOMMEND_RESCHEDULE: An important service provided by the factory monitor is one which assesses whether or not a new schedule is required. Thus, this service analyses whether or not the progress of the schedule is within acceptable limits, and if significant deviations occur, a reschedule activity is recommended to the factory scheduler.
5 FAC_MON$SHOW_SCHEDULE_STATUS: It is important that any supervisor or production planner can utilize the information within the factory monitor. The following four services are examples of what

we believe are the more relevant useful segments of information which people would be interested in. Displaying the schedule status for each of the cells within the plant shows immediately which jobs are not on time, and corrective action may then be taken. The basis of this display is to compare the planned start and finish times with the actual start and finish times.

6 FAC_MON$SHOW_QUALITY_PROBLEMS: This service highlights where the location and possible cause of current quality problems in the plant. It is based on the aggregated information from the cell monitors.

7 FAC_MON$SHOW_UNUSUAL_STATISTICS: The context of unusual statistics will more than likely very depending on the manufacturing environment. However, these could be identified in advanced and programmed into the software system. For example, reduced output from a normally reliable cell could be classified as an 'unusual statistic', which the FC layer would be made aware of this.

8 FAC_MON$SHOW_OVERALL_TREND: The overall trend of a manufacturing facility shows aggregated figures such as overall equipment utilization, percentage of jobs completed on time, percentage quality levels etc. These are the kind of measures that would be relevant to other departments with a manufacturing organization, such as Finance, Marketing and Materials Planning.

Thus, the factory layer controls the cell layer by providing it with the guidelines necessary to achieve the overall manufacturing goals. The cell layer consists of the main planning and control tasks of PAC, and together with the workstation layer it forms the complete PAC architecture in our IT Reference Model. The overall goal of the cell layer is to take the schedule guidelines from the FC layer and ensure that these are adhered to as much as possible. This involves generating a detailed plan for each of the manufacturing resources within a particular cell, and then implementing this plan based on the state of the cell resources. Implementation of the plan requires direct communication of commands, and feedback from the workstation layer. Events which occur that have an overall significance to the FC task are communicated upwards to the factory layer.

Within the cell layer, there are three distinct entities:

- The PAC scheduler takes the guidelines provided by the factory layer, and generates a specific work plan for each resource within the cell without violating these guidelines. The services provided by the PAC scheduler include:

1 CEL_SCH$DEVELOP_SCHEDULE: Once the scheduling strategy has been selected, the schedule will be generated based on the guidelines received from the factory layer.
2 CEL_SCH$RELEASE_SCHEDULE: This service then releases the schedule to the cell dispatcher and the cell monitor. This schedule is then implemented by the cell dispatcher, which sends instructions to the device control layer of our IT Reference Model.
3 CEL_SCH$PERFORM_RESCHEDULE: If a request for a reschedule arrives from the cell monitor, this service then performs this task. The resulting schedule is then released using the CEL_SCH$RELEASE_SCHEDULE service described above.

- The PAC dispatcher receives the detailed plan from the scheduler and then attempts to implement that plan. It issues specific commands to the workstation layer on when and where to commence operations on jobs. The services provided by the PAC dispatcher include:

1 CEL_DIS$INTERPRET_SCHEDULE: The dispatcher receives the schedule for the cell from the scheduler and its primary task then is to ensure that that schedule is implemented.
2 CEL_DIS$COMMAND_WORKSTATIONS: When implementing the schedule, the PAC dispatcher issues commands to the workstation layer so that the products flow effectively through the cell.

- The PAC monitor keeps track of the important production system information and then provides decision support to the PAC scheduler and PAC dispatcher, as well as entities in the factory layer. The services provided by the PAC monitor include:

1 CEL_MON$ANALYSE_DEVICE_DATA: The primary aim of the PAC monitor is to make sense of the mass of data emanating from each of the workstations. Essentially, this service transforms the raw manufacturing data into useful information which may then be used by other monitor services to advise and inform other PAC modules and activities.
2 CEL_MON$ADVISE_MODULES: Based on the data collected and aggregated, the PAC monitor may advise certain modules on necessary courses of action. For example, this service could be used to advise the PAC dispatcher to reorder more of a certain component once it has been depleted below its reorder point.
3 CEL_MON$GENERATE_REPORTS: This is a most useful feature of the PAC monitor, and the reports generated may be used by all levels of management at some further stage for reflective analysis.
4 CEL_MON$SHOW_SCHEDULE_STATUS: During the course of a days production, the following four services provide both supervisors and

operators with the capability of viewing important trends and informations within their cell. The first service shows the progress of the schedule, and basically compares the planned start times with the actual start times. These measures may then be used to calculate whether or not the schedule implementation is within satisfactory control limits.

5 CEL_MON$SHOW_DEVICE_STATUS: This service enables the status of each device within the cell to be viewed. Relevant statistics such as average utilization, number of breakdowns, etc. are key attributes of this service.

6 CEL_MON$SHOW_COMPONENTS_STATUS: The consumption of components/raw materials is a regular occurrence within production which may require close attention. This service provides the means for viewing components usage.

7 CEL_MON$RECOMMEND_RESCHEDULE: The cell monitor may assess whether or not a new cell schedule is required. Thus, this service analyses whether or not the progress of the schedule is within acceptable limits, and if significant deviations occur, a reschedule activity is requested from the cell scheduler.

Thus, the cell layer receives guidelines from the factory layer. In turn, it then controls the workstation layer by issuing specific instructions on when and where to commence operations on jobs. The role of the workstation layer is to translate the cell layer commands into specific instructions which will then control the appropriate device.

Within the workstation layer, there are two distinct entities:

- The PAC producer is the process control system which ensures that the command from the dispatcher is translated into specific operation steps for a particular job. The services provided by the PAC producer include:

  1 WST_PRO$DOWNLOAD_PART_PROGRAMS: An essential service of the PAC producer is to retrieve the appropriate part program file from its source and download it to the device.

  2 WST_PRO$REPORT_TO_DISPATCHER: The PAC producer also returns the status of the device to the dispatcher after a significant operation has taken place.

- The PAC mover coordinates the materials handling activities within the cell. Essentially, it is the interface between the PAC dispatcher and the physical transportation and storage devices in the cell. The services provided by the PAC mover include:

  1 WST_MOV$CONTROL_MOVING_DEVICES: This service issues specific instructions to the moving devices, based on the initial commands received from the PAC dispatcher.

2 WST_MOV$REPORT_TO_DISPATCHER: The mover sends important information from the moving devices to the dispatcher indicating information such as the location and the destination of the mover.

Thus, we have described the three layers of the IT reference architecture for shop floor control systems, the distinct entities that belong to these layers, and a selection of core services provided by each entity. However, a description of entities and services alone does not provide the basis of a reference model. It is also necessary to consider the protocols which determine the basis of communication between each of the entities of our model. Since a complete protocol definition is outside the scope of this book, we concentrate on two entities within out model and outline the basis of one such protocol between these entities. Prior to describing the protocol, we give a descriptive summary of the graphical modeling technique, Petri nets, upon which we based our protocol model.

4.4 A descriptive summary of Petri nets

Petri nets were designed specifically to model systems with interacting concurrent components (i.e. activities of one component may occur simultaneously with the activities of other components). They developed from the early work of Carl Petri in 1962 (Peterson, 1977), who formulated the basis for a theory of communication between asynchronous components of a computer system. He was particularly concerned with the description of the causal relationships between events. The practical application of Petri nets to the design and analysis of systems can be accomplished in several ways. One approach considers Petri nets as an auxiliary analysis tool and in using this approach conventional design techniques are used to specify a system. This system is then modelled as a Petri net and the model may then be analysed. Any problems encountered in the analysis of the Petri net point to flaws in the original design. A process of revision of the design and analysis of the Petri net model continues until an acceptable design is formulated.

A Petri net is composed of the four basic elements:

1 A set of places P.
2 A set of transitions T.
3 An input function I.
4 An output function O.

The input and output functions relate transitions and places. Formally, a Petri net C is defined as the four-tuple C=(P,T,I,O). As an illustration, consider the following example (Table 4.1) of a Petri net structure defined as a four-tuple.

The simple Petri net view of a system concentrates on two concepts, namely

$C=\{P,T,I,O\}$
$P=\{p_1,p_2,p_3,p_4,p_5\}$
$T=\{t_1,t_2,t_3,t_4\}$

$I(t_1)=\{p_1\}$	$O(t_1)=\{p_2,p_3,p_5\}$
$I(t_2)=\{p_2,p_3,p_5\}$	$O(t_2)=\{p_5\}$
$I(t_3)=\{p_3\}$	$O(t_3)=\{p_4\}$
$I(t_4)=\{p_4\}$	$O(t_4)=\{p_2,p_3\}$

Table 4.1. *Formal Petri net structure*

transitions and places (places are also known as conditions). Transitions are actions which take place in the system and the occurrence of these transitions is controlled by the state of the system. The state of the system may be described as a set of places. For a transition to occur, it may be necessary for certain conditions to hold, and these are termed the preconditions of the transition. The occurrence of the transition may cause the precondition to cease to hold and may cause other conditions, called postconditions, to become true.

The above definition is useful for formal work with Petri nets. However, it is not suitable to illustrate concepts of Petri nets in an informal and concise manner. For this purpose, a different representation of a Petri net is more useful, that is, the Petri net graph. Fig. 4.3 shows the corresponding graph to the structure defined in Table 4.1.

A Petri net structure consists of places and transitions with input and output functions. In a Petri net graph there are two types of nodes corresponding to the places and the transitions of the Petri net structure. A place is represented by a circle and a transition is represented by a bar. The input and output functions are represented by directed arcs from the places to the transitions and from the transitions to the places.

A marking μ of a Petri net is an assignment of tokens to the places in that net. The execution of a Petri net is controlled by the position and movement of these tokens. The tokens reside in the places of the Petri net and the number and position of tokens in a net may change during its execution. The vector $\mu=(\mu_1, \mu_2,....\mu_n)$ gives, for each place within the Petri net, the number of tokens in that place. On the Petri net graph, tokens are represented by small solid dots inside the circles (places) of the net. This is illustrated in Fig. 4.4, which is an example of a marked Petri net. It represents the structure described in the previous section, with the marking μ=(1,0,1,0,2). A Petri net C=(P,T,I,O) with a marking μ becomes the marked Petri net M=(P,T,I,O, μ).

A Petri net executes by firing transitions. A transition may fire only after it has been enabled. A transition is enabled when each of its input places

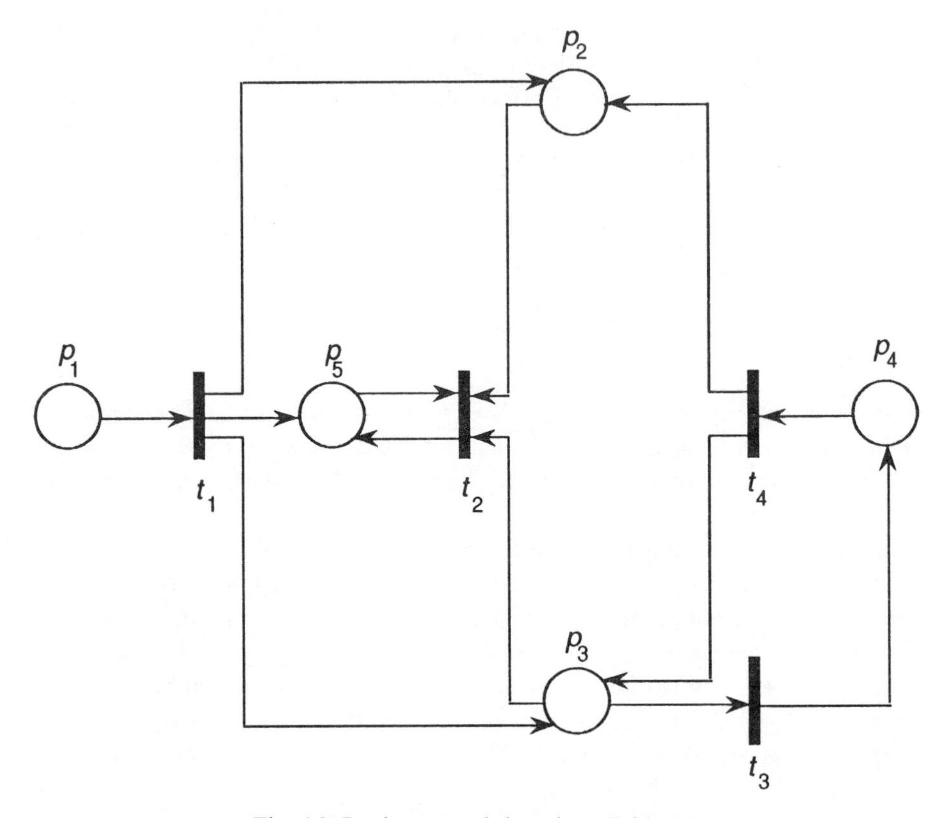

Fig. 4.3. Petri net graph based on Table 4.1.

contains at least one token. Considering Fig. 4.4, transition one (t_1) is enabled because place one (p_1) contains a token. However, t_2 is not enabled because p_2 does not contain a token, although the other input arcs to the transition (p_3 and p_5) each contain at least one token. A transition fires by removing one token from each of its input places, and deposits one token in each of the output places. Firing a transition changes the marking of the Petri net μ to a new marking μ'. Note that since only enabled tokens may fire, the number of tokens in each place always remains non-negative when a transition has been fired.

To illustrate the concept of firing, consider the marked Petri net as shown in Fig. 4.4. The initial marking is μ =(1,0,1,0,2) (i.e. places p_1, p_3 and p_5 contain tokens). At this stage, the transitions of t_1 and t_3 are enabled since each of their input places contain at least one token. If transition t_1 is fired, the token in place p_1 is removed and the number of tokens in places p_2, p_3 and p_5 are incremented by one. The resulting Petri net graph is shown in Fig. 4.5.

A consequence of firing transition t_1 is that this transition may not now be

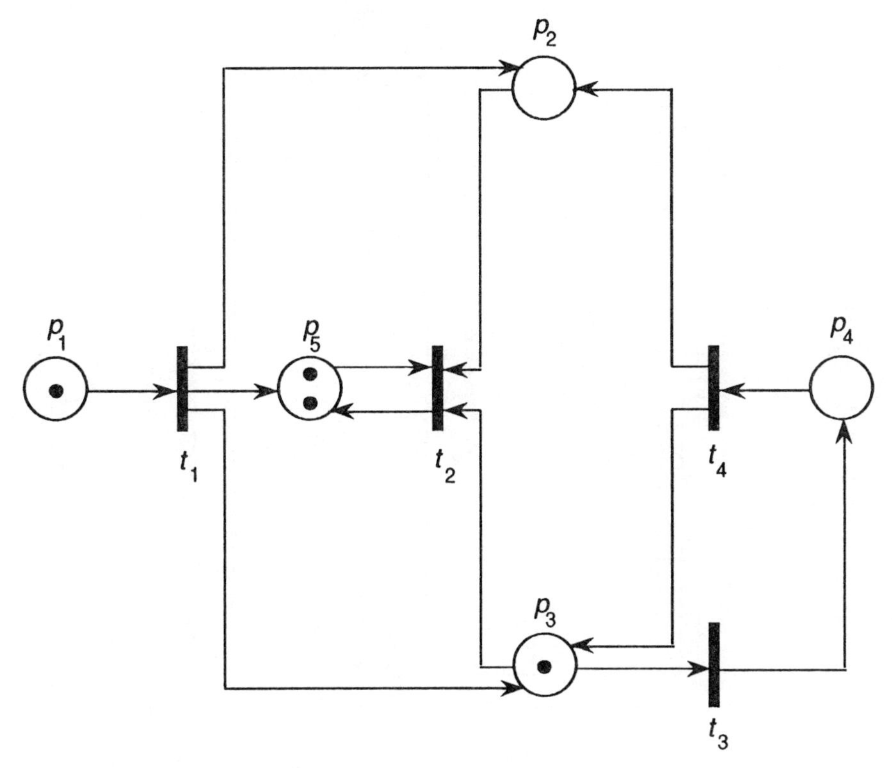

Fig. 4.4. A marked Petri net.

enabled again since there is no token present in place p_1. A further result of firing t_1 is that transition t_2 is enabled as there are now at least one token in each of its input places.

The state of a Petri net is defined by its marking, and thus the firing of a marking represents a change in the state of the Petri net. Figs 4.4 and 4.5 represent different states of a Petri net.

Timing in a Petri net may be represented in the following manner. Consider a Petri net representation of an activity which takes a known duration to complete (say n time units). The activity may be modelled using two transitions and one condition. The time factor may be taken into consideration by marking the starting time of the activity (x time units) and then scheduling the end of the activity for $(x + n)$ time units. Therefore the first transition (start activity) happens at time x and schedules the second transition (end activity) to fire at a time $x + n$ (Fig. 4.6).

As a result transition t_2 now must meet two constraints before it can fire. The first of these is that there must be at least one token in its input place, and the second is that the transition must be scheduled for the current clock time.

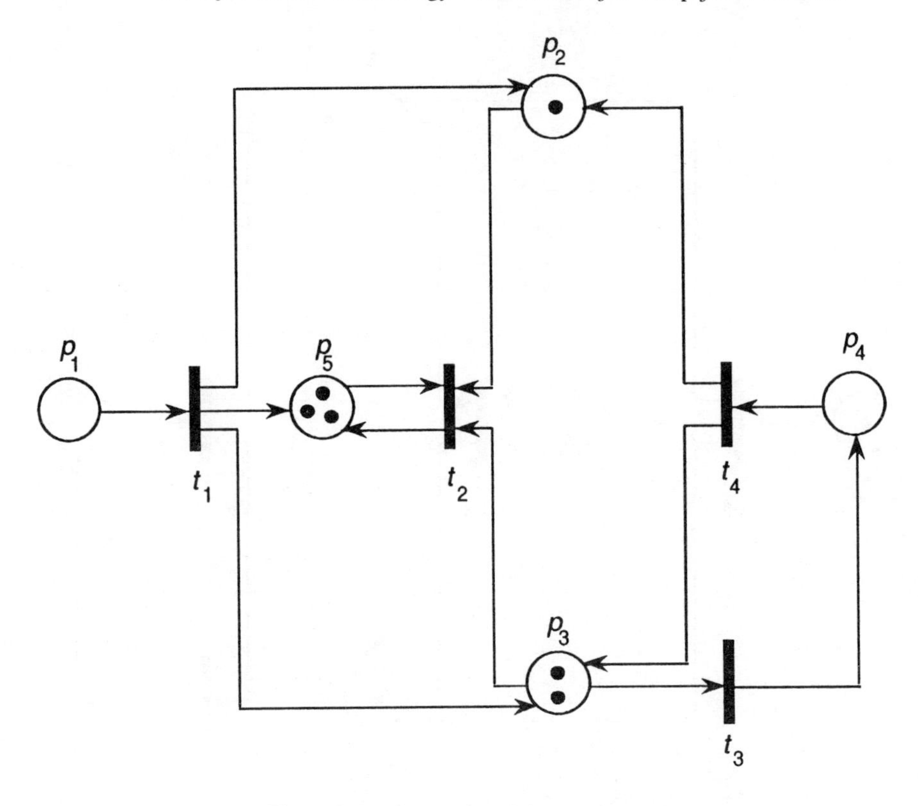

Fig. 4.5. Petri net after firing transition 1.

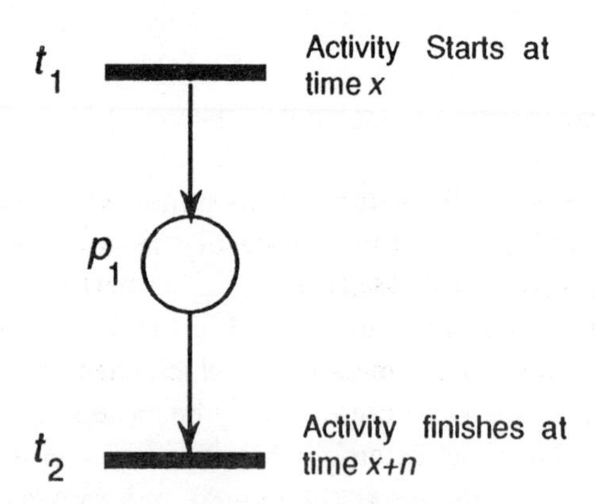

Fig. 4.6. Timing of a Petri net.

The clock time is incremented to the next scheduled firing of a transition when all of the transitions scheduled for the current clock time have been fired.

A timed Petri net is an example of an extension to the original Petri net model proposed by the work of Carl Petri. Tzafestas (1989) describes other relevant extensions to the Petri net model including references to coloured, structured adaptive, structured coloured adaptive, modified, coloured stochastic and fuzzy Petri nets.

Thus, Petri nets may be viewed as a valuable tool for analysing systems, and the in section following we illustrate their applicability to modelling the information flow between two distinct building blocks of a PAC system.

4.5 A sample protocol of the reference architecture

In this, the final section of the chapter, we examine a sample protocol between two entities of the IT reference architecture. A protocol is a set of rules and formats which determines the communication behaviour of entities. The modular nature of FC and PAC requires a number of difficult protocols to enable effective communication between the different entities. A complete protocol definition for FC and PAC is outside the scope of this book. We have decided to offer a model of one protocol out of a choice of many. The sample protocol is between two distinct building blocks which exist within two separate layers of the reference architecture, the cell layer and the workstation layer, and are the dispatcher and the producer.

There are a number of distinct messages which are passed between the dispatcher at the cell layer and the producer at the workstation layer. These include:

1. DIS$PRO_GIVE_STATUS: The PAC dispatcher sends a request to the producer to return the producer status.
2. DIS$PRO_STARTUP: When the PAC dispatcher requests the producer to startup, this message is sent.
3. DIS$PRO_START_JOB: This is the most common command the PAC dispatcher sends under normal manufacturing circumstances. The parameters sent along with this message includes the part identification and the quantity to be operated on.
4. DIS$PRO_SHUTDOWN: This message is sent when the PAC dispatcher wants the PAC producer to cease operations. The reasons might include, planned maintenance or the end of a daily shift.
5. DIS$PRO_HALT_JOB: If for some reason a job has to be stopped during its operation, this message will be invoked. It is rare that such a situation arises within manufacturing, but a possible reason could be that a job with a very high priority needs to use the producer.
6. PRO$DIS_RETURN_STATUS: In response to the PAC dispatcher's request for status information, this message returns the information the PAC dispatcher requires.

7 PRO$DIS_NOTIFY_STARTUP: When startup has been achieved, this message is sent.
8 PRO$DIS_NOTIFY_START_JOB: Once the manufacturing operation has been started on the job, the PAC dispatcher must be notified.
9 PRO$DIS_NOTIFY_FINISH_JOB: The finishing of an operation releases the producer to perform subsequent tasks, and this message informs the dispatcher that the producer is once again available for work.
10 PRO$DIS_NOTIFY_HALT: When the dispatcher commands the producer to halt, this message is returned.
11 PRO$DIS_NOTIFY_BREAKDOWN: If for some reason the device to which the producer is connected fails, the producer must inform the dispatcher that a breakdown has occurred. The cause of the breakdown should also be transmitted in this protocol.
12 PRO$DIS_NOTIFY_SHUTDOWN: Finally, when the producer performs an orderly shutdown, this message is sent.

Thus, these messages form the basis of the communication between the PAC dispatcher and the PAC producer. We now describe the interaction of the dispatcher and producer resulting from the DIS$PRO_START_JOB message from the dispatcher. The other messages which result from this initial command are:

- PRO$DIS_NOTIFY_START_JOB
- PRO$DIS_NOTIFY_FINISH_JOB

The Petri net model highlights the information flow between the dispatcher and producer, which belong to two distinct layers of our reference architecture. The model does presents the essential properties which explain the basis of the message passing between the dispatcher at the cell layer and the producer at the cell layer.

Essentially, this model shows the relationship between the controller, which in this case is the dispatcher, and the controlled, the producer. The Petri net model is composed of four transitions and seven conditions. Table 4.2 describes the relationships between the transitions and the conditions, Table 4.3 gives details of the conditions, and the completed graphical Petri net model is illustrated in Fig. 4.7.

Analysis of the initial Petri net model reveals that only one transition can fire, and all other transition events follow on from this activity. The flow of control within the Petri net model is now considered, and the relationships between the transitions are described.

1 As we can see from the initial diagram in Fig. 4.7, the first transition to be enabled is t_1, since all of its three input conditions (p_1, p_2 and p_3) contain one token. (Note that the shaded area of the Petri net diagram highlights those transitions which model the producer actions). This means that if a

Transition	Description	Pre-Conditions	Post-Conditions
t_1	dispatcher commands producer	p_1,p_2, p_3	p_4, p_5
t_2	producer starts operation	p_3, p_4	p_6, p_7
t_3	dispatcher resumes	p_5,p_6,	p_2,
t_4	producer finishes operation	p_7	p_1, p_3

Table 4.2. *Table of Transitions for Petri net model*

Condition	Description
p_1	A job is waiting for an operation
p_2	The dispatcher is available
p_3	The producer is available
p_4	The producer has received a produce command
p_5	The dispatcher awaits producer acknowledgement
p_6	The dispatcher has received producer acknowledgement
p_7	The producer is processing a job

Table 4.3. *Table of descriptions of conditions*

job is available for processing p_1, and the dispatcher is available to make a decision c_2, and the producer is available to perform that operation, then the dispatcher will command the producer to start the operation. When this transition fires, we see in our model that tokens in the input places (p_1, p_2, p_3) are all removed, and tokens are inserted into the output places (p_3, p_4, p_5). The insertion of a token into place p_4 models the inter process communication which occurs, namely the message that is passed from the dispatcher to the producer.

2 We now consider what happens to our model after the first transition

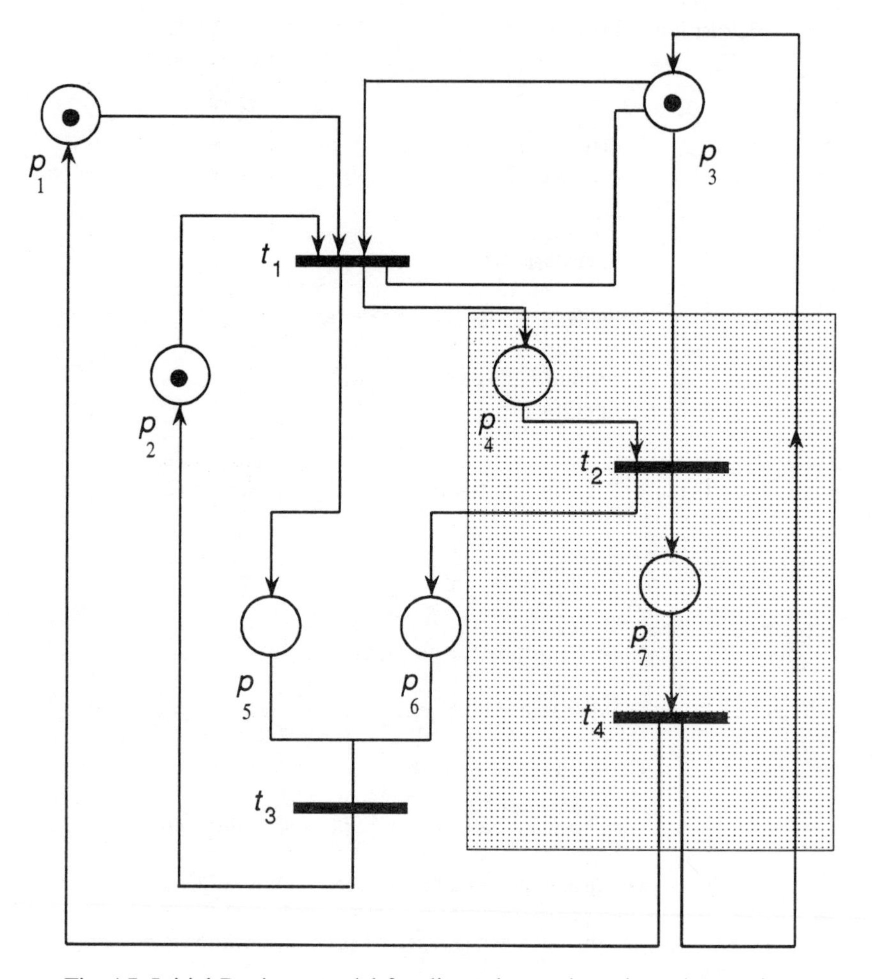

Fig. 4.7. Initial Petri net model for dispatcher and producer interaction.

has been fired, and this is illustrated in Fig. 4.8. Remembering that a transition can only fire if a token is present in each of its input places, we examine the Petri net to find the enabled transitions. Notice that transition t_1 can no longer fire because there are not any tokens in each of its input places. The only transition that can now fire is t_2, which is the event that models the producer starting the operation. Thus all the conditions are present for the producer to start working on the job, namely tokens exist in places p_3 and p_4. Transition t_2 now fires, thereby changing the state of our Petri net model once more. As a result, the tokens in c_3 and c_4 are removed, thereby making the producer unavailable for the present time. A token is also entered into place c_6, and this models

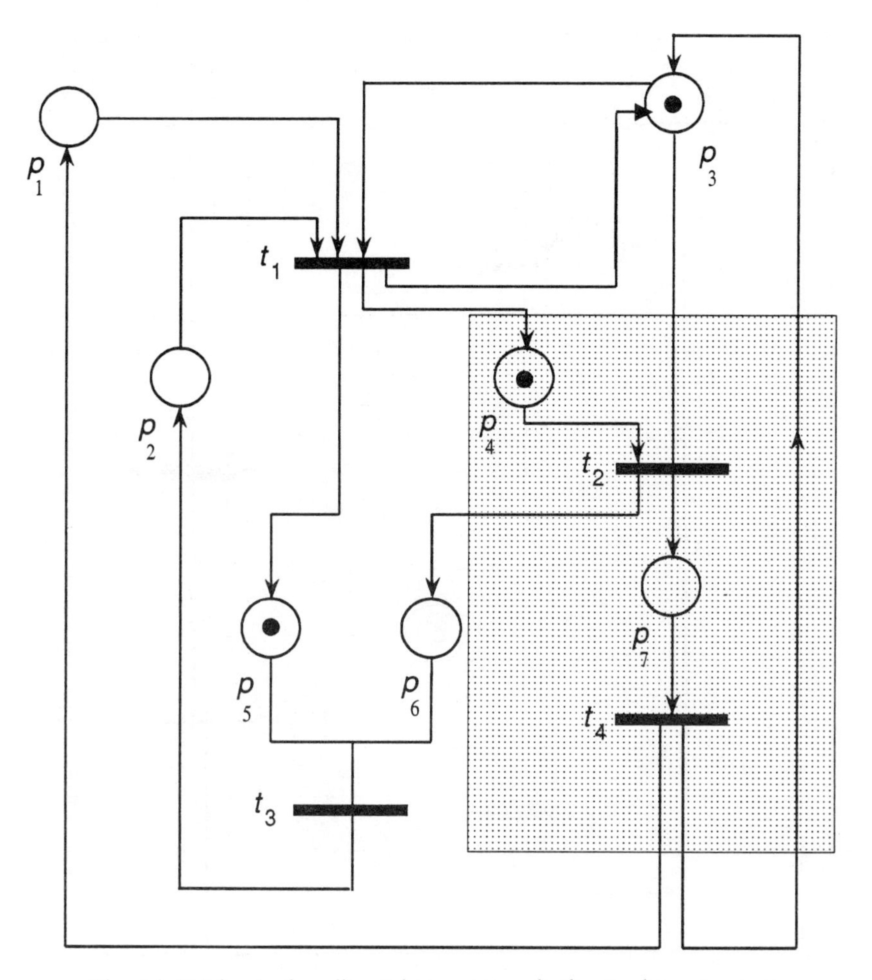

Fig. 4.8. Petri net after dispatcher commands the producer to start.

the flow of information from the producer back to the dispatcher. This data transfer unit informs the dispatcher that the operation on the job has successfully started, and now the dispatcher can resume controlling the other entities within the PAC system.

3 Fig. 4.9 shows the state of the Petri net after the producer has commenced the processing operation on the job. From this model we can see that two transitions are now enabled. Assuming that the processing time for the job on the producer is greater than the time taken for the data transfer unit to make its way to the dispatcher (a reasonable assumption!), the next transition to fire will be t_3. The firing of this transition will then release the dispatcher once more so it can control other jobs and producers in the system.

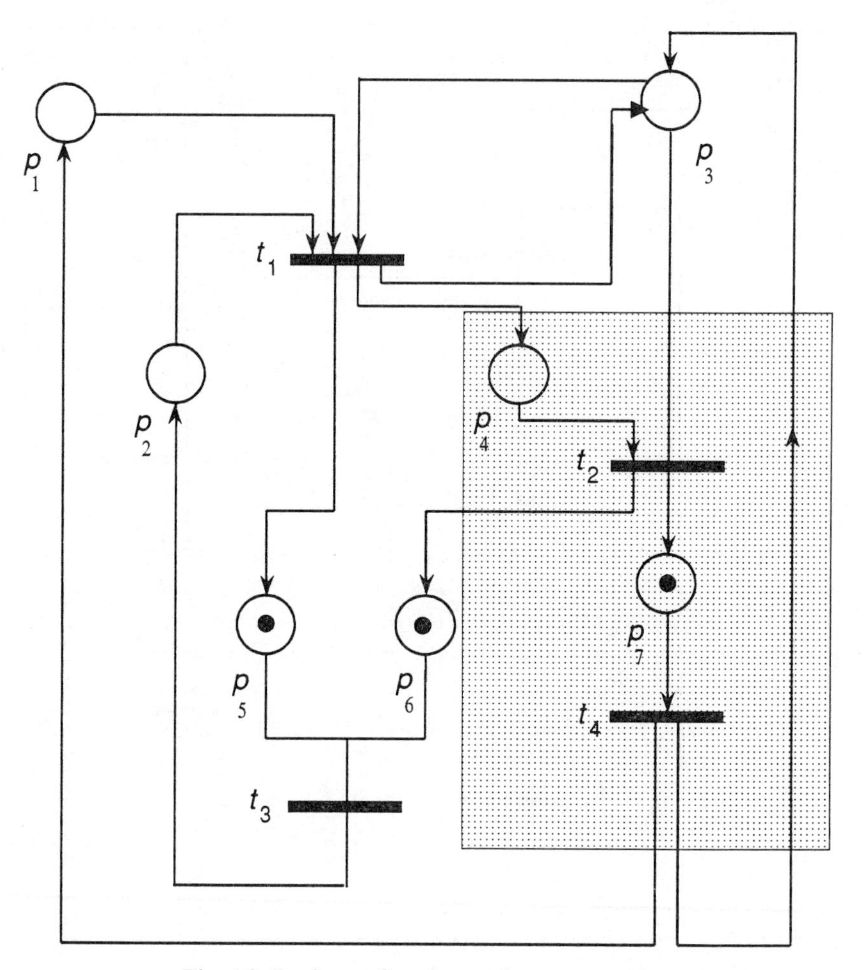

Fig. 4.9. Petri net after the producer commences.

4 Finally, the last transition to fire in our modelling sequence is transition t_4, which models the completion of the producers task on the job. As a result, two places have tokens returned to them, c_3 and c_1. This means that a job is again made available for an operation, and the producer is freed to perform an operation task on any job passing through the system. This act of placing a token in condition c_3 is again another example of how messages pass within the PAC system. The data transfer unit this time is notify finish.

The procedure the dispatcher uses to select a particular job is not shown, as this in itself is a complex activity which will be looked at in detail in our review of scheduling strategies in Chapter 6. It is important to realize that this model is a simplification of one of the protocols which occur between the dispatcher and the producer. Other protocols which could have been considered are:

- The dispatcher orders the producer to shut down;
- the producer informs the dispatcher of a process problem which halts the operation in progress;
- the dispatcher asks the producer for its status;
- the dispatcher orders the producer to start up;
- the producer requests a job from the dispatcher.

It is evident that a complete formal protocol specification for the dispatcher and producer would require a substantial effort in terms of analysis and documentation. We believe that this is outside the scope of this book.

4.6 Conclusions

In this chapter, we have presented a basic information technology reference model for FC and PAC. The development of this model was based on the layered approach, and three distinct layers were identified for our proposed software implementation of shop floor control. These layers consisted of:

1. A factory layer, containing four entities which perform the tasks of FC.
2. A cell layer, containing three of the PAC building blocks, the scheduler, dispatcher and monitor.
3. A workstation layer, containing the remaining two PAC building blocks, the mover and producer, which translate cell layer commands into specific device instructions.

In order to implement this model of FC and PAC, it is clear that a reference model in itself is not sufficient. What is also required is a set of IT tools, both hardware and software, which can realise this model of shop floor control. Clearly there is a need for a communications mechanism between entities and layers, a data management facility to store the manufacturing data, and software applications capable of performing the countless manipulations of each of the individual entities.

Our aim in Chapter 5 is to analyse the requirements of the IT reference architecture presented here, and suggest the most suitable IT tools to realise the threefold requirement of communications, data management and applications development.

5

Implementation technologies for shop floor control systems

5.1 Introduction

In the previous chapter we presented an information technology reference architecture for FC and PAC. We now describe some Information Technology tools and standards, currently available and under development, which we believe may be important in realizing successful implementations of FC and PAC. Our primary objective in this chapter is to examine our information technology reference architecture for shop floor control systems, and associate with it specific hardware and software tools. We briefly review the most important information technology tools and standards, and give examples of how these tools can be applied in real-life shop floor control systems. Furthermore, we stress the importance of portability, flexibility and inter-operability for shop floor control systems, attributes which we believe are vitally important in a multi-vendor computing environment. The structure of this chapter is as follows:

1 We present an initial overview of information technology, highlighting the importance of flexibility, inter-operability and portability of IT systems. A general computing model is then introduced, and this contains four essential elements required to realize an information technology implementation of shop floor control. These elements are communication systems, data management systems, processing systems and user-interface systems.

2 The main body of the chapter describes the relevant state-of-the-art technologies of the four elements included in the general computing model. At the conclusion of each section, a synthesis of what we believe are the most suitable technologies is presented, in the form a shop floor control model for each particular computing element. Thus, we also present a communication systems model, a data management model, a

processing model and an user-interface model for the implementation of FC and PAC systems.

3 Finally, we focus our attention on the emerging object-oriented technology, which holds the promise of highly portable, flexible and inter-operable solutions within the domain of shop floor control.

5.2 An overview of information technology

The overall goal of information technology within a manufacturing environment is to help achieve the business goals of an enterprise. Information technology may be viewed as an adaptive mechanism for following the frequent changes and evolution of the technological environment, the market, and organizational structures. Three key aspects of any information technology system are:

1 flexibility;
2 portability;
3 inter-operability.

Each of these key features will now be discussed in turn.

1 **Flexibility**
In a production environment, new products have to be introduced with minimal disruption to existing manufacturing applications. A shop floor control system implementation must also be able to adapt quickly to the changes brought about be the introduction of new products and processes. Flexibility may be achieved through distribution and decentralization, and by placing the decision making at the point where the changes are likely to be introduced. For example, decisions in terms of production planning should not be made centrally in a top down approach, but more responsibility should transferred to the individual production units.

The trend in production systems is towards decentralization, and this places a high demand on distributing the databases, user-interfaces, and the individual functionality of each application. We believe that the importance attached to flexibility will result in a growing demand for configurable shop floor control systems, and consequently a corresponding demand for the required software tools.

2 **Inter-operability**
The typical information technology based manufacturing environment is characterized by a multitude of hardware and operating system platforms from different vendors. The primary goal of inter-operability is to integrate these various platforms, and enable designers to build coherent shop floor control systems utilizing the most suitable base hardware

and software components. One of the characteristics of a multi-vendor computing environment should be that the location of information or processes is totally transparent to the user. Ideally, a user of a computerised system should not need to know whether, for example, a scheduler runs on a local computer, or remotely on a different computer. The Open System Interconnect (OSI) model, discussed in the communications section, defines standards for inter-operable systems. The OSI standard will eventually provide most of the standards for software tools necessary to bridge various hardware and software platforms.

3 **Portability**
Applications must be portable across various computer platforms. Portability implies that application software developed on a given hardware or operating system platform can easily be transported to another platform if desired. This requires the use of standard programming language syntax, as well as standard calls to the operating system.

Thus, we have identified the three most important criteria, flexibility, interoperability, and portability, in selecting the software tools and standards for a successful shop floor control system implementation. The software tools and standards outlined in this chapter are selected bearing in mind these important criteria.

Now we discuss briefly the software tools and standards currently available and emerging that can be used for successful shop floor control system implementations. We introduce a general computing model, illustrated in Fig. 5.1. This computing model is used as a framework in order to categorize the software tools and standards to be discussed. In this computing model we can identify four basic components, namely:

1 Communication;
2 data management;
3 processing;
4 user interface.

Ideally, each of these components should be distributable in a heterogeneous multi-vendor environment. For instance, the scheduler building block might be constructed so that the user-interface (for example, the task of displaying and modifying Gantt charts) is executed on a workstation while, the actual computing of the schedule is performed on an alternative, more powerful, computer. This represents a typical requirement in a manufacturing environment with heterogeneous computing resources in place. The communication component in Fig. 5.1 is concerned with the exchange of messages within a shop floor control system. The data management component is responsible for handling access to data for each of the building blocks. Processing performs the actual application functionality of the FC and PAC building blocks.

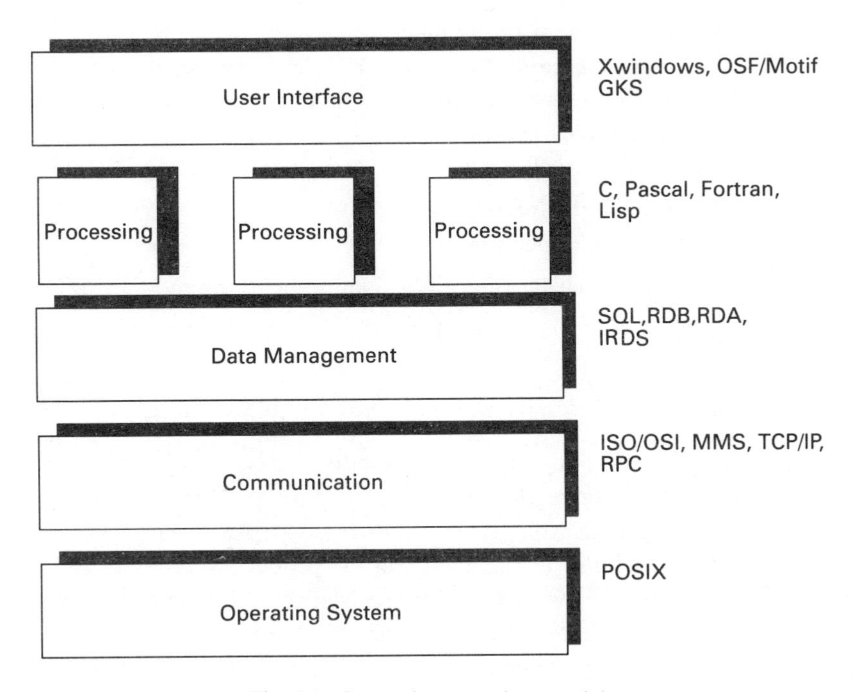

Fig. 5.1. General computing model.

Now we present an brief overview of the software tools and emerging standards under the topics of data management, communication, user interface, and processing. We have chosen these software tools and standards based on the selection criteria (flexibility, portability, and inter-operability) identified earlier in this chapter. We also present possible scenarios for using these software tools for FC and PAC implementations.

5.3 Communication systems

Today CIM is expanding beyond the boundaries of the manufacturing company to the total manufacturing enterprise. An important goal of an enterprise is to integrate the various internal and external business functions, which includes direct links with the supplier chain on the supply side and links forward into distribution channels and to the end customers on the market side. In a CIM environment, networks provide the means to communicate, and have become increasingly important for leading manufacturers. Standards provide the enabling technologies for implementing global networks in a cost-effective manner. In this section we present an overview of a selection of common network technologies used within manufacturing, namely:

1 ISO/OSI (Open System Interconnect);
2 CNMA (Communications Network for Manufacturing Applications);
3 MAP (Manufacturing Automation Protocol);
4 MMS (Manufacturing Messaging System);
5 RPC (Remote Procedure Call);
6 EDI (Electronic Data Interchange);
7 Low Cost Device Connection.

We conclude this review by offering a communications model for FC and PAC, which is based upon a synthesis of the forementioned technologies.

Open System Interconnect – OSI

The emerging industry-wide standard architecture for computer networks is the Open System Interconnection (OSI) architecture developed by the International Standards Organization (ISO). ISO has created a seven-layer architectural model, illustrated in Fig. 5.2, and is in the process of defining, endorsing, and approving protocols for each layer. The basic structuring technique in the OSI reference model is that of layering. According to this technique, each open system is viewed as being logically composed of an ordered set of subsystems. Information is transferred in various types of data-units between peer-entities across a multi-vendor environment.

As the highest layer in the OSI reference model, the application layer provides a set of services for the application-processes to access the OSI environment. Examples of application layer services include file transfer, database access, job transfer and order entry. Taking each layer in turn, we can make the following observations:

1 The purpose of the *application layer* is to facilitate the exchange of information between correspondent application processes across different nodes in a multi-vendor environment.
2 The *presentation layer* may transform data before it is presented to the application layer.
3 The *session layer* organises and synchronises the dialogue and manages data exchange.
4 The transport layer provides transparent transfer of data between session-entities and relieves them of any involvement in the detailed way in which reliable and cost effective transfer of data is achieved.
5 The network layer establishes, maintains and terminates network connections between open systems. Furthermore it performs the routing and relaying of network connections.
6 The data link layer provides functional and procedural means to establish, maintain and release data link connections.

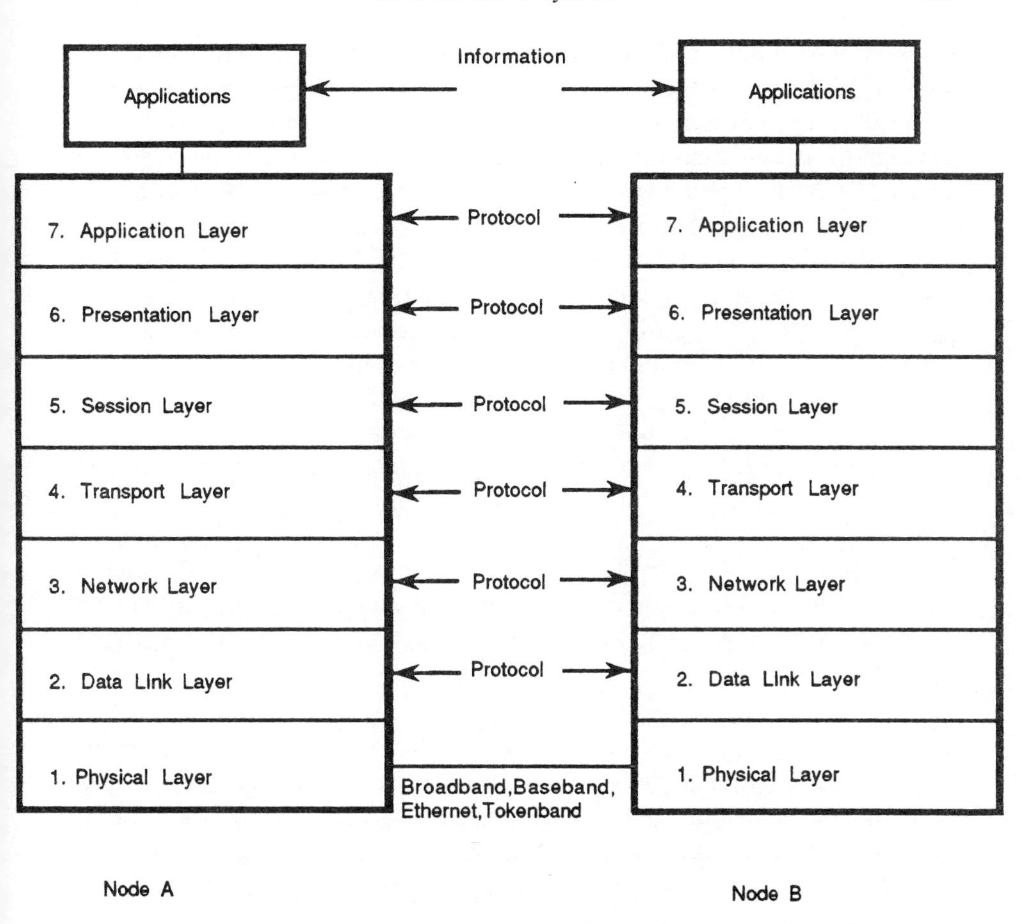

Fig. 5.2. OSI reference model.

7 The physical layer provides mechanical, electrical, functional and procedural means to activate, maintain and deactivate physical connections for bit transmissions.

Thus, the OSI model forms a framework for the development of communication protocol standards. We now discuss several communications products designed and developed to adhere to the OSI standards.

TCP/IP This is a set of protocols developed to allow cooperating computers to share resources across a network. It was developed by a community of researchers sponsored by the U.S. Advanced Research Projects Agency (ARPA) in 1974 (Cerf and Kahn, 1974). TCP/IP has become most popular in the UNIX^{TM*} operating system environment. Today however most of the com-

* UNIX is a trademark of UNIX Systems Laboratories, Inc.

puter vendors support TCP/IP on various kinds of operating systems including OSF/1[TM†] from the Open Software Foundation. TCP/IP is a 'connectionless' protocol. Information is transferred in 'packets' Each of these packets is sent through the network individually. Similar to the previously described ISO/OSI Model, TCP/IP is also based on layered protocol structure. IP (the 'internet protocol') is responsible for routing individual packets. IP functionally fits in the network layer of the ISO/OSI reference model. It is mainly responsible for routing messages in a complex network of computers. The TCP (transmission control protocol) is layered on top of IP. It is responsible for breaking up the message into packets, reassembling them at the other end, resending anything that gets lost, and putting things back in the right order. On top of the TCP there are a number of application protocols performing specific tasks, e.g. transferring files (ftp) between computers, sending mail (mail), remote login (telnet), and remote procedure call (rpc) and more. TCP/IP can communicate over Ethernet, ISO/CCITT X.25 (public packet-switching) and simple serial line networks. The internet standard is continuously extended by the Internet Activities Board (IAB). As OSI becomes more widely implemented there will be need for interoperating TCP/IP with ISO/OSI protocols (Open System Handbook, 1991).

CNMA Communications Network for Manufacturing Applications (Kreppel, 1988) is a European initiative, supported by the Commission of European Communities under the ESPRIT (European Strategic Program for Research and Development in Information Technology) programme. CNMA aims to specify, implement, validate and promote communication standards for manufacturing which are emerging within the framework of the ISO seven-layer model for open systems interconnection. The goal of this activity is to make it easier for vendors of computer, communication and control equipment to quickly develop and market equipment which is compatible, thereby meeting the user requirement of interoperability. Probably the most significant goal of CNMA is to define protocol profiles through close cooperation amongst vendors and users. CNMA offers a choice of different protocols, including Manufacturing Messaging Specification, and local area networks are employed as a means to link devices. CNMA has been developed to ensure compatibility with MAP and ISO standards.

The Manufacturing Automation Protocol – MAP MAP is based upon the seven-layer OSI standard, and is designed to meet the requirements of manufacturers dealing with multi-vendor device connections on the shop floor. MAP

† OSF/1 and Open Software Foundation are trademarks of Open Software Foundation.

has chosen the broadband, token passing bus topology as the physical carrier. An important component of MAP V3.0 is the MMS (Manufacturing Messaging Service), which is a protocol for communications between computers and plant-floor devices. Currently, vendors and users of robots, Programmable Logic Controllers (PLC), Numerical Control (NC) machines, and process controllers are writing companion standards for MMS to define how each class of the devices will implement MMS protocols. When these companion standards are completed, the task of integrating applications with devices from multiple vendors will be greatly simplified. MMS can be regarded as a level seven (application layer) service which is likely to become available also in other OSI implementations using Ethernet, Baseband, etc.

The Manufacturing Messaging Specification – MMS MMS supports communications to and from programmable devices such as Numerical Controllers and Programmable Logic Controllers, and can be positioned as an OSI layer seven protocol. MMS originated from the work within the Electronic Industry Association (EIA) in defining a communication standard for NC equipment. MMS provides a set of services to application programs. In the case of FC and PAC, it would enable the producer to talk to the plant floor device controllers. MMS allows the programmer to define objects which can be identified in plant floor devices. Examples of MMS objects are variables, events, alarms, and other objects of interest to application programs. Furthermore MMS provides a set of services to manipulate these objects, and to read/write to variables located within the device controllers.

MMS provides a set of convenient services to the application program for interfacing to device controllers. MMS creates an additional level of portability across different device vendors thus increasing the user's flexibility in procuring shop floor devices.

Remote Procedure Call – RPC The aim of the RPC (Remote Procedural Call) (ECMA, 1987) standard is to facilitate the implementations of distributed applications. It enables programmers familiar with simple procedural calls to call programs on remote computers without being concerned with communication issues. RPC enables application programs to be written in different programming languages and executed in different operating system environments. Applications can be written in a local environment and later become distributed with little or no change. RPC is supported by networks conforming to the emerging OSI standards.

Electronic Data Interchange – EDI In modern manufacturing there is a growing demand from trading partners for design engineers to communicate from a Computer Aided Design (CAD) workstation to vendors' CAD systems.

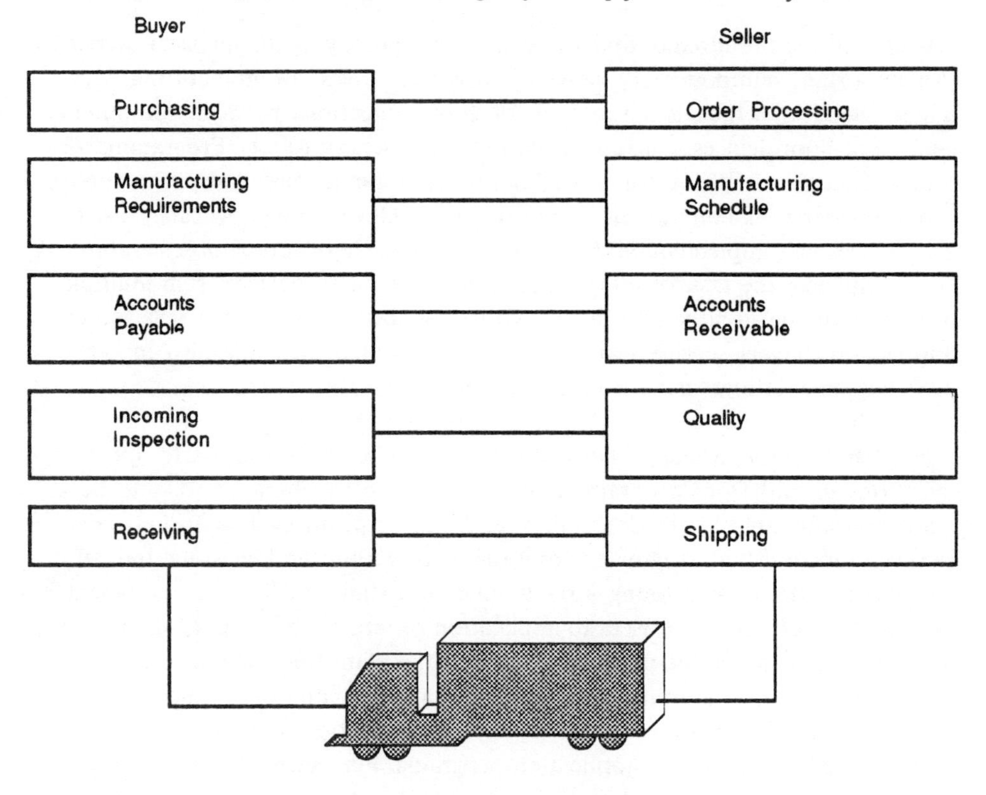

Fig. 5.3. EDI concept.

This can improve design quality, reduce design time, and monitor product performance. EDI is the computer-to-computer exchange of inter- and intra-company business and technical data, based on the use of agreed standards.

Aerospace and automotive companies have already implemented such schemes, and in many cases these are being extended to include their sub-contractors. Today many manufacturers are already exchanging electronic data on orders and shipping details. These companies have created EDI trading links to reduce their paper flows, and to remove postal and processing delays.

Alongside techniques such as Just In Time, EDI enables companies to work more closely together, to their mutual benefit. Planning, production, and communications are coordinated into a form of partnership.

Low Cost Device Communication RS232 was introduced in 1962 and is widely used in manufacturing to connecting PLC and other intelligent devices. RS232 was later replaced by the RS423 which implements a data transfer rate of

100 kbits/s over a distance of 100 m. RS232 and RS423 facilitate low cost solutions to device connection problems.

The Fieldbus standard is currently being developed for controller to device connections. It will replace the 33-mA, and RS232 serial line standard commonly used today at the plant floor. The Fieldbus implements a digital replacement for the older analog technology providing higher reliability.

The BITBUS is a high-speed serial interconnect developed by Intel Corporation to support communication between PLC and plant-floor computers that monitor and control the manufacturing process. The BITBUS message protocol addresses up to 250 nodes within a single network, and implements a transfer-rate of 62.5 kbits/s at a distance of 13km over a twisted pair line.

5.3.1 A communications model for shop floor control

Fig. 5.4 illustrates an example of the use of communication software tools and standards for FC and PAC. The main task of the various networks in the model is to connect the building blocks and allow them to exchange messages between factory, cell, workstation and the device level.

We recommend the use of Open System Interconnect (OSI) standard compliant networks for the implementation of shop floor control systems in order achieve a high degree of inter-operatability. Remote Procedural Call (RPC), which will be described in section 5.5, is also suitable for implementing communication between the building blocks. MMS (Manufacturing Messaging System) is best suited for the communication between the producer and mover to device controller on the plant floor. It can perform the tasks of downline loading part programs to CNC machines, monitoring, and control of the devices.

We have outlined the most important communication technologies that support a multi-vendor environment. Of course, we acknowledge that there are alternative propriety networks that can perform the function of communicating between building blocks in a shop floor control system implementation.

5.4 Data management systems

Bray (1988), defined a database management system as ‘a set of system software to define, retrieve and modify data stored in a database’. An important objective of database management system is to give access to information in a timely and concise manner, and to manage and effectively use information as a corporate asset. Database management systems allow information to be distributed across various computing platforms on a network and be presented as a compact unit of information to a user or an application.

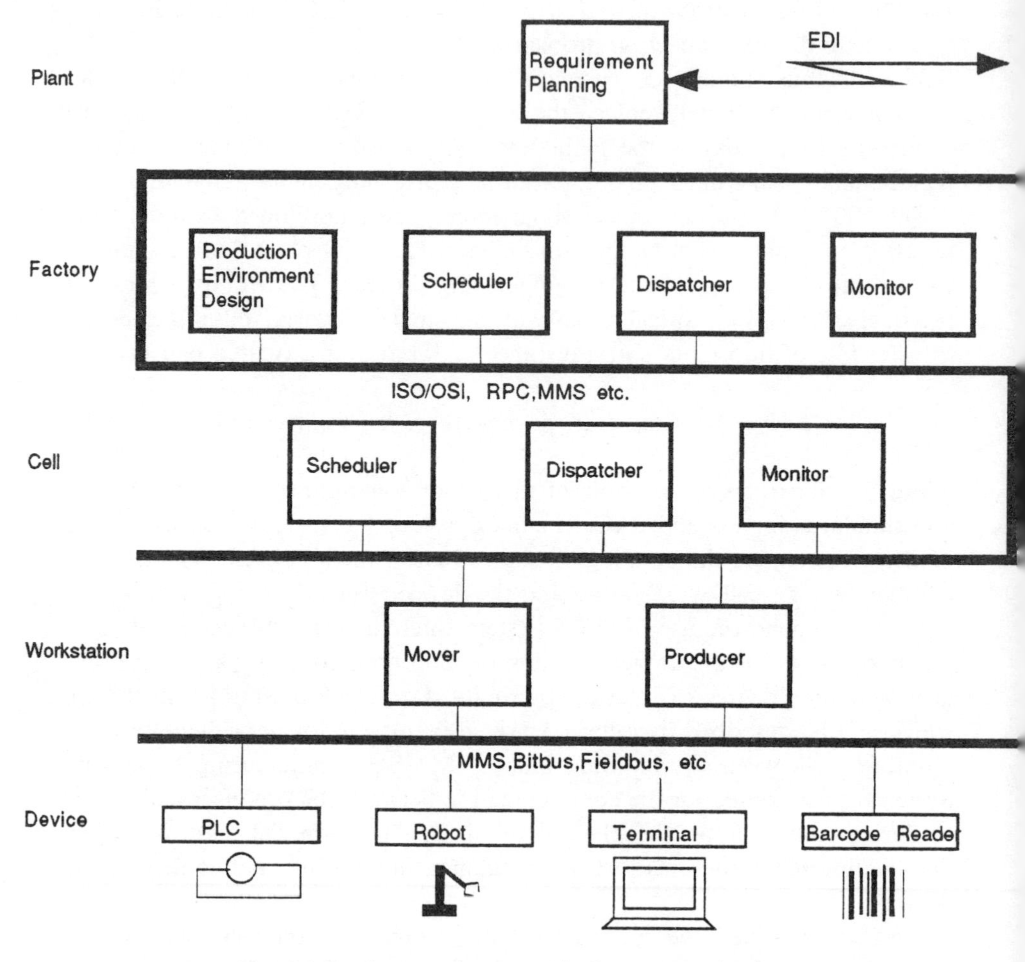

Fig. 5.4. Implementation scenario for communication.

Within this section, we review the three major database management systems:

1 Hierarchical;
2 network;
3 relational.

We also consider systems which are available to create, retrieve, modify and store data in databases, namely:

1 Standard Query Language (SQL);
2 Information Resource Data Dictionary (IRDS).

We then conclude this review by offering a data model for shop floor

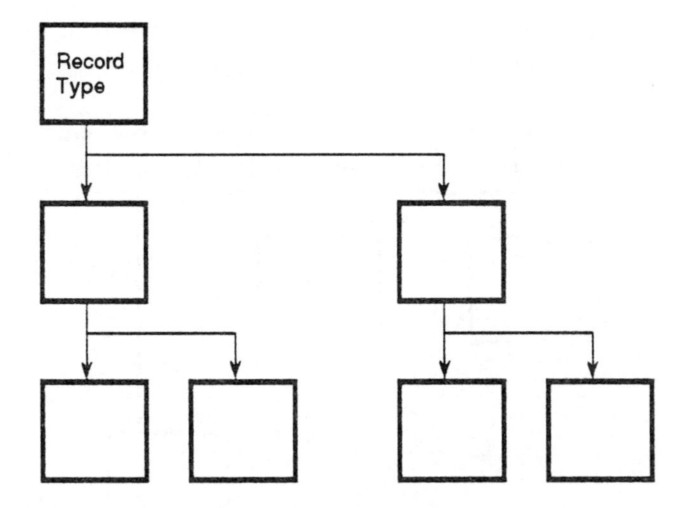

Fig. 5.5. Hierarchical data model.

control, which is based upon a synthesis of a selection of the reviewed database technologies.

Hierarchical databases In a hierarchical database management system records are related to each other in a hierarchy or tree structure, as indicated in Fig. 5.5. Each record can have zero, one, or more subordinate records (children) in the hierarchy, but may have at most one superior record (parent). The hierarchical approach emerged out of the principle of sequential file structures. It has a major disadvantage compared to the more modern database management systems, in that access to data is biased towards only one access path, and this may result in a considerable amount of redundancy in data models.

Hierarchical databases were one of the first data management systems that become commercially available. However, according to Smith and Barnes (1987), this data model is of historical significance only, and is not likely to be the basis of many future data management systems. Readers interested in a full discussion on hierarchical data models should refer to Date (1988).

Network databases Network database management systems may be regarded as an extended version of the hierarchical data model. The principal difference is that a child record can have any number of parent records, and this is illustrated in Fig. 5.6. The network data model approach to database management systems allows more flexibility in modelling data relationships than the hierarchical approach. Network database management systems consist of three major components for definition and querying of the data.

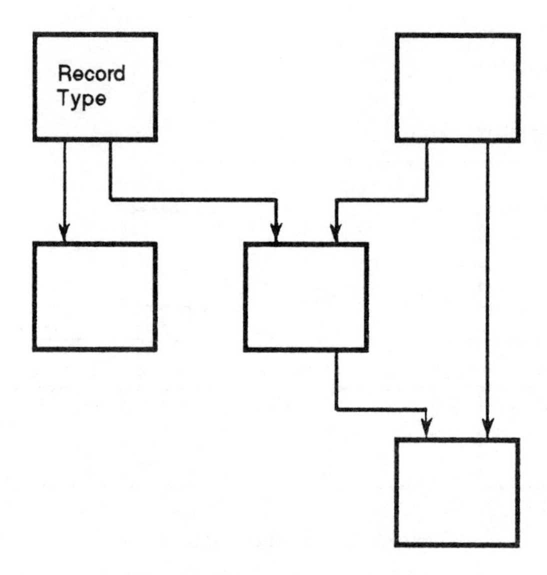

Fig. 5.6. Network data model.

1 A schema Data Description Language (DDL) describes the records and network structure.
2 A subschema Data Description Language (subschema DDL) allows the external views of the database to be defined.
3 A Data Manipulation Language (DML) provides a set of operators for retrieving and modifying data in the network database.

Network database management systems have been commonly applied in commercial systems, although the current trend is towards applying relational database management systems. Network database management systems are suitable for an environment where the data model remains reasonably stable over a long period of time, and where large amounts of data must be handled. The major disadvantage of the network data model is its complexity and structure. From an application programmers perspective, detailed knowledge of the logical structure of the database is required. Smith and Barnes (1987) acknowledge that this places a heavy burden on the programmer, and application programs written for network databases tend to be complex and not easily modified.

Relational databases Relational database management systems have been the most popular database management approach since the late 1970s. Since then, they have substantially grown in functionality and performance. In the field of distributed databases, the relational model is considered the most advanced technology in current use. The principles of the relational model were originally

Product_Name	Operation	Workstation	Time
Board_X	1	Station_A	20
Board_X	2	Station_B	10
Board_X	3	Station_C	5
Board_Y	1	Station_A	15
Board_Y	2	Station_B	25
Board_Y	3	Station_C	5

Table 5.1. *The process relation*

laid down by Codd (1970), and the relational approach is somewhat different from the hierarchical and network data model approaches. In the network model, for example, pointers embedded in the each data record are used to link the records, thus establishing the desired logical relationships among records. The relational data model approach does not employ embedded pointers.

The relational data model can be thought of as two dimensional tables consisting of columns and rows. A row is similar to a record and a column consists of all values for a given data item for a record. For a given database, many relations can be defined which together comprise the database. In order to access the database and generate logical relationships, a series of operations are performed on the different relations, using a language (Structured Query Language) based on the relational algebra.

A relation is a mathematical term for a table. In the relational data model the data structure is composed of relations, domains, and tuples. A sample relation for manufacturing process information is illustrated in Table 5.1, and this example will now be used to illuminate some of the important ideas of the relational model.

Relational data models use the terms domain, attribute, tuple, primary key, and relation itself. For further clarification we draw an analogy to other terms typically used in some of the available database management systems.

1. A relation is comparable to a traditional file with one record-type.
2. A tuple corresponds to a row within a table or relation and it can be compared with one specific record of a traditional file systems.
3. A domain corresponds to a column within a table or relation. For example, a domain is all occurrences of operation in Table 5.1.
4. A primary key uniquely identifies a tuple or record within a table or relation.
5. A foreign key links two relations or tables. Domains of two relations are linked if the values of the tuples match.

There are three main relational operations that can be applied to manipulate data within the relation data model. These are selection, projection and join.

Product_Name	Operation	Workstation	Time
Board_X	1	Station_A	20
Board_X	2	Station_B	10
Board_X	3	Station_C	5

Table 5.2. *Selection operation in the relational model (Board_X)*

Product_Name	Operation	Workstation
Board_X	1	Station_A
Board_X	2	Station_B
Board_X	3	Station_B
Board_Y	1	Station_A
Board_Y	2	Station_B
Board_Y	3	Station_C

Table 5.3. *Projection operation in the relational model*

Firstly, selection is the simplest of the three operations, and it may be viewed as selecting data from one or more rows of a particular relation. The general form of this operation, as described by Smith and Barnes (1987), is:

relation2 = SELECT relation1 WHERE condition

This concept is illustrated through the simple example in Table 5.2. In this case, relation2 is the output shown in the table and relation1 is the data in Table 5.1. The important condition is to scan relation1 for all data on the product Board_X.

Secondly, projection may be viewed as selecting data from particular column(s) of the relation. The general form of this operation, as described by Smith and Barnes (1987), is:

relation2 = PROJECT relation1 ON [attribute,attribute...]

Projection is illustrated through the example in Table 5.3. For this operation, relation2 is the output shown in the Table, and relation1 is the data in Table 5.1. The attributes isolated for this particular projection operation are the product, operation and workstation, and all data pertaining to these three attributes are presented in Table 5.3.

Finally, join is a dyadic operator which amalgamates related data from different relations, and presents the resulting attributes in one table. The most general form of this operation, as described by Smith and Barnes (1987), is:

relation3 = JOIN relation1,relation2 WHEN condition

To illustrate this join operation, it is necessary to define a new relation which is related to our existing relation. This new relation, illustrated in Table 5.4,

Product_Name	Lead_Time
Board_X	10
Board_Y	21

Table 5.4. *The product relation*

Product_Name	Operation	Workstation	Time	Lead Time
Board_X	1	Station_A	20	10
Board_Y	1	Station_A	15	21

Table 5.5. *Join operation in the relational data model*

is one which describes the product, and it shares the common attribute of product_name with the previously described process relation.

Table 5.5 illustrates the join operation, and in this case the join occurs only where the following two conditions are met:

product.product_name = process.product_name, i.e. search the two relations for the same product_name;

process.operation = 1, i.e. only join data with an operation number = 1.

Hence, all the data from the two relations with the same product names and where the process operation number equals one, are presented in the new table.

Data within relational databases may be accessed through Structured Query Language (SQL). We believe that relational database management systems are the most suitable technology for implementing the data management function of shop floor control systems. This is mainly due to their inherent flexibility, and their potential for distributiveness. Relational database management systems are today available on all ranges of computers, from personal computers to large mainframes.

Comparison of hierarchical, network and relational data models

From a performance viewpoint, the hierarchical and network data models are more or less equivalent in terms of their space efficiency. The storage space is well utilized, as each of these models implement data associations through the use of pointers rather than by data values. This common feature reduces the transaction time. However, transactions in the relational model take longer to process, because the associations between the individual data structures must be derived dynamically.

From an application programmer's viewpoint, the use of the data manipulation language for each data model also varies. The hierarchical and network models are similar, and both of these are significantly more complex to manipulate than the relational model. The elegance and simplicity of the relational model, illustrated through the examples in the previous section, has facilitated its use for modern day applications.

Finally, in a relational data model a user can dynamically build new relationships to restructure the data model. This makes the relational data model approach much more flexible for coping with unanticipated changes. Network and hierarchical database management systems require the data model to be predefined in the database definition, and hence if an application needs a new data structure, these database management systems often must be redefined and converted. This is an important factor within the context of shop floor control systems, as the inevitable introduction of new product and process technologies make corresponding demands for flexibility in the data management system. Thus, we believe that the relational model is the most suitable to handle the data management aspect of FC and PAC systems.

SQL – Structured Query Language SQL (Structured Query Language), defined originally by Chamberlin (1974), is a data definition and data manipulation language for relational databases. Using SQL, a database may be created, populated with data, and its data and structured modified. The SQL language can be used interactively by the user to query the database, or alternatively it can be used directly from most common programming languages. SQL offers the select, project and join operations described previously. Joins are the most powerful feature of relational database management systems, as they allow a user to dynamically traverse and determine the structure of the database.

Various standard organizations such as the International Standards Organization (ISO) and the American National Standards Institute (ANSI) are now actively pursuing standards in relation access to relational database management systems. The remote data access (RDA) project is an initiative within ISO that has chosen SQL as the basis for defining a standard to access remote databases.

IRDS – Information Resource Data Dictionary Data dictionary systems provide the ability to create, analyse and administer metadata. Metadata describes data, how this data is represented, and how it is used. It includes the location, type, format, size, and change history of the data. Through the use of a data dictionaries, the integrity of metadata used in a distributed database environment is ensured. This is an important aspect of data management, since it is vital that all databases use the same definitions of an attribute, e.g. 'part'. The Information Resource Data Dictionary (IRDS) is a proposed ANSI/ISO standard for the definition of data dictionary interfaces. The adoption of this standard enables standardization on the data dictionary product across multiple system platforms.

Thus far, we have reviewed the four classes of available data management systems, and also SQL and IRDS. We now return to our IT reference architecture for shop floor control, and suggest the most appropriate tools to fulfill the data management aspects of FC and PAC.

5.4.1 A data management model for shop floor control

One of the key requirements for the implemention of shop floor control systems is that the building blocks remain autonomous components, which can be configured according to the need of a specific business problem. We distinguish between static and dynamic information. Dynamic information should be kept local to the individual building blocks and is required to perform the function of a given building block. Static information common to all building blocks should be contained in a shared manufacturing database. The five databases shown in Fig. 5.7 can also be contained in one database with separate logical units; alternatively they can be physically distributed to the computer where the building block is running. The software tools shown in Fig. 5.7 allow for a flexible database environment in shop floor control systems. SQL is a consistent interface for building blocks to access databases. As discussed earlier we recommend the use of Relational Database Management Systems (R-DBMS) because of their inherent flexibility and distributiveness. The data dictionary function of shop floor control systems should be handled by IRDS compliant data dictionaries. Finally, access to remote databases should be achieved through the RDA mechanism discussed earlier.

5.5 Processing systems

Processing systems allow each of the individual entities of the FC and PAC systems to be implemented, as they enable the development of coded instructions required to perform the many different tasks of shop floor control. Within this section we review three approaches to the development of the processing element of the reference architecture, namely:

1 Programming languages;
2 Remote Procedure Call (RPC);
3 POSIX.

We conclude this review by offering an implementation model for processing within FC and PAC systems.

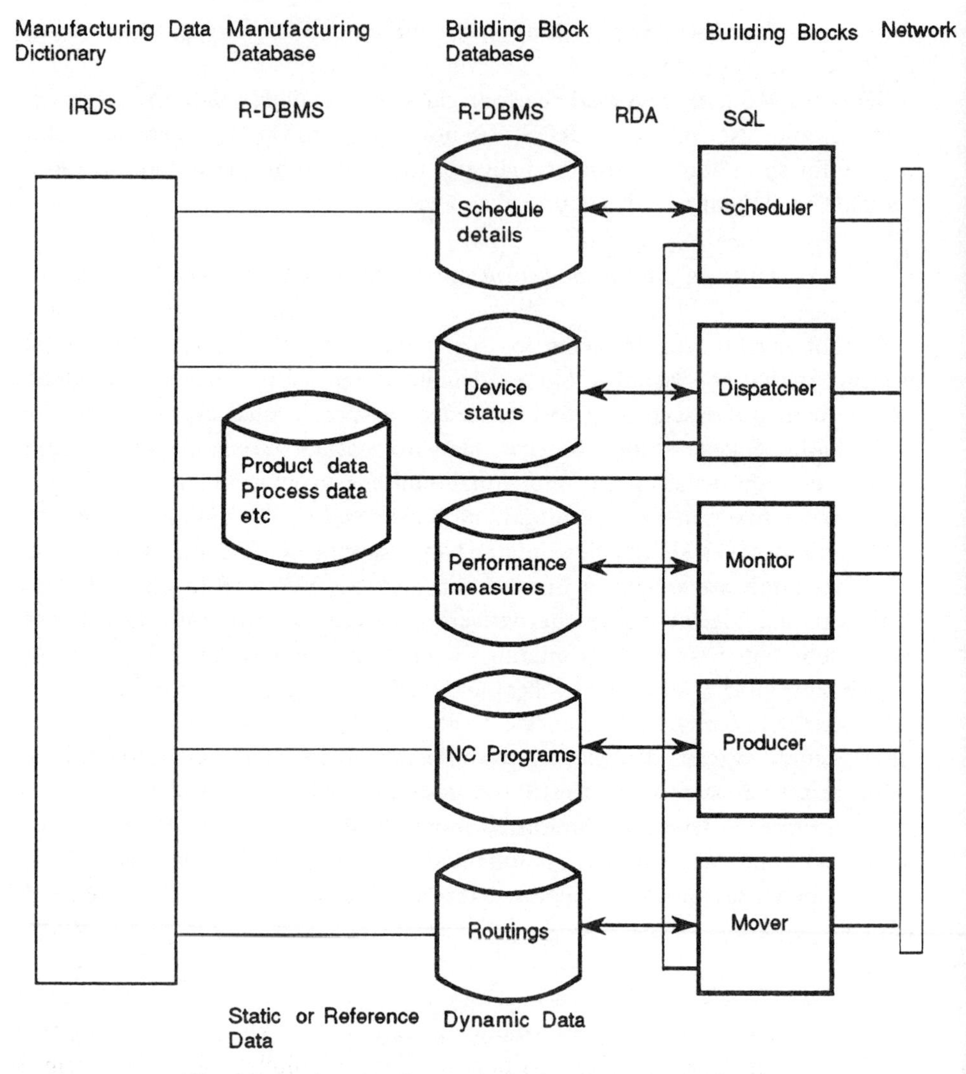

Fig. 5.7. Implementation scenario for data management

Programming languages There are a wide variety of programming languages available to implement the functional requirements for the individual FC and PAC applications. The most widely known and used conventional programming languages are Pascal, C, FORTRAN, ADA and COBOL. Baron (1986) suggests that present day computer languages may be viewed in terms of six distinct approaches to problem solving.

1 **Imperative/Algorithmic**

The terms imperative and algorithmic refer to the same set of computer languages. Imperative indicates that these programming languages are issuing instructions as to what values should be placed in the memory locations of the computer. Algorithmic tells us how these instructions are actually given: namely, through algorithmic statements telling us how to carry out procedures. Imperative/algorithmic are the best known and most widely applied of all computer languages, and examples include C, BASIC and Pascal.

2 **Functional/Applicative**

An alternative conception to program design is functional, or applicative, languages. These terms derive from their mathematical meaning, and in mathematics, a function is a way of relating elements of one set (the domain) to elements from another (the range). In a functional language, these functions are applied in order to solve a specific problem, hence the name applicative. Three separate attempts have been made at developing computer languages in terms of mathematical functions. The earliest of these was LISP (LISt Processing), developed in the late 1950s and early 1960s at MIT, Boston, in the United States. LISP is an example of an artificial intelligence language. Other functionally conceived languages are APL (A Programming Language), developed at Harvard in 1960, and SNOBOL (StriNg Oriented SymBOlic Language), developed at Bell laboratories in the mid-1960s.

3 **Object-Oriented**

Object-oriented languages use a different perspective on programming. These languages are structured in terms of objects, where an object is defined as an area in computer memory that serves as a basic structural unit of analysis. The object-oriented approach is discussed in more detail in the the final section of this chapter.

4 **FORTH**

FORTH is known for its application in the fields of astronomy, robotics and graphics. FORTH is more of a grammar than a language, in that it allows users to write a new language each time a new program is written. Thus, in theory, FORTH is potentially an infinitely extensible system. It is made up of a small set of units, known as words, and these are used for constructing new words. According to Baron, FORTH is just now coming into the public eye, although she believes that its potential for widespread use is not yet clear.

5 **Logic Programming**

Another conceptual approach to programming is logic programming, the best known example of which is PROLOG (PROgramming in LOGic), a language developed in the early 1970s at the University of Marseilles. Logic programming is a form a theorem proving, and a PROLOG program essentially evaluates theorems to determine whether or not they are true. Logic programming has attracted considerable interest since PROLOG was adopted by the Japanese for programming the fifth generation machines they have planned to design. PROLOG is an example of an artificial intelligence language.

6 **Query**
Finally, there are query languages, which require little or no knowledge of formal programming conventions. Users specify what problem they wish to solve and what data may be relevant, and the language translator itself determines how to solve the problem.

Thus, these programming languages provide a possible means of implementing the functional aspects of the individual PAC and FC building blocks. We now go on to consider one further approach POSIX.

POSIX (POSIX, 1988), an acronym for portable operating system interface for computer environments, seeks to standardize the interface to operating systems for application programmers. It allows application programs to be implemented on one operating system, and be moved to another operating system with minimum or no effort involved. POSIX defines a set of standard library routines for accessing basic operating system services such as the creation of processes, creation/deletion of files, primitive input/output functions, and security features, etc. The use of POSIX in the development of building blocks provides an additional level of independence from operating systems.

5.5.1 A processing model for shop floor control

Fig. 5.8 shows the four basic interfaces of a given FC and PAC building block. Three of these have already been discussed, the fourth is the topic of the following user interface section. We do not recommend a specific programming language to implement the building blocks of FC and PAC, as in principle any of the modern programming languages discussed earlier may be applied. The choice of programming language often depends on the preferred programming language of the local manufacturing information system department.

We recommend the use of POSIX as the interface to operating system services, as this creates a high degree of portability among building blocks across various operating system platforms. Building blocks should use SQL as an interface to the database environment, since SQL leaves the decision open on whether to access data locally or remotely. The user interface, discussed in detail in the next section, may be realized through Xwindows, since this enables the user interface to be distributed over a network. For example, a workstation and the user interface can be situated on the plant floor, while the computing task itself is performed centrally on a different computing machine. The user interface is now discussed in more detail.

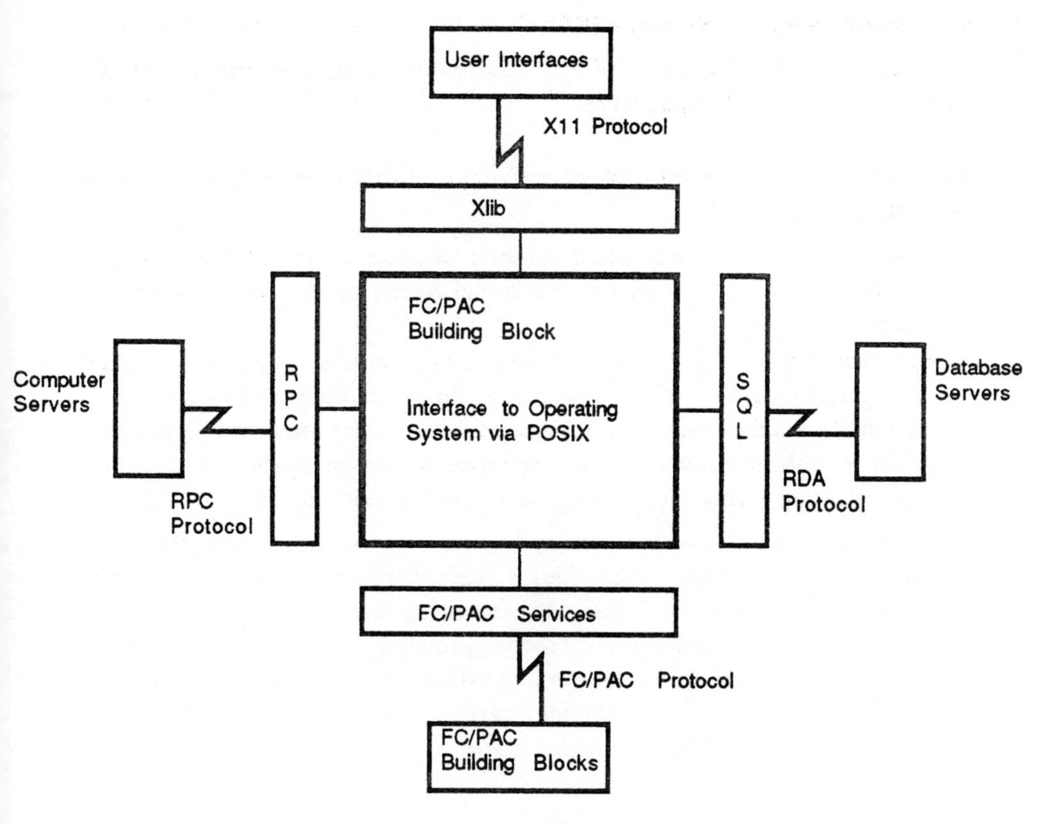

Fig. 5.8. Implementation scenario for processing.

5.6 User interfaces

The user interface has become a major topic of research in recent years. A good user interface aims to facilitate synergy between man and computer. Late in the 1980s graphical user interaction become widely available due to the advance in workstation technologies. A workstation provides graphical interactions with the user through windows and direct manipulation dialogues. From an information technology perspective, a user interface can be defined as a piece of software that handles the interaction between the application and the terminal, or workstation. It is responsible for displaying windows, forms, menus, icons, etc.

Here, we discuss the four user interface related topics, namely:

1 User interface classification, describing the two main types of user interface systems;
2 Xwindows, an example of an emerging user interface technology;
3 OSF/Motif;

4 Graphics Kernel System (GKS).

We conclude the discussion with a description of an user interface model for shop floor control applications.

User Interface Classification In general we can identify two different types of user interface software:

- A kernel based system, where the user interface is part of the operating system. The application and the kernel based system must run on the same computer.
- A client/server system, where the graphic system is a server process which is typically running on a workstation computer. Calls to the system are made by sending messages using a well defined protocol to this process. The advantage of the client/server model is that the application does not need to run on the same computer as the graphics system.

Client/server based systems can work in a heterogeneous computing environment. For example, a graphics system can run on a UNIX workstation displaying windows, menus etc of an application which is running on VAX/VMS‡. The client/server model provides a high degree of interoperability because the server can be located on any computer on a network implementing the client/server communication protocol. Xwindows is an example of a client/server based implementation.

Xwindows

Xwindows is a windowing system based on the client/server model developed by Massachusctts Institutc of Tcchnology (MIT). It provides user transparent network access to graphic systems. The Xwindows system architecture consists of four main components:

1 The X-server runs on the hardware to which the display and keyboard are attached. It allows for low-level graphics, windowing and other input functions. It also relies on a low-level interface that is supplied for each type of supported workstation, consisting of device-dependent components.

2 The X11 protocol provides for communication between the application and the server. It defines the format of requests between client applications and display servers over the network. The X11 protocol also allows the server to run on different hardware, while allowing the client to be unconcerned with the network connection.

‡ VAX/VMS is a trademark of Digital Equipment Corporation

3 The library Xlib provides a procedural interface to the X11 protocol. It converts procedure calls into requests that are transmitted to the server and translates messages from the server into return values and events for the application. Xlib is the interface normally used by applications.
4 The X-toolkit consists of general-purpose routines which create a set of user-interface objects on the screen.

Xwindows is an emerging standard pursued by several national and international standardization bodies, hence an increase in its use for applications is likely.

OSF/Motif OSF/Motif § is a graphical user interface which presents a single user interface across various computer platforms. OSF/Motif was developed by the Open System Foundation (OSF), a non-profit organization with members from many of the major computer manufacturers. OSF/Motif provides a high degree of portability for applications in manufacturing. It contains a toolkit which is based on Xwindows, a language for the presentation of graphics, text, windows, etc, a window manager, and a style guide. OSF/Motif could be used to implement the graphical user-interface of the FC and PAC building blocks. The client/server model implemented within OSF/Motif allows the tasks of the user interface function to be distributed on workstations, while the actual computing is performed on another computer.

GKS The GKS (Graphical Kernel System) standard defines a set of functions for interactive and non-interactive computer graphics applications that need to define and display computer generated pictures using a variety of computer graphic equipment. The GKS standard is documented by the International Standards Organization. GKS makes the graphical interface of applications portable across machines that implement the standard. GKS, as the name suggests, is kernel based graphics system.

5.6.1 A user-interface model for shop floor control

Here, we give an example of a shop floor control system implementation using the Xwindows technology explained previously. Fig. 5.9 shows a multi-vendor computing environment with three of the building blocks (scheduler, dispatcher, and monitor) running on a computer of vendor A, and the monitor and producer being executed on another computer. Any of the building blocks shown in Fig. 5.9 can display windows on any workstation. Xwindows also provides the capability to combine two different operating systems into

§ OSF/MOTIF is a trademark of Open System Foundation

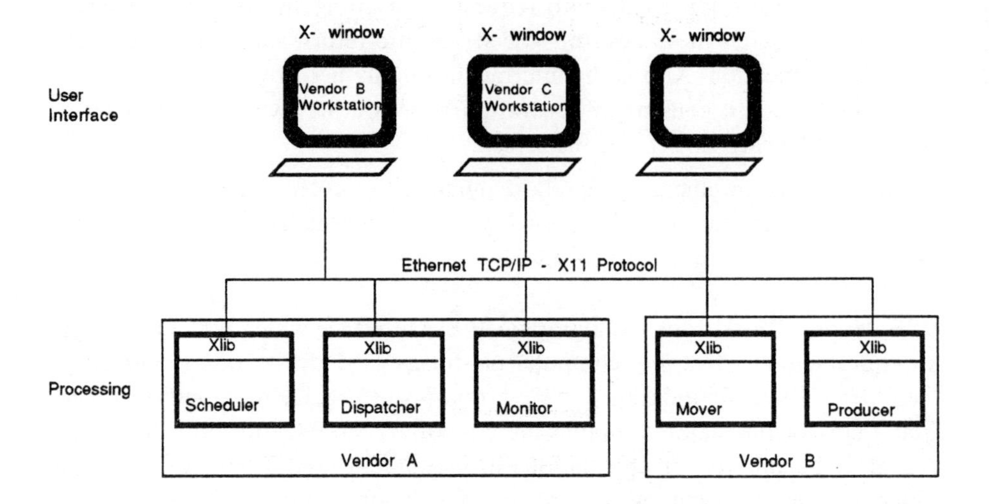

Fig. 5.9. Implementation scenario for the user interface.

a single view on a given graphical workstation. For instance, a workstation can display windows originating from UNIX™¶ and from the VAX/VMS operating system simultaneously.

Up to now, we have developed what we believe to be useful and applicable IT reference models for FC and PAC implementation. These models have been based on available technologies in the areas of communications, data management, processing and user-interface. We believe that it is essential to look into the future, to examine the likely effect of emerging technologies which could have significant impact in the area of shop floor control. One such technology is the object oriented approach, and its main ideas are now briefly presented.

5.7 The object oriented approach

The object oriented approach promotes the idea of combining the structural data and behavioural functions into one entity known as an object. This is a fundamentally different approach to the traditional design of software applications. In traditional systems design we define the informational (data) and behavioural (functional) element separately, and in doing so clearly distinguish between information and functions. Information is typically stored in a shared

¶ UNIX is a trademark of American Telephone and Telegraph Company.

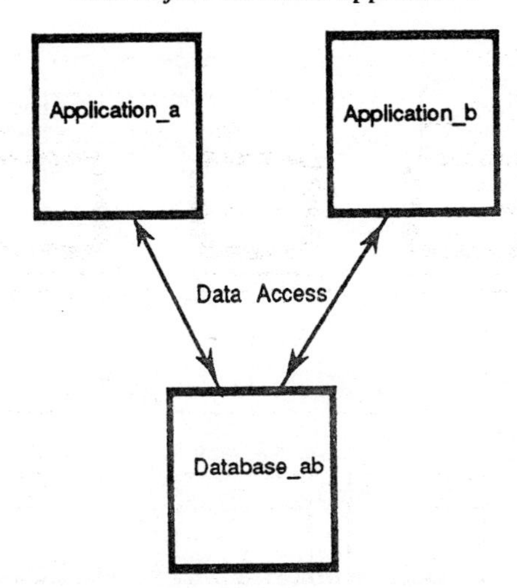

Fig. 5.10. Traditional design of software systems.

or distributed database which can be accessed by applications. Applications are implemented in individual programs. The traditional approach of building applications creates a number functional dependencies, thus reducing flexibility for the introduction of changes.

Consider the illustration in Fig. 5.10, which contains two distinct applications: *application_a* and *application_b*. Each of these have a common data source, namely *database_ab*. The main problem of the traditional approach to software systems design, is the potential disruption that may be caused through the introduction of enhanced applications. For instance, consider the effect of introducing a new version of application_a, and we may assume that this new version requires a new form of data structure in database_ab. This may be due to the enhanced functionality of the new application. A change in the database brought about through the introduction of application_a will more than likely require a further change in the structure of application_b. While changing two applications may not seem like a great deal of effort, consider the effect on a much larger software system, perhaps containing up to a hundred different applications operating from the same data source. The object oriented approach offers a more elegant and time saving solution to this problem.

Fig. 5.11 illustrates in very simple terms the object oriented system design approach. In place of two applications, a and b, we have two objects, object_a and object_b. Because one of the unique features of an object is that it combines both the data and functional aspects, there is no need for the common data source as required in the traditional approach. Similar to the previous example, we assume that object_a needs to be replaced by an enhanced version. This

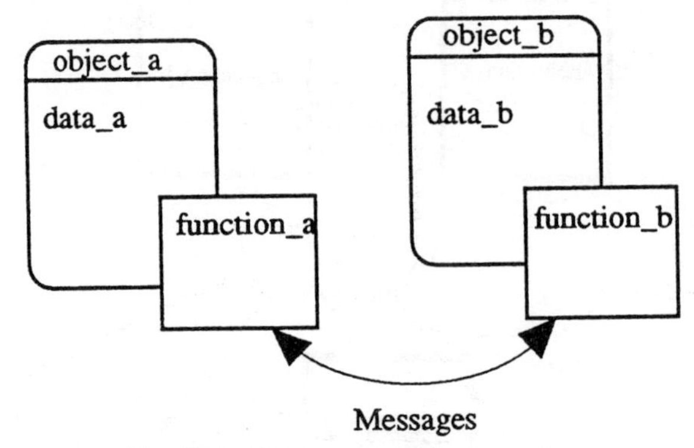

Fig. 5.11. Object oriented approach.

can be achieved with a version of object_a which uses the same messages as expected by object_b. Unlike in the previous example no other entities are indirectly affected by the change. Thus, with the object oriented approach, if an object is to be replaced then only the objects which exchange messages with that object are affected. In a traditionally designed system, a change affects many applications, which is not always apparent at the time when the change is planned and introduced.

Objects can represent organizations, incidents, or roles which individuals or organizations play. Object oriented technologies also employ the principle of inheriting characteristics or attributes from super class objects. The inheritance mechanism of object oriented technology supports reusability of software, and simplifies the design. In the object oriented approach, objects are only allowed to communicate via messages.

Fig. 5.12 illustrates the concept of inheritance in the object oriented approach. It shows two manufacturing objects, a robot and an Automated Guided Vehicle (AGV), each of which inherits attributes from its superclass object, namely a device with its attributes speed and capacity. Operations are associated with object such as start and produce for the robot, and start and move for the AGV. The instances of each object are actual implementations of robots and AGV, such as the well known PUMA and SMART robots.

From an historical perspective, the notion of an object as a programming construct first appeared in a product known as Simula. Simula (Birtwistle 1973) is a language designed for computer simulations. In 1983 Goldberg and Robinson defined Smalltalk‖ (Goldberg, 1983) as an object oriented language based upon the concepts of object classes introduced by Simula. Smalltalk is now available on various hardware and operating platforms. In fact, there

‖ Smalltalk is a trademark of Xerox Corporation

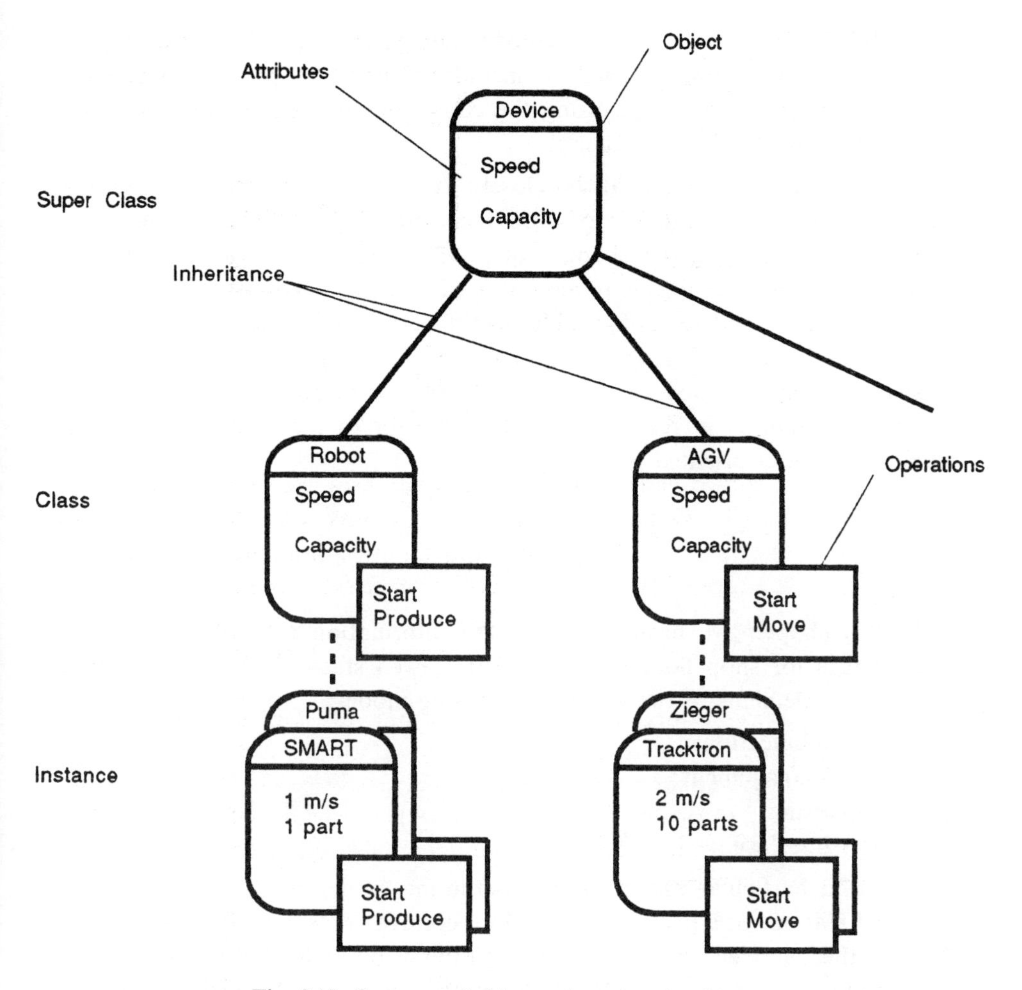

Fig. 5.12. Concept of object oriented technology.

are variety of object oriented programming languages available on the market today. Below we list some commonly used object oriented languages:

- Smalltalk-80 was originally developed by XEROX Palo Alto Research Center. It offers a graphical interactive environment with object management features and system environment objects: file system, display, handling text, keyboard and pointing devices.
- C++ (Wiener and Pinson, 1988) is an extension to the commonly used C programming language offering object oriented programming features. C++ was developed at AT&T Bell Laboratories in the early 1980s and is still evolving. C++ might well replace the C programming language in the future.

- *Trellis/Owl*** is an object oriented language developed by Digital Equipment Corporation in 1985. It includes a graphical programming environment and language, using the conventional programming language syntax of the Pascal language.

Object oriented systems can be closely associated with real-world objects, such as devices, operators, etc. and therefore they can provide increased flexibility in following the changes of in the physical environment. Object oriented systems lead to highly modularized systems which are a prerequisite for configurable and re-configurable manufacturing systems. Object oriented methodologies are emerging and are supported by requirements analysis and design, as demonstrated by Bals (1989). In the future, we can expect an increased emphasis in research and development of manufacturing systems employing object oriented technologies.

5.8 Conclusions

In this chapter we have examined the information technology reference architecture for shop floor control, and reviewed state-of-the-art technology within four elements of the general computing model:

1 Communications;
2 data management;
3 processing;
4 user interface.

For each particular aspect of the computing model, we have outlined the essential features of a particular shop floor control implementation model, and each of these models can be viewed as an initial blueprint for a technological implementation of FC and PAC. The importance of each particular model, and the technology recommended therein, is that emerging and established IT standards are applied. Hence, the models which we have outlined, support the important IT goals of inter-operability, flexiblity and portability. We have also examined and reviewed what we believe is the most important recent development in IT systems, namely the object oriented approach.

** Trellis is a trademark of Digital Equipment Corporation

Part Four
STATE-OF-THE-ART-REVIEW

Overview

In our discussion on the architecture of shop floor control systems, we identified individual functional building blocks concerned with issues such as production environment design, scheduling, dispatching etc. Within production environment design we talked about process planning, and Group Technology tools to define product families and product based production cells. Clearly a tremendous amount of effort, in terms of research, development and implementation, has been expended on those areas over the past number of years. In this part of the book, we take an overview of this work, and try to position it in terms of our understanding of shop floor control systems.

The structure of this part is as follows. In Chapter 6 we review scheduling strategies. We distinguish what we term the 'traditional' approach to scheduling and the 'modern' approach, and try to position our own ideas on scheduling / dispatching in this framework. We argue that the traditional research approach has been concerned primarily with seeking elegant and optimal solutions to the scheduling problem, while in many cases making assumptions which render the results invalid in a factory environment. More recently, researchers using computer based tools, in particular data representation and solution techniques from the field of artificial intelligence, are offering more realistic approaches.

In Chapter 7 we consider the techniques used to support the production environment task in Factory Coordination. We review the state-of-the-art in terms of Group Technology, and discuss various techniques available to support the identification of product families and associated product based manufacturing cells. We consider the various approaches to the development of process planning systems and conclude the chapter with a short discussion on manufacturing systems analysis techniques.

In essence we can say that this part of our book is concerned with the 'how' of FC and PAC, whereas Part Two in particular was concerned with the 'what'.

6

A review of scheduling strategies

6.1 Introduction

Scheduling is an important activity within FC and PAC. Research in scheduling has absorbed the creative talents of many academics over the past thirty years or so. Despite the numerous developments in scheduling, a common complaint from shop floor personnel is that scheduling techniques and solutions developed by academic researchers are 'out of touch' with the reality of the shop floor. In a sense this perception of a wide gap between research and practice is accurate, and we will suggest that many scheduling techniques developed over the years are restricted in their application. Yet recent developments in computer based scheduling solutions suggest that the gap between the research world and the shop floor is being slowly bridged.

Our review presents two viewpoints on scheduling. The first of these we term the traditional approach, which may be viewed as employing the ideas and methods from operations research. The second we term the modern approach, which proposes a synthesis of Operations Research (OR), Artificial Intelligence (AI), discrete-event simulation and ideas from control theory. Our argument for this modern perspective is supported by a recent publication by Solberg (1989), who argued that 'there is every reason to explore the opportunities for merging and blending various points of view to obtain the best attributes of each'. The 'various points of view' he refers to include the different approaches to scheduling employed by OR, AI, control theory and discrete-event simulation approaches.

The overall structure of the chapter is as follows:

- We review the *traditional* approach to scheduling;
- We propose a *modern* approach to solving the scheduling problem, one which we believe is a blueprint for future scheduling solutions and involves the synthesis of the forementioned 'various points of view';

- Finally, we review the latest reported developments in scheduling research which may be classified as being in line with this modern approach. Within this, we also include our own opinion on an effective approach to scheduling.

6.2 Traditional scheduling approaches

A formal mathematical notation has been developed to classify scheduling problems, and we adhere to this notation throughout this chapter. The notation consists of four parameters: $n/m/A/B$, where:

1. n is the number of jobs.
2. m is the number of machines.
3. A describes the flow pattern or discipline within the factory, i.e.:
   - F for the flow shop case which assumes that all tasks are to be processed on the same set of machines with an identical ordering of the processing steps.
   - P for the permutation flow-shop case. Here the search for a schedule is restricted to the case where the job order is the same for each machine.
   - G is the general job-shop case where there are no restrictions on the processing steps for each task.
   - When there is only one machine (i.e. $m = 1$), A is left blank.
4. B describes the performance measure by which the schedule is to be evaluated.

The approaches to developing possible scheduling solutions vary, and in this section we attempt to categorise the traditional scheduling approaches. These range from manual computational methods developed by operational researchers, to the latest computer based technology approaches which promise more realistic and viable solutions. The areas addressed by this review of scheduling approaches include:

1. Single machine solutions;
2. algorithmic solutions;
3. enumeration methods;
4. scheduling heuristics;
5. project scheduling and control approach;
6. discrete-event simulation;
7. artificial intelligence approaches.

Single Machine Solutions Single machine scheduling problems arise in practice more often than one might expect. For example, there are job-shops where one machine may act as a bottleneck, and it may make sense to schedule these shops as a single machine problem (French, 1982). The format of this type of problem is $n/1//B$ (n jobs, one machine). The following results are well known:

1 The shortest processing time heuristic minimizes the mean flow time: $n/1//\bar{F}$, (note that $\bar{F}$ is used to represent the minimization of the mean flow time, and the mean flow time is another way of describing the average throughput time of jobs through the manufacturing system).

2 The earliest due date heuristic minimizes the maximum lateness: $n/1//L_{max}$.

An extension to the single machine problem is the addition of precedence constraints that require that certain subsets of jobs be processed in a given order. An example of a precedence constraints might be one job being given a higher priority because of the importance of a particular customer. With this in mind an algorithm was developed by Lawler (1973), which minimized the maximum cost of processing a job based on the constraint to process certain jobs before, but not necessarily immediately before, others. Smith (1956) developed an algorithm based on shortest processing time sequencing subject to due-date constraints. We refer readers interested in the detailed explanation of these algorithms and other aspects of scheduling theory to French (1982), Baker (1974) and Rinnoy Kan (1976).

Algorithmic Solutions Here we present three algorithms, based on an initial algorithm developed by Johnson (1954). Each of these develops an optimal solution for a specific scheduling problem, based on the manufacturing data of the problem, by following a simple set of rules which determine an exact order for processing. The objective of Johnson's Algorithm is to minimize the maximum flow time (i.e. F_{max}), and the three problems now described are:

- $n/2/F/F_{max}$ (an n job, two machine, flow-shop problem)
- $n/1/G/F_{max}$ (an n job, one machine, job-shop problem)
- $n/3/F/F_{max}$ (an n job three machine, special case flow-shop problem)

1 $n/2/F/F_{max}$

In this problem, n jobs must be processed through two machines, so that the maximum flow time is minimized. The Algorithm proceeds as follows:

(a) Start processing with the job having the shortest processing time on machine 1.

(b) Finish processing with the job having the shortest processing time on machine 2 (since this is the time that machine 1 has to be idle).

If the smallest time is the same for two jobs, then an arbitrary selection is made. Consider the following simple illustration of Johnson's Algorithm

Job	Machine 1	Machine 2
1	10 mins.	6 mins.
2	3 mins.	11 mins.
3	12 mins.	2 mins.
4	7 mins.	13 mins.
5	8 mins.	9 mins.

Table 6.1. *Process information for the $5/2/F/F_{max}$ problem*

Iteration No.	Action Taken	Latest Schedule
1	Job 2 Scheduled	2 * * * *
2	Job 3 Scheduled	2 * * * 3
3	Job 4 Scheduled	2 4 * * 3
4	Job 1 Scheduled	2 4 * 1 3
5	Job 5 Scheduled	2 4 5 1 3

Table 6.2. *Resultant schedule for the $5/2/F/F_{max}$ problem*

(Table 6.1). The solution is illustrated in Table 6.2, and it is derived by following the two simple steps we have just described. Thus the overall solution to the problem which minimizes the maximum flow time, is the job sequence (2,4,5,1,3).

2 $n/2/G/F_{max}$

For this problem we drop the assumption that made the previous problem a flow-show type. Thus, this solution is for a job-shop and each job need not be processed in the same order through each of the two machines. With this in mind, we can divide the jobs into four separate categories; A, B, C and D.

A: Those jobs to be processed on machine M_1 only.

B: Those jobs to be processed on machine M_2 only.

C: Those jobs to be processed on both machines, in the order M_1 followed by M_2.

D: Those jobs to be processed on both machines, in the order M_2 followed by M_1.

The development of an optimal schedule proceeds as follows:

1. Schedule the jobs of type A in any order to give the sequence S_A.
2. Schedule the jobs of type B in any order to give the sequence S_B.
3. Schedule the jobs of type C using Johnson's Algorithm for $n/2/F/F_{max}$ to give the sequence S_C.
4. Schedule the jobs of type D using Johnson's Algorithm for $n/2/F/F_{max}$ to give the sequence S_D (in this case M_2 is the first machine).

The resulting optimal schedules for each machine are as follows:

- Machine M_1 : (S_C,S_A,S_D)
- Machine M_2 : (S_D,S_B,S_C)

3 $n/3/F/F_{max}$
Johnson's Algorithm for the $n/2/F/F_{max}$ may be extended to a special case of the $n/3/F/F_{max}$ (a three machine problem). This algorithm is applicable for a flow shop with three machines and assuming one condition is met. This condition is: the maximum processing time on the second machine is no greater than the minimum time on either the first or the third machine.

Thus Johnson's Algorithm, and variations on it, present optimal solutions to the three cases discussed. The obvious drawback of this approach is that the majority of scheduling problems cannot be approximated to any of these three models. We now discuss the development of enumeration methods to search for optimal solutions to larger scheduling problems.

Enumeration Methods We have just described algorithms which exist for a small class of scheduling problems, and these algorithms can be used to obtain optimal scheduling solutions. However, it is widely acknowledged that larger scheduling problems are Non-Polynomial (NP) complete, in other words, that the number of possible solutions increases NP exponentially with the size of the scheduling problem. Cunningham (1986) showed that for a simple n jobs and m machines case, the upper bound on the number of solutions is $(n!)^m$, which is almost 25 billion for a five machine-five job problem.

Because of the enormity of the potential solution space, viable solution techniques were developed with the objective of eliminating large groups of non-optimal solutions. Two such enumeration methods, are dynamic programming and branch and bound. An enumeration method lists, or enumerates all possible schedules and then eliminates the non-optimal possibilities from the list, leaving those which are optimal.

Dynamic programming originates from Bellman (1957) and can be applied to many optimization problems outside of the scheduling domain. Held and Karp (1962) applied dynamic programming ideas to sequencing problems and derived solutions to the single machine problems. In French's (1982) study of mathematical solutions to scheduling, a solution to a four job, one machine ($4/1//\ \bar{T}$) problem is generated, but a major drawback is the amount of calculation required. French also showed that gross estimates on the computer time needed to solve dynamic programming scheduling problems is one minute for twenty jobs on a one machine problem ($20/1//\ \bar{T}$), and four years generate a solution for forty jobs ($40/1//\ \bar{T}$). Thus, he concluded that dynamic programming may only be suitable for solving relatively small problems.

Branch and Bound is a form of implicit enumeration. It involves the

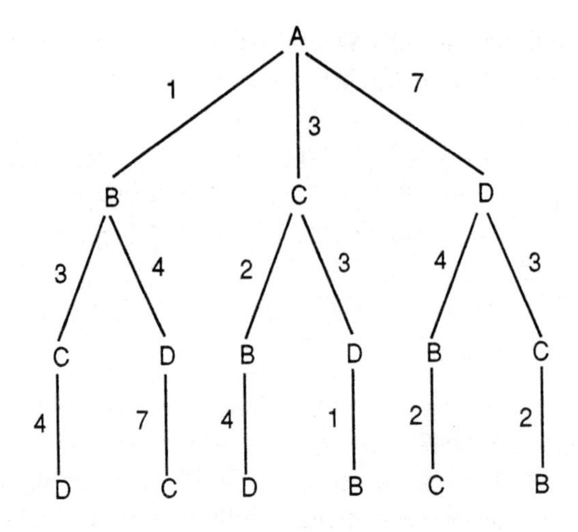

Fig. 6.1. Possible sequences of 4 jobs A, B, C and D (Cunningham and Browne, 1986)

formation of an elimination tree, which lists possible permutations. Branches in this tree may then be eliminated if it is evident that their solution will not approach the optimal. In theory, as with the dynamic programming approach, an optimal solution is found, but this can be costly in terms of computation time. Winston (1977), gave the following definition of this useful technique:

'During the search there are many incomplete paths contending for further consideration. The shortest one is extended one level, creating as many incomplete paths as there are branches. These new paths are then considered along with the remaining old ones, and again the shortest is extended. This repeats until a destination is reached along one path. To ensure that none of the incomplete paths yields a better solution they are all extended as before until each is longer than the proposed solution, or extends it.'

To illustrate the application of the branch and bound approach, consider the following example taken from Cunningham and Browne (1986) (Fig. 6.1), which lists the possible sequence for four jobs A,B,C and D on one machine. The assumption for this example is that job A is selected first. For this small problem, we can expect $(4!)^1$, or 24 possible schedules to choose from. Since we are assuming A to be the first job, the number of alternatives are reduced to six, and this is illustrated in the branch and bound diagram for the scheduling problem. In this diagram, the numbers shown represent the cost of scheduling jobs in that order, and this cost is based on the number of set-up changes that are involved. The branch and bound procedure is now described in more detail.

The solution procedure is as follows. Firstly, 'branch' on AB which results in the possibilities ABC and ABD. Since these are both longer than AC, we then branch on AC to produce ACB and ACD. Again, we evaluate the shortest permutation, which in this case is ABC and then produce ABCD. We can then eliminate ACBD and ACBD since they are both longer than ABCD. After that, ACDB is preferable to ABCD, and AD is eliminated because of its length. Finally, ACDB is accepted as the best solution. A branch and bound procedure was developed by the authors, implemented in LISP, and a sequencing problem on an automatic insertion machine for electronics assembly was tackled. While a ten job problem required ten minutes processing time on a mini-computer, it was calculated that 15 jobs would take almost seven years. This 'combinatorial explosion' in the amount of computation time needed reflects accurately the problem of 'NP completeness', which was referred to earlier.

To overcome this computation time problem, a heuristic scheduler was developed, which reduced the possible solution space. The resulting schedule was still found to be near optimal. The heuristic strategy deployed concentrated on analysing the variety within the scheduling problem. Cunningham and Browne discovered that the problem became more manageable if the list of jobs to be sequenced were divided into groups containing similar jobs (similar to the group technology approach), and then the branch and bound scheduler was used to schedule individual groups. This successful application of a heuristic significantly increased the applicability of the scheduler to more complex problems. According to the authors, using the grouping heuristic with 25 jobs reduced that problem to the same size as the original ten job problem.

Scheduling Heuristics From a practical viewpoint, work on the scheduling problem has clearly highlighted the need for heuristic (non-optimizing) approaches to scheduling problems. Barr (1981) described a heuristic as 'a rule of thumb, strategy, trick, simplification, or any other kind of device which drastically limits the search for solutions in large problem spaces ... all that can be said for a useful heuristic is that it offers solutions which are good enough most of the time.'

A large number of single stage heuristics have been developed. Single stage heuristics (also known as dispatching rules, scheduling rules and priority rules) select a job to be processed on a machine based upon some easily computed parameter of the jobs, operations or machines. Thus, a dispatching rule is a method used to select a job from a queue of jobs at a processing workstation. The type of dispatching rule ranges from the simple first-in first-out (FIFO) rule, to more complex rules requiring a range of manufacturing variables. In their review of dispatching rules, Blackstone (1982) compared several rules based on results of published studies. They concluded that it is not reasonable to identify any single rule as the best in all circumstances. However, they identified several rules that exhibited good performance in general (a total of 34 different rules are described in their paper), and these will now be presented in terms of:

1 Rules involving processing times;
2 dynamic scheduling rules;

Job	Entry Time	Process Time	Due Date	Selection Order
1	11.11	23 mins.	11-NOV	3
2	10.24	15 mins.	14-NOV	1
3	12.19	18 mins.	10-NOV	2
4	9.13	45 mins.	12-NOV	5
5	11.25	29 mins.	15-NOV	4

Table 6.3. *Job selected on the basis of the shortest imminent operation rule*

3 rules involving due dates;
4 simple rules.

1 **Rules involving processing times**
The two main rules here are based on the shortest imminent operation time of the jobs awaiting processing in a workstation queue.

- SI: the shortest imminent operation rule, selects for processing that job for which the operation can be completed in the least time. Conway and Maxwell (1962) found that within a multiple machine environment this rule retained the advantages of throughput maximization it had exhibited in the single server environment, and that imperfect information about processing times had little effect on the operation of the SI rule. The principal difficulty with the SI rule is that jobs with long processing times can be very late, and to overcome this, the truncated SI rule was developed. Table 6.3 shows how the effect of applying the SI rule to five jobs in a queue.
- Truncated SI: this rule, a modified version of the SI rule, ensures that jobs with long process times which have been waiting for a certain time period, are given priority so that they may meet their due dates. In practice, this involves examining the due dates as well as the processing time when selecting jobs from a queue.

2 **Dynamic scheduling rules**
With dynamic scheduling heuristics, jobs are scheduled in real-time based on the current state of the system. A widely-known dynamic heuristic is the minimum slack time heuristic, which calculates the amount of slack time remaining for each job in the system. This calculation is based on the due date of the job (d), the current date (p) and the sum of the remaining processing times (sum).

Minimum slack time $= d - p - sum$

Another effective dynamic rules is the least slack per operation rule, which is based on the ratio of the minimum slack time to the total number of remaining operations.

Job	Entry Time	Process Time	Due Date	Selection Order
1	11.11	23 mins.	11-NOV	2
2	10.24	15 mins.	14-NOV	4
3	12.19	18 mins.	10-NOV	1
4	9.13	45 mins.	12-NOV	3
5	11.25	29 mins.	15-NOV	5

Table 6.4. *Job selected on the basis of the earliest due date*

Least slack per operation $= [d - p - sum]/[no\_rem\_ops]$

3 **Rules involving due dates**

The principle advantage of having due-date based rules over processing time based rules is a smaller variance of job lateness, and often a smaller number of tardy jobs. Conway (1967) has shown that this advantage is especially manifest when due dates are established as some multiple of total processing time. Conway also studied three due date based rules: earliest due date, least slack and least slack per operation, and of these three it was found that least slack per operation produced a smaller variance of job lateness and a smaller number of tardy jobs than the other two measures. The method for selecting by the due date criterion is illustrated in Table 6.4, and in this case it is interesting to note how the selection order differs with that generated using the SI rule in the previous table. It is worth mentioning that when the authors introduced the concept of formal scheduling to an electronics assembly plant, the due date heuristic was the one most favoured by the shop floor personnel. Another member of the due date family of rules is the critical ratio, which is (due date minus date now) divided by (lead time remaining).

4 *Simple rules*

The most commonly used simple heuristic is First-In-First-Out (FIFO), and a number of researchers have found that this rule performs substantially the same as random selection with respect to mean flowtime or mean lateness, although FIFO produces a lower variance of performance measures than random selection. In a practical sense, FIFO is an attractive alternative due to its simplicity of definition and usage. The priority rule is another simple method of selecting jobs from a queue. The use of the priority rule requires that a priority value be assigned to each job, and those with the highest priority are processed first. The FIFO dispatching rule is illustrated in Table 6.5, and in this case the entry time into the queue is the important attribute.

Blackstone (1982) concludes that SI seems to be the best alternative when:

1 The shop does not set due dates;

Job	Entry Time	Process Time	Due Date	Selection Order
1	11.11	23 mins.	11-NOV	3
2	10.24	15 mins.	14-NOV	2
3	12.19	18 mins.	10-NOV	5
4	09.13	45 mins.	12-NOV	1
5	11.25	29 mins.	15-NOV	4

Table 6.5. *Job selected on the basis of first in first out*

2 the shop sets very 'tight' due dates;
3 the shop sets 'loose' due date during periods of great congestion.

This view of the SI heuristic is supported by Kiran and Smith (1984b) who state that 'the SI rule is superior to other simple priority rules for job-completion times based on jobs-based process criteria'. They concluded that many dispatching rules are remarkably simple and can be easily applied with a minimum of computation. For a comprehensive review of dispatching rules, we refer the interested reader to Panwalker and Iskander (1977), whose work references 113 different dispatching rules.

Project Scheduling and Control Theory Project Scheduling involves two methods (Taha, 1982, Daellenbach, 1983), namely PERT (Program Evaluation and Review Technique) and CPM (Critical Path Method). A project is defined as a collection of interdependent activities which must be accomplished in a certain sequence, the similarity between this and production of a product is apparent. Project scheduling using PERT/CPM consists of three phases: planning, scheduling, and controlling. The planning function is the determination of the projects activities and their time estimates, with the result being a precedence network diagram with directed arcs highlighting the relationships between activities. Scheduling then uses this network of nodes and arcs to determine critical paths, the slack-time in non-critical activities, and finally the start and finish times of each activity. Rickel (1988) suggests that 'although project scheduling techniques are not general enough to be used directly for most scheduling problems, the techniques which are used to generate the activity network and extract useful temporal information from it have potential applicability'.

Solberg (1989) and Rodammer (1988) point to developments in control theory applications for scheduling. The view of scheduling taken by the control theory approach is that it is a dynamic activity, and the essential problem of scheduling is understanding how to reschedule. This idea gives expression to the common phrase among scheduling personnel that: 'there is no scheduling problem, only a rescheduling problem'. The usefulness of the control approach is that it accepts the dynamic nature of the manufacturing environment, and attempts to calculate schedules which are robust and flexible in adapting to the inevitable environmental changes. However, the main problem with applying control theory is that the mathematics and techniques of control theory apply to continuous- and discrete-time systems, but are not well adapted to discrete event systems.

Discrete-Event Simulation Pritsker (1984) described digital simulation as 'the establishment of a mathematical logical model of a system and the experimental manipulation of it on the computer'. As a result of experimentation with a simulation model, inferences may be drawn about the modelled system. The main advantage of the simulation approach is that information may be obtained about system performance prior to implementation, thus enabling the system designer(s) to pinpoint the risks and benefits of a proposed system. The five main steps in developing a simulation model of a system are:

1. System definition and problem formulation, which seeks to ensure that the problem and its boundaries are well defined, and the assessment criteria identified. Thus, the main requirement of this step should be the explicit identification of the purpose of the simulation.
2. Collection of the data is the next step in the development cycle. A valid simulation model requires highly accurate data.
3. Construction of a computer model. This may be achieved using a simulation language, a simulation package, or a general purpose third generation computer language. We will discuss some of these packages later on in this chapter.
4. Verification and validation of the model. Once the computer model has been developed, it must then be verified and validated for accuracy and completeness. Verification is the process of confirming that the computer program performs as intended, while the task of validation consists of building an acceptable level of confidence that any inferences or conclusions from experiments are realistic for the real system.
5. Experimentation with the model, which occurs at the final stage in the development cycle. Here, the simulation model is used to evaluate the potential effectiveness of various alternatives. The results may then be analysed so that the best possible course of action can be pursued.

Coll *et al.* (1985) identified two distinct roles for simulation within a manufacturing environment. Firstly, they concluded that simulation may be used as a tool to evaluate production system designs; secondly, they suggested that it may also be applied as a control mechanism to support functions such as scheduling and monitoring. This second conclusion is supported by Kiran and Smith (1984a), who recognize that in complex problems such as dynamic job-shop scheduling, simulation may be an indispensable tool.

Law *et al.* (1989) gave a comprehensive survey of the available simulation systems which have applications in manufacturing, and a selection of these are now briefly discussed under the headings simulation languages, simulation packages and conventional computer languages. In general, simulation languages are more flexible than simulation packages, however they do require both simulation and programming experience. Standard statistical output is also provided by these types of languages, and this output forms the basis of the evaluation and experimentation stages of model development. Examples of simulation languages are SLAM II, SIMAN, GPSS, and GASP IV.

In order to make simulation more accessible to people with a limited programming and simulation background, a number of data driven simulation packages have been developed. These packages are likely to be less flexible than general purpose simulation languages and they require practically no simulation and programming experience. It is also likely that the overall development time for a simulation model will be less than the time taken using a conventional simulation language. Essentially, simulation packages involve a menu driven user interface which allows the developer to describe a manufacturing system in detail. When the system has been described, simulation experiments can be performed and animation and graphical facilities are used to present the simulation output. Example include SIMFACTORY and Pcmodel (Law, 1989).

Finally, simulation models may be developed using conventional programming languages (e.g. C, Pascal, FORTRAN), although it must be said that this type of activity is usually only carried out by those having an intimate knowledge of simulation and a conventional language. Thus, people involved in research activities are most likely to develop simulation models using this approach, and a simulation tool to be discussed later on in Chapter 9, was developed using the Pascal computer language.

Thus, any (or a combination) of these simulation approaches may be used to analyse the effectiveness of scheduling systems. Potential plant bottlenecks and other problems of material flow in a plant can be evaluated and a solution proposed. To date, the main use of discrete-event simulation has been to test fixed scheduling heuristics and dispatching rules. However, it is important to realize that using simulation models to generate schedules can be costly, both in computer time used in generating the schedule and the human effort required to build an accurate model of the system.

Artificial Intelligence Approaches Artificial Intelligence (AI) approaches typically depict the scheduling problem as the determination and satisfaction of a large number of constraints that are found in the scheduling domain (Rodammer, 1988). The use of expert systems seems appropriate when applied to the scheduling problem, because scheduling is heuristic in nature, and, as we have indicated earlier, often requires the use of rules of thumb to achieve acceptable solutions. However, one of the difficulties with the use of AI is that expert systems are expensive and time consuming to develop. Also, for reasonably sized problems, there is often a problem with computation speeds. McKay *et al.* (1988) even go so far as to suggest that 'conclusive results have not been presented that would indicate that the first generation of AI based scheduling systems generate better schedules than would a traditional heuristic approach'.

Despite these problems, considerable research effort has been devoted to the use of AI to the scheduling domain, and the latter sections of this chapter describe some of the recent work (both research and practical implementations). Before we discuss current AI- and discrete event simulation based scheduling systems, we summarize what we believe are the main drawbacks of the traditional approach.

6.2.1 Drawbacks of the traditional approach

We have just described scheduling solutions which may be categorized as being part of the traditional approach to scheduling. We believe that these approaches, with the notable exception of AI and discrete-event simulation, have failed to bridge the gap between the theory and practice of scheduling. The reason for their obvious lack of success can be attributed to their impracticability, and this argument is supported in many academic scheduling papers, including Browne *et al.* (1981), who articulated the view that 'the scheduling problem as it is posed in the literature, does not exist in real factories'. Hadavi (1987) echoed this view by stating that 'analytical approaches aiming at optimal solutions have proven to be feasible for a small subset of scheduling problems, namely those which assume highly idealized conditions and small number of jobs and machines'. In his criticism of the analytical approaches, Hadavi listed some of these 'highly idealized conditions' required, including the following:

1 Each machine resource is continuously available;
2 there is no rework;
3 each operation can be performed by only one machine on the shop floor.

These assumptions are not really valid in the context of modern manufacturing systems, since, for example, alternative routes are often used to distribute production loads on a shop floor. Thus, in the pursuit of optimality, one could argue that reality is lost along the way, and it is our firm opinion that an approach solely concerned with the view that scheduling is an optimization problem is likely to fail. Further criticisms of optimization were outlined by Ackoff (1977), who pointed out that the major problem with this approach is that the 'preoccupation with optimization leads to a withdrawal from reality'. Ackoff argued that optimal solutions deteriorate because the system and environment change, and that more effort should be placed on designing models that adapt well rather than optimize. In our view, this is the key to the future of successful scheduling systems. Any successful implementation of a scheduling system must adapt well to the inevitable variability in the manufacturing environment.

We now present a modern view of the scheduling problem, one which promises to bridge the gap between research and the reality of modern manufacturing. There are three important features of the following section:

1. We present an interesting perspective on scheduling, articulated recently by a distinguished member of the scheduling research community. Essentially, this view argues for a modern approach, because of the evident failure of past scheduling solutions.
2. We present a selection of scheduling solutions which we believe may be classified as being part of the modern approach to scheduling.
3. Finally, we summarize our approach to scheduling, which we believe presents a coherent and flexible strategy for ultimately implementing shop floor control systems.

6.3 Modern scheduling approaches

Up to now we have outlined some of the approaches taken to develop solutions to the scheduling problem over the last thirty years or so. Solberg (1989) presented an interesting perspective to categorize these various approaches, and to do this he used the concept of a paradigm. The notion of a paradigm was articulated by Thomas Kuhn (1970), and it may be described as the world view taken at some period of history by a significant number of scientists within a certain domain of research. Kuhn himself described it as '. . . the entire constellation of beliefs, values, techniques and so on shared by members of a given community.' Each paradigm has a conceptual framework and the community of researchers accept this framework without conscious evaluation. The community also shares a literature, informal communication networks, common terminology and so on. Solberg recognizes three distinct 'world views', or paradigms, of the scheduling research community, namely:

- The optimization paradigm;
- the data processing paradigm;
- the control paradigm.

The optimization paradigm takes the view that production planning and scheduling problems are problems which require optimum solutions. Many of the techniques described so far in this chapter fit within this paradigm, although Solberg identifies discrete event simulation and artificial intelligence approaches as 'important variations' to the optimization world view. The important variation of simulation is that it can capture as much of the complexity of a real scheduling problem as the user wishes to contend with; however the resulting model cannot guarantee an optimal solution. The approach of artificial intelligence allows attractive alternatives to formal optimization constraints since artificial intelligence can accommodate qualitative rules such as 'never start a big job on Friday'. Despite these variations, the two approaches are essentially seen by the scheduling research community as a means to move towards optimal schedules.

The data processing paradigm reflects the view of an entirely different community of people who see manufacturing and scheduling problems fundamentally in terms of data management issues. Planning and scheduling are performed through calculations of the net effects of customer orders upon the manufacturing system. An obvious example of this approach is Materials Requirements Planning (MRP). It is interesting to note that much of the research work conducted within this world view is commercial in nature, and probably identifies with aspects of applied computer science (e.g. databases and management information systems).

Finally, the control paradigm reflects the view of a community of researchers who see production planning and scheduling problems as control issues. These researchers focus on aspects necessary to keep a production system 'under control'. Solberg describes the problem as 'one of following a desired trajectory in time or maintaining equilibrium in the presence of disturbances'. In recent years this paradigm has focused on hierarchical control where commands are issued downwards and status information is sent upwards through layers of responsibility. Thus the emphasis of the control paradigm is to monitor a system to ensure that it is 'going in the right direction.'

Solberg points out that his 'brief and over-simplified characterizations' of the three paradigms are not intended to be disrespectful to any particular community. He points out that what is important is that three separate research communities have adopted different approaches to what is essentially the same problem. He suggests that when one takes a global view of the three different approaches, it seems obvious that each is dealing only with some aspects of the real scheduling problem. He agrees with Graves (1981) by acknowledging that real world manufacturing practices rarely exploit any of the methods offered by these three paradigms, relying instead on informal methods of production scheduling. He concludes that 'these paradigms do not adequately address some of the important aspects of real production planning and scheduling', and the aspect that most concerns him is that these paradigms do not take due cognisance of uncertainty. The problem is that it is in the nature of manufacturing that events do not go strictly to plan, machines break down, raw materials are unavailable etc. The reason for this lies with our basic inability to accurately predict what the future holds, and as Solberg says

'.. even if better models were available, a reliability problem would remain'. Thus Solberg argues for a new approach to scheduling and outlines the main characteristics of such an approach as:

- Having a decentralized control structure, so that complexity is reduced and each sub-system works within its own parameters and guidelines to plan and control the flow of work through the manufacturing system.
- Having 'real time negotiation of resource assignments which fits well with the dynamically varying and uncertain conditions that characterize modern manufacturing systems', in other words, designing control systems that can react effectively to changing circumstances on the shop floor.
- Having independent software 'objects', and each of these objects should contain well defined functions, be modular and present a simple standardized interface to any user of the production control system.

We now present reported production scheduling systems which could be generally classified amongst the modern types of scheduling systems. These incorporate different 'points of view' which have been described so far in this chapter. The systems we discuss are:

1 DISPATCHER (Acock and Zemel, 1986);
2 FACTOR (Factrol, 1986);
3 ESPRIT Research Projects (Tiemersa, 1988, Meyer *et al.*, 1988);
4 PLATO-Z (O'Grady, 1988);
5 MADEMA (Chryssolouris, 1987);
6 ISIS (Fox *et al.*, 1983);
7 SONIA (Collinot *et al.*, 1988);
8 LMS (Fordyce, 1989);
9 The Leitstand Approach (Haverman, 1991, Doumeingts, 1991).

DISPATCHER This is a specific expert system which controls and monitors automated materials handling systems, and this system has been implemented within two of Digital Equipment Corporations manufacturing systems in the USA by Carnegie Group Inc., Pittsburgh, PA. The environment for which the DISPATCHER system was initially developed was the plant's Printed Circuit Board (PCB) assembly area, consisting of a conveyor transport system which carries work in progress to over a hundred workstations, linking each station with a central dispatch and storage area.

This system was designed to replace a human dispatcher whose task it was to search through a file of index cards to find the correct information for a returning tote, decide whether or not the tote could be sent to another workstation, and update the information on the card. As the typical number of totes flowing through the system on any one day might number up to five hundred, this task of searching for information, making rapid decisions, and tracking the totes was a reasonably frantic one. This inevitably increased the probability of human error, and as a result the usefulness of an automated system which would provide speed and consistency was investigated, resulting in the DISPATCHER system.

The developers attribute the success of DISPATCHER mainly to the fact that it was not an over-ambitious project and that there was a clear need for the system. Before its development began, it was known that the human operator alone would not be able to control the automated materials handling system efficiently. Thus, in order to justify the cost of the materials handling system, a system such as DISPATCHER was required. Another advantage for this implementation was that the functional requirements from the system were well defined. Therefore, the DISPATCHER project did not suffer from some of the problems which have plagued previous AI development tasks: the lack of a real need for the system, the lack of constraints on the compass of the system, and a vague definition of the system's requirements.

According to Acock and Zemel (1986), another major factor which contributed to DISPATCHER's success was its phased introduction to production. In the first phase of the implementation process the new system imitated the functions of the previous system. Thus, users gained confidence in the new system since it performed tasks adequately. Following acceptance at this stage, the full system was then implemented in incremental steps, each one unfolding new layers of functionality. The importance of this incremental approach of installing DISPATCHER was twofold. First, each stage dispelled fears about, and built confidence for, the following stage. Secondly, the installation sites had the assurance that at any point they could abandon some new functionality, and retreat to the level of the previous stage which they were familiar with. This common sense approach to implementing new ideas on a piecemeal basis recognizes that people need time to adjust to new ideas. The topic of implementing technological systems in manufacturing environments will be dealt with in some detail in Chapter 8.

FACTOR This (FActory Control Through Operations Replication) is a computer based scheduling package designed for on-line generation of detailed schedules for the factory floor (Factrol, 1986). FACTOR uses discrete-event simulation technology to generate schedules. According to the developers of the system, FACTOR supports the implementation and control of manufacturing systems. Using MRP due dates and current order status for initialization, FACTOR produces reports highlighting shop order performance and resource utilization, and then produces the best possible schedules of production resources. As well as providing schedules for daily use on the shop floor, FACTOR facilitates disciplined experimentation through its powerful simulation features. For instance, if a rush order is received, or a machine goes down, or raw materials are unavailable, FACTOR has the capability to evaluate various scheduling alternatives and predict possible outcomes for each particular course of action. Overall potential benefits accruing from this simulation facility include reduced work in progress, improved on-time order deliveries and reduced labour costs.

Grant (1987) describes manufacturing locations which have used the FACTOR software system. One such facility was at a plant in the USA, which was at the time involved in an overall CIM project that the corporation embarked on to address what it considered were the three dominant themes in its marketplace: cost control, higher quality levels and the requirement for shorter lead and cycle times and delivery dates. This plant produces resistor networks for the data processing, telecommunications, automotive, and assorted electronics market. As part of their CIM solution, they recognized the need for a tool that could schedule efficiently. The FACTOR system was installed to take the customer order requirements from their MRP system, and generate a schedule based on these orders and the manufacturing data. This detailed schedule is then passed down to the supervisory system, which disseminates the information to each of the machines on the shop floor (each machine operator having his/her own computer terminal). FACTOR may be viewed as a finite scheduling system, and thus it could be applied as the scheduling building block in our proposed PAC architecture.

ESPRIT Project Research The European Strategic Program for Research and Development in Information Technology is a European Community initiative to support the development of information technology within the European Economic Community (EEC). The ten year ESPRIT programme, launched in February 1984, has three major objectives:

1. To provide the advanced industrial technology base needed to meet the competitive requirements of the 1990s in the key information technology areas of: micro-electronics, information processing systems and application technologies, particularly office systems and computer integrated manufacturing.
2. To promote European industrial cooperation in information technology across borders and across the industrial/academic divide. With the new phase of ESPRIT (ESPRIT II), the European dimension was extended to include participants from EFTA (European Free Trade Association) countries.
3. To contribute to the development of internationally accepted standards, helping European companies to become leaders in information technology standardization.

Research in shop floor control comes under the umbrella of the CIM section of ESPRIT, and there are three projects in the scheduling/production control domain which are of interest in this chapter: project 809, project 932 and project 477. Project 477, entitled COSIMA (Control Systems for Integrated Manufacturing), involved the authors of this book, and is described in detail in Chapter 9. The three other research efforts are now briefly described.

ESPRIT project 809 (Tiemersa, 1988) deals with the development of a system aimed at improving the production control capabilities in small batch part manufacturing. This is achieved by a system which integrates the functions of scheduling, workstation control and monitoring, and which is connected online to the production equipment on the shop floor. The resulting production control system (PCS) makes use of an expert system and the knowledge and intelligence of the operators. A prototype has been implemented at a pilot plant.

PCS was developed for a completely automated flexible manufacturing cell (FMC), consisting of:

- several pieces of equipment (machining centres, milling machines, lathes etc),
- transport systems for tools and products, loading and unloading devices,
- a cell computer on which the PCS runs, which connects with the different items of equipment and devices in the manufacturing cell by a communications network.

This PCS system includes four main modules:

1 A scheduling module;
2 a dispatching module;
3 a workstation control module;
4 a monitoring module.

The input of the PCS from the system level consists of production orders and the necessary process plans. Based on the up-to-date availability of the equipment and production times, these production orders are scheduled into a workplan by the scheduling module. This workplan contains the sequence of jobs to be executed, and for each job the start and finish time is calculated. The output of the scheduling module serves as a command for the dispatching module, which then transforms the workplan into commands for the different jobs at each process step. In circumstances where a scheduled job cannot be started, the dispatching module passes status information to the monitoring module. The output of the dispatching module serves as a command for a workstation control module. This workstation control module collects the status information of the different items of equipment and passes it on to the monitoring module. The monitoring module subsequently performs the task of feeding back information within the PCS. Status information is collected and processed in the form required by the different modules and stored in the database of the PCS. This monitoring module also reports the performance of the manufacturing cell to system and factory level. This system was implemented on a pilot basis at Morskate Transmissions, B.V., Hengelo, The Netherlands, a company which produces transmission components such as spindles, clutches, gears, parts of gear boxes and special high precision products.

ESPRIT project number 932 (Meyer *et al.*, 1988) developed a methodology to analyse and design an intelligent cell controller. This intelligent cell controller was developed using AI tools and the purpose of the controller is to plan and control activities within a cell. The prototype cell controller was realized for a Surface Mount Device (SMD) cell of a car radio factory. The controller supports the lowest levels of the planning function, by bridging the time gap between the three days planning period of the higher planning functions and the shift operations on the factory floor. Prior to the installation of this cell controller, the task of manually sequencing orders was a complex one, the main problems being:

1 The plan from the higher planning system was not appropriate for short term production planning;
2 there were a large amount of process parameters to deal which increased the complexity of the scheduling task.

The solution to this task is a flexible planning support system based on heuristic or approximative algorithms and rules. The core strategy is to determine the bottleneck workstation, and optimise the throughput for this bottleneck in the manufacturing system.

PLATO-Z PLATO-Z (Production Logistics and Timings OrganiZer) is an application of the artificial intelligence techniques of blackboard and actor based systems to develop an intelligent cell control system. PLATO-Z includes an architecture for an Intelligent Cell Control System (ICCS) which may be categorized as part of the planning and scheduling system. It also includes other functions such as error handling and monitoring. O'Grady argues that for such an intelligent control system, a decentralized control structure is preferred since the decisions are made as low as possible in the hierarchy and the cell level can take over much of the responsibility for running the cell. Thus the major functions of the ICCS of PLATO-Z is to schedule jobs, machines and other resources in the cell so as to achieve the goals set at the shop level by efficiently using the resources within the cell. Instructions are then passed to the equipment within the cell, and a monitoring function keeps track of the processed operations based on feedback from equipment.

Prior to describing each of the separate functions of the ICCS, it is important to clarify what is meant by blackboard and actor based systems. A blackboard is a shared data region surrounded by knowledge sources and by a blackboard controller (Erman *et al.*, 1980, Hayes-Roth, 1985, Nii, 1986a,b). The knowledge sources may be heuristics, optimizing techniques or complete expert systems. According to O'Grady, problems placed onto a blackboard can be solved through the cooperative efforts of the knowledge sources. The actor based methodology, first proposed by Hewitt (1977), consists of autonomous data and processing entities called actors which communicate through message passing. This methodology maps naturally onto multi-processor computers and has the advantage of low computational time. From these two methodologies, O'Grady proposes a hybrid multi-blackboard and actor based framework, which seeks 'the advantage of computational efficiency under steady conditions but flexibility in changing environments'.

Thus, PLATO-Z ICCS has been implemented using the multi-blackboard/ actor-based framework contains several blackboard sub-systems, each of which performs major cell control functions (i.e. scheduling, dispatching etc). This ICCS system provides an architecture in which many different control functions can be performed in a modular fashion and where overall control can be achieved by passing appropriate messages between blackboard systems. The four blackboard systems which reflect the different functions of the ICCS are:

1 The scheduling blackboard which schedules resources within the cell so as to achieve the goals set by the higher shop level function;
2 The operation dispatching blackboard which generates detailed operation requests to the equipment level;
3 The monitoring blackboard which filters and classifies the information gathered from the equipment level of the cell;
4 The error handling blackboard which recognizes and analyses the errors and problems occurring in the cell and provides possible corrective actions to these problems.

In addition to these four main blackboards there are also three support functions which facilitate initialization and termination of cell activities, communication and networking, and a user interface. Each of the main blackboards are now described.

1 The scheduling blackboard

The ICCS scheduling blackboard receives goals or commands from the shop level and must manage the cell resources in order to achieve these higher level goals. The goals are in the form of jobs to be completed and their required completion dates. Thus, the major functions of the scheduling blackboard include resource availability checking, resource assignment, sequencing, interacting with external systems, and responding to the feedback information from the monitoring and error handling systems. The control functions of the scheduling blackboard can be divided into two main activities: regular scheduling and rescheduling.

The regular scheduling generates the sequence in which jobs are to be released to the dispatching blackboard. Rescheduling incorporates the ability to react to any error conditions occurring in the cell environment.

2 The operation dispatching blackboard
This blackboard takes commands from the scheduling blackboard and generates a detailed operation request for the equipment level. The operation dispatching blackboard reports any major changes in the status of the system to the scheduling blackboard. The main functions of the operation dispatching blackboard are:

- To generate and disseminate detailed operations to the equipment within short time frames;
- to respond to requests for 'corrective action' from the error handling blackboard;
- to update the cell status (e.g. part status, machine status, robot status, etc), and;
- to report major status changes of jobs, machines, and other resources to the scheduling blackboard.

3 Monitoring blackboard
The monitoring blackboard constantly keeps track of the status of the operations and resources through regular feedback information sent from equipment level. It filters and classifies the feedback information and reports relevant information to the higher level blackboard systems. The major functions of the monitoring blackboard are:

- To track the status of operations, parts, and resources;
- to filter and classifying feedback information;
- to report status changes to the higher level systems;
- to keep statistics on the cell components such as utilization of machine tools, down times, set up times, etc.

4 Error handling blackboard
This blackboard is responsible for dealing with error problems or exceptional conditions within the cell, such as machine failures and tool breakages. It recognizes and analyses the errors and provides possible corrective actions to these problems. The suggested actions are passed to either the scheduling or dispatching blackboard. The main functions of the error handling blackboard are:

(a) To interpret and classifying error conditions reported from the monitoring blackboard;
(b) to generate corrective actions to an error;
(c) to request corrective actions from the operation dispatching blackboard for minor deviations;

(d) to request rescheduling actions to the scheduling blackboard for the major disruptions;

(e) to report major failures to the shop floor control system through the scheduling blackboard when they are beyond the control of the ICCS.

To summarize, the PLATO-Z system consists of four main blackboard modules and three support functions which are combined using artificial intelligence techniques, to provide an intelligent cell controller. This control system is conceptually similar to our architecture for shop floor control, since it recognizes the individual components of scheduling, dispatching and monitoring. According to O'Grady, the latest implementation is in LISP on a Symbolics 3645 machine, and this multi-processing environment simulates the concurrent operation of the four main blackboards.

MADEMA MADEMA, an acronym for 'MAnufacturing DEcision MAking', was initially developed at the Massachusetts Institute of Technology (MIT), USA, under a contract from Computer Aided Manufacturing -International Inc (CAM-I). It is based on a concept for manufacturing decision making at the work centre level which involves the assignment of resources (machine tools, handling devices, operators, robots etc.) to production tasks.

The MADEMA approach recognizes two primary factors which must be considered in order to create adequate decision-making procedures: flexibility of systems and the nature of existing scheduling techniques. Flexibility, a primary aim of any modern manufacturing system, may be compromised and restricted if the scheduling techniques are not capable of responding to changes in the manufacturing environment. The MADEMA methodology also considers that most modern scheduling techniques optimize based on a single criterion, and that such an approach requires all aspects of a system to be reduced to the single criterion in a predetermined manner. The drawback of the one criterion approach is that this approach assumes that the manufacturing environment is static when, in reality, priorities, requirements and conditions are constantly changing. Thus, MADEMA is designed to support multi-criteria decision making in the shop floor environment. It also attempts to establish a decision making framework for manufacturing systems that integrates traditional techniques for process planning and scheduling.

Today, MADEMA is a fully operational software system designed for intelligent control of the factory floor. It is based on three main elements:

1 A simulated manufacturing environment for testing the response of the manufacturing system to different decision making procedures. The input information describing the factory is used to create data structures which describe jobs, work centres, and resources that make up the

factory. These data structures, when aggregated, defined the overall manufacturing system. This view of the system is simulated and relevant data changed to reflect changes in the system during the duration of the simulation run.

2 A decision making module that can make decisions according to five separate steps which ultimately selects the most appropriate alternative. This module is called on by the simulator at certain intervals of the simulation run. When called, the five steps perform the tasks of firstly determining the alternatives, then evaluating the relevant decision criteria before finally applying the decision rule with a view to selecting the alternative with the maximum expected utility.

3 A database that provides and manages the data needed for the decision making processes. In addition to these initial data requirements for MADEMA and the simulator, MADEMA also requires extensive data and support to continually update decisions according to a variety of new information.

An important feature is the ability to simulate production operations through the accommodation of a wide range of 'what if' evaluations. It is designed to provide optimal allocation of resources and serves also serves as a management and educational tool. Bunce (1988) describes a major future enhancement of the MADEMA system as the accommodation of dynamic scheduling within the manufacturing environment.

ISIS The ISIS system is based upon the notion that scheduling is is concerned primarily with the determination and satisfaction of a large number and variety of constraints. Hence, ISIS is an artificial intelligence based scheduling system capable of incorporating many constraints in the construction of job shop schedules. The developers of this system have categorized manufacturing constraints into five separate categories:

1 The first of these encountered in the factory is an organizational goal. An organizational goal constraint can be viewed as an expected value of some organizational variable (e.g. due dates, work in progress levels, resource levels, costs, production levels and shop stability).

2 The second category of constraints are physical constraints, and these specify the technological constraints which limit the functions of each physical resource.

3 Causal restrictions constitute a third category of constraint, and these define what conditions must be satisfied before initiating an operation (e.g. precedence and resource requirements for operations).

4 Preference constraints can be viewed as an abstraction of other types of constraints, and the reason for the preference may be due to cost or quality, but sufficient information does not exist to derive actual costs.

5 Availability of resources is the final category. During the generation of a schedule, as each resource is assigned to an operation, it has attached to it a constraint that defines its unavailability for other jobs/operations during the relevant time period.

A key task of scheduling, according to the developers of ISIS, is to obtain feasible schedules taking constraints into account. Often, there may be many different and conflicting constraints within a given production system. ISIS uses a knowledge representation system as the repository for the knowledge necessary to schedule production. It recognizes the different types of constraints and allows inter-relationships between the constraints to be explored. For example, if a conflict between two constraints cannot be solved, one or both of the constraints is relaxed.

In developing a suitable schedule, ISIS first performs capacity based scheduling before then generating a detailed schedule. ISIS also provides the user with the capability to interactively construct and alter schedules. The ISIS system has been designed for practical use in job shop production management and control. The ISIS model also supports organizational modeling, reactive plant monitoring and simulation.

SONIA This, described by Collinot *et al.* (1988), is 'an on-line reactive scheduler which deals with real-time disturbances'. It integrates scheduling, dispatching and monitoring functions using an AI based approach to guide production and process engineers. It is organized into a schedule maintenance module and a set of analysis and problem solving modules divided into three categories:

1 Predictive and reactive scheduling modules which use a heuristic search approach to generate acceptable schedules.
2 Analysis modules that refine the available information and provide reports about the quality of possible solutions. These reports are then interpreted to direct the problem solving effort, and provide possible answers to the questions of what action to take and how to take it.
3 Control modules which are used to coordinate the activities of the other modules of the SONIA system.

The schedule maintenance module aims at maintaining the consistency of the scheduling decisions which may be taken by either predictive (planned mode) and reactive modules (real-time mode), or by the user in an interactive manner.

In all, there are five modules in the SONIA system:

1 A global selection module is used to prioritize activities and to select them with respect to the available shop capacity. Its primary inputs are manufacturing orders characterized by MRP constraints concerning required materials and associated due dates.

2 A product selection module selects the successive operations of the process routeings of one product, and this selection is dependent on the status of operations and the type of product flow.
3 A detailed scheduling module sequences the activities allocated to a given workstation, while satisfying sharing resource constraints and interruption constraints.
4 A global re-scheduling module modifies the whole plan forward from the current date, the advantage of this strategy being that it can rapidly generate a new admissible plan by relaxing due dates.
5 Some local methods are used when conflicts concern a particular manufacturing order or a work area.

The SONIA system includes a simulator which is used to test the possibilities of disturbances to the original plan. The simulator provides the capability to execute the planned actions as they are sequenced and timed by SONIA, and can also use its own decision making with the help of dispatching rules. The developers of SONIA also report that the predictive components of the system have already been tested on actual data from a sheet-metal workshop. Future plans for further development of SONIA includes the design of 'complementary reactive algorithms' and the validation of various strategies to generate and modify scheduling plans.

Logistics Management System (LMS) The LMS system, developed at IBM's semi-conductor facility in Essex Junction, Vermont, is described as an 'integrated decision support and knowledge based expert system to monitor and control manufacturing flow'. It serves as a monitoring and control function with the aim of improving manufacturing performance in tool utilization, serviceability and cycle time. Significant success has been achieved in these areas, and the LMS system is now a critical component in running major areas of the manufacturing facility.

According to Fordyce (1989), LMS makes use of two key principles:

1 Secure reliable and timely data sources are an essential element of effective shop floor control.
2 A manager would rather live with a problem he cannot adequately solve, than accept a solution he cannot understand.

The first requirement of accurate and timely data was achieved within the IBM facility since the Essex Junction plant had built a set of very reliable (strong data integrity, hardware backup, etc) real-time data collection computer systems to record activities on the shop floor (movements of lots, down machines, change in operators etc.) However, these systems had limited decision support facilities, and as the size and activity level of the plant increased, the support systems were no longer sufficient to meet the needs of the business. As a result, the facility remained data rich but information poor.

The LMS system was then developed to overcome the obstacle of poor decision support, and it included a monitoring facility with a scheduler. The monitoring system made sense of all of the data being captured using a knowledge based expert system. The scheduler was developed with the view of improving the lead time of products and enhancing tool availability.

The Leitstand Approach According to Haverman (1991) a Leitstand is a computer aided decision support system for interactive production planning and control. A number of such tools have been developed, particularly in Germany. A tool called SIPA PLUS, developed at the GRAI Laboratory of the University of Bordeaux can also be categorised as a Leitstand. A Leitstand is essentially a finite scheduling tool which is positioned within the PAC or the Factory Coordination system. In terms of the architecture presented in this book it functions as a scheduler and a partial monitor. SIPA PLUS as described by Doumeingts (1991) is a computer based interactive Gantt chart, which allows a workshop manager to create a detailed capacity sensitive schedule which takes due account of the local knowledge of the user and the actual capacity of each resource in the cell or factory. Based on a simulation model of the cell or factory, SIPA PLUS generates a schedule using a user selected dispatching rule or scheduling heuristic. The expectation is that the user would go through a number of scheduling iterations, perhaps modifying the capacity of individual workstations (overtime), or the priority of individual jobs before accepting a particular schedule. A very important requirement of such Leitstands in practice is that they integrate easily with exising MRP systems and production databases. In this way the planned orders output from an MRP style system can be translated into actual orders for the shop floor.

6.3.1 Scheduling for shop floor control (FC and PAC)

We have completed our review of scheduling which encompassed the modern and traditional approaches, and now we present a summary of our own opinions on the most suitable solution to the scheduling problem. Our approach may be viewed as a synergy of the following five essential elements:

1. A well defined control architecture is an essential prerequisite to any implementable scheduling solution. We believe this architecture should incorporate both flexibility and decentralization, and decision making should be made at the lowest possible level of the hierarchy. This is why we have developed the architecture for FC and PAC, to ensure both flexibility and decentralization in shop floor control related activities.
2. Information technology has an important role to play in implementing such a scheduling solution. An information technology reference model for FC and PAC has already been described in Chapter 4, and the most

appropriate tools available to implement this reference model reviewed in Chapter 5.

3 A sound strategy is required to develop detailed schedules for any particular shop floor. As was pointed out earlier, no one scheduling heuristic can be viewed as the most appropriate in all cases. Hence, we recognize the need to construct, through the use of IT tools, what we term a scheduling library, which contains a wide range of possible scheduling solutions. This can enable the selection of the most appropriate scheduling strategy, which can subsequently be used to develop a plan for the shop floor.

4 As outlined in Chapter 2, we distinguish clearly between finite and infinite scheduling approaches. We believe that finite scheduling should be used at the PAC level of the production management hierarchy. The finite scheduling approach is also preferrable at the FC level, but in reality this may not always be appropriate.

5 Finally, it is important to realize that in most scenarios, scheduling occurs in a human environment, and thus the scheduling system should take full cognisance of the important role that humans have to play in intelligent decision making. We believe that scheduling systems must be designed to achieve a worthwhile balance between available scheduling heuristics and the experience of humans. For this purpose, we propose an interactive scheduler, which allows humans to use their experience to enhance possible scheduling solutions.

Thus, we have summarized five essential components for our solution to the scheduling problem. Each of these elements are included in our design tool for shop floor control systems, which is described in detail in Chapter 9.

6.4 Conclusions

In this chapter, we have presented a literature review of the many different approaches to the scheduling problem, highlighting the strengths and weaknesses of these methods. The drawbacks of the optimization approach were highlighted and the future direction of scheduling and control systems was discussed. We also presented our own approach to the solution of the scheduling problem, and this approach is further illustrated through the design tool to be described in Chapter 9. A further important aspect of our overall approach to shop floor control, is the recognition of the importance of production environment design strategies. The following chapter presents a state-of-the-art review of a selection of production environment design strategies.

7

A review of production environment design strategies

7.1 Introduction

The factory coordination task which we described in Chapter 2 earlier, involves a production environment design element. The production environment design element reorganizes a product based layout to accommodate new products and to operate more efficiently. We argued that the production environment task should be regarded as a selection of techniques which support manufacturing personnel in maintaining a product based layout within their manufacturing system. We identified three procedures within this element as being significant in the organization of a product based manufacturing system; product based manufacturing, process planning, and manufacturing systems analysis.

In this chapter we present current thinking in each of the three areas associated with the production environment design task. Firstly, in section 7.2 we examine product based manufacturing. Within this section we also look at the methodology of Group Technology and its use in relation to the maintenance of product based manufacturing. This is followed by a discussion on five different methods of product family and cell formation. We conclude the section on product based manufacturing by examining potential problems with such a mode of manufacturing from an organizational and social perspective. Secondly, we introduce the reader to the task of process planning by describing some of the important issues involved in process planning in section 7.3. We continue the discussion on planning by looking at current planning strategies and at current developments in planning systems involving variant and generative process planning systems. Thirdly in section 7.4, we examine different techniques for simplifying the manufacturing system.

From the material contained in this chapter, the reader can familiarise himself/herself with the details and procedures involved in product based manufacturing, process planning and manufacturing systems analysis. The reader can also use the material to examine the particular uses of product based manufacturing, process planning and manufacturing systems analysis within the operation planning and control architecture, with other similar applications in the area of production management.

7.2 Product based manufacturing

Gunasingh and Lashkari (1989) describe product based manufacturing as an organizational approach to produce products with similar processing requirements within groups of resources known as cells. In a process based layout, the various different resources are arranged together in groups according to their function. For example in a machine shop organized by process, all lathes are grouped together in one department, while all drills are arranged together in another department. In a product based layout all resources which are required for the manufacture of a product family are grouped together into one or more cells. The changeover to a product based layout can lead to a more organized layout. Burbidge (1989) lists four main advantages which accrue from the use of product based layout:

1. A decrease in throughput times;
2. a reduction in the complexity of the delegation of tasks;
3. an increase in the range of skills of supervisors and operators;
4. the provision of a good base for developing automation.

Throughput time refers to the time taken to completely produce a product in a factory. By decreasing throughput times, work in progress stock levels and stock holding costs are lowered. Since throughput time is a component of the total time of getting product to market, increased sales can result from faster market introduction of new products, especially in 'make to order' environments. By organizing the production of each family of products among a small number of cells, it is possible to delegate responsibility to each team of operators for managing all aspects of the production of each product family.

With a process based layout, no one supervisor has complete responsibility for the production of a product or family of products, and therefore the decision making in relation to the production of a family of products has to be centralized. With a centralized decision making system, expensive bureaucratic controls can develop, which can encourage inefficiencies. In a product based layout, each team of operators and their supervisor have greater responsibility for organizing all aspects of the production of a product family. This distributed responsibility increases the levels and range of skills of the operators and supervisors, thus ensuring good production performance. In order to ensure the maximum possible benefits from the introduction of product based manufacturing, it is important that the organization and the production environment in particular has certain characteristics. We shall discuss these characteristics in section 7.2.2.

A process based organization may hinder the introduction of automation to a factory because:

- it gives no indications as to which products are most suitable for manufacture by automation;
- unlike in a product based layout, no one supervisor or manager has complete responsibility over all the required resources for a family of products;
- any automation development in a factory can affect several different departments. These departments cannot be isolated into a single grouping under one supervisor or manager, as in a product based layout.

A methodology is required to ensure that the evolution towards product based manufacturing proceeds efficiently. One suitable approach for the development of a product based layout is Group Technology, which we shall now discuss.

7.2.1 Group Technology

Groover (1980) described Group Technology (GT) as a 'manufacturing philosophy in which similar parts are identified and grouped together to take advantage of their similarities in manufacturing and design'. Group Technology can also be regarded as a means of not only organizing common products, but also common concepts, principles, problems and tasks with the overall goal of improving productivity (Greene and Sadowski, 1983). With the formation of product families and cells, the production environment is organized on a more efficient basis. This efficient organization generates benefits in production, process planning, materials handling and employee satisfaction (DeVries *et al.*, 1976). Gallagher and Knight (1973) argued that the production related benefits such as reduced work in progress levels, and reduced set-up times can be achieved in a product based layout, because of the similarity between the manufacturing and/or design characteristics of the range of products in a family. Gunn (1987) outlined the following results from a survey of thirty five product based manufacturing systems (Table 7.1).

7.2.2 Suitable conditions for the use of group technology

Group Technology is generally recognized as being useful in batch manufacturing environments, but particular batch manufacturing environments can be more suitable for the use of Group Technology than others. Leonard and Koenigsberger (1972) have identified a number of production features which influence the feasibility of using Group Technology in a production

Production Characteristics	Percent Reduction
Manufacturing Lead Time	55 %
Set-up Time	17 %
Average Batch Travel Distance	79 %
On-time deliveries increase	61 %
Average Work in Progress	43 %

Table 7.1. *Demonstrated production benefits of product based manufacturing (Gunn (1987))*

environment and a range of different production problems which the Group Technology methodology can help to solve. The overall conclusion from their work is that all production features of a batch manufacturing environment must be considered before a suitable environment for group production can be developed. We shall now discuss each of these manufacturing features in turn.

The product flow within a factory should consist of a large number of small batches, in order for the application of Group Technology to prove successful. Mass production techniques are appropriate in a manufacturing environment where products are manufactured in a small number of large batches. Group Technology can be of great benefit in a manufacturing environment, where the fulfilment of customer order due dates is important. Connolly *et al*, (1970) argued that the use of Group Technology helps to reduce production lead times through better organization of the production process, thus making the production control task relatively easier. In an organization which is having problems with the control of raw materials, the application of Group Technology can help to increase control through standardizing the range of raw materials being used in production.

Another factor which is helpful in the process of product family and cell formation using Group Technology techniques is a large degree of similarity of components and manufacturing operations. The introduction of a design for manufacture and assembly methodology into a manufacturing environment can help to standardize the range of components being used in the production of a collection of final products.

In batch manufacturing environments which use capital intensive equipment, management may feel that the use of Group Technology will lead to under utilisation of these resources. However with an efficient analysis procedure during the formation of product families and cells, appropriate levels of resource utilization can be achieved. Suitable analysis procedures include the use of simulation or simple statistical techniques. Purcheck and Oliva-Lopez (1978) illustrated the use of statistical techniques to help identify potential problems with under utilised resources using Pareto-type histograms.

To help a product based manufacturing environment to cope with fluctuating demand, a factory should have a high degree of labour flexibility and a range of techniques for determining suitable production loads for each cell. Operator training is essential in order to develop high labour flexibility particularly with highly skilled tasks. El-Essaway (1971) argued that in some manufacturing situations, high labour flexibility may not be enough and that cell production loading techniques should also be used to plan a near even distribution of production requirements, among all cells.

The practice of inspecting products after each process step does not help the development of product based manufacturing because it may require movement to a separate inspection area. This movement can cause unnecessary disruption to product flow within a product based layout. A more efficient inspection procedure proposed by Marklew (1970) is the use of a 'travelling inspector' procedure which involves assigning one group operator in each cell to be responsible for product inspection. This group member can move around the cell, checking products without causing any unnecessary disruptions. Another feasible approach towards inspection is to adopt total quality control principles from Just In Time, and to assign responsibility for quality control at each process step to the operator who completes that particular process step.

This range of production conditions and production problems, that we have just discussed, illustrate the necessity of giving consideration to a wide variety of features within a production environment prior to the introduction of product based manufacturing. Once a suitable production environment has been developed and the decision to proceed with the introduction of group production has been made, the next step is the formation of product families and cells. We shall discuss a range of methods for product family and cell formation.

7.2.3 *Product family formation*

Product family formation involves examining the range of products and based on certain criteria, placing them into groups. Groups of resources are formed into cells, which cater for the production requirements of each product family. Waghodekar and Shau (1984) list five methods for forming product families:

- Visual inspection;
- classification and coding systems;
- classification algorithms;
- production flow analysis;
- composite part.

All five methods use the basic Group Technology principles of common manufacturing processes and/or common components and design attributes, as the main criteria for grouping products together into families. We shall now look at the main ideas of each of the formation techniques, with a particular emphasis on the particular benefits and drawbacks associated with each method.

Visual Inspection Of the five formation methods, visual inspection is the least sophisticated and least expensive. Parts are classified by examining the actual physical parts, drawings of parts or photographs of them. Based on these observations parts are grouped. The main drawback to this method is that it can be very subjective. However there are cases where companies have made the changeover to group production using visual inspection (Holtz, 1978).

Classification and Coding Systems Classification and coding systems allow the formation of product families and cells to proceed in a systematic manner. Chang and Wysk (1985) describe coding as 'a process for establishing symbols to be used for meaningful communication' and classification as 'a separate process in which items are separated into groups based on the existence or absence of characteristic attributes'. The coding process serves as a prelude to the classification of products into groups. Products are classified according to interpretation of each product's code, which depends on the coding scheme being used. A code differs from a product number in that it is not unique to each product and it contains a string of characters which hold information about the product. There are three main types of codes, depending on the variation in the assignment of symbols; monocodes, polycodes and mixed mode codes (Hyer and Wemmerlov, 1985). Monocodes otherwise known as hierarchical codes consist of a string of symbols, where the interpretation of each symbol depends on the meaning of the subsequent symbol. For example using the approach of Fig. 7.1, a rotational unmachined part with a length to diameter ratio of less than 0.5 is described by a monocode 010. The interrelationship of symbols in a monocode can make the code difficult to interpret and to construct, but a wide range of detail can be contained in a relatively short length of code. The symbols in a polycode are independent of each other, i.e. the interpretation of each symbol does not depend on the meaning of any other symbols in the code. The independence of symbols makes a polycode relatively easy to read and to develop. However, a polycode can be quite long, if it is to represent and contain a large amount of information. A mixed mode code is a mixture of a monocode and a polycode. Product characteristics which are accessed frequently are contained in the polycode section, while characteristics which have a wide variety of values are stored in the monocode section. Interpreting a mixed mode code is not as difficult as

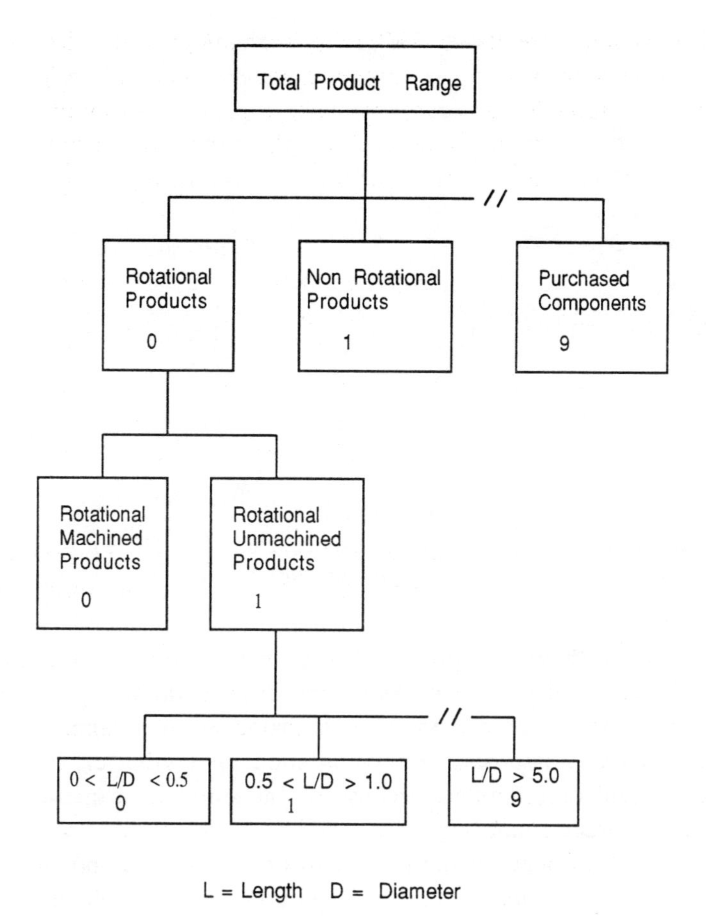

Fig. 7.1. An example of a monocode coding system.

reading a monocode and it can contain a large amount of information in a relatively short code.

When all products have been coded, Hyer and Wemmerlov (1985) recommended the following steps in the formation of product families:

- Define the objective of the family formation exercise;
- based on this objective, decide on the specific attributes that are associated with each product family members;
- having defined each set of attributes, determine the necessary code symbols which match the set of attributes;
- use each matching code to retrieve products with similar attributes from the database of coded products.

Membership of product families depends on the objectives of the formation exercise, whether they are connected with efficiencies in manufacturing or with

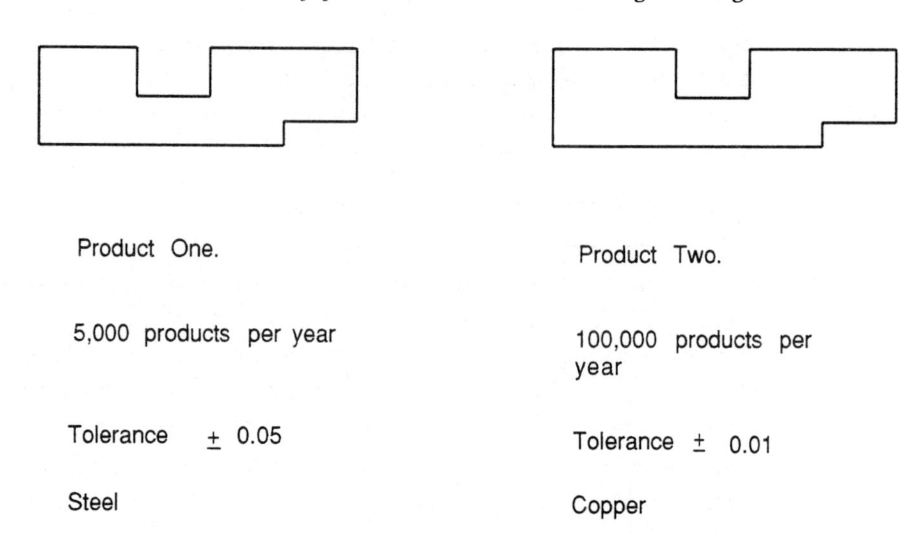

Fig. 7.2. Two components of similar shape and size, but different manufacturing characteristics.

improvements in the design process. For example, the two parts illustrated in Fig. 7.2 are identical in general shape and size. Using design characteristics as a criterion, they would belong to the same product family. However if manufacturing characteristics are being used as a criterion, the two parts might be in different families because of the differences in tolerances, raw materials and order quantities.

The relationship between the objectives of the exercise and the chosen attributes to distinguish each family is important. For example, if the objective of the formation exercise is to group together products using similar manufacturing processes, then a specific attribute for a family can be that it contains all machined products. Then the code symbols which describe a machined surface on a product are determined and used to develop a list of machined products from the total product range. Brankamp (1969) found that the complexity of the retrieval process is dependent on the number of attributes included in the family definition and the type of coding system being used. The family formation process should be regarded as an iterative one, because in some situations, a product family may be too small or perhaps become too large.

Classification and coding systems are very helpful in forming product families, but Burbidge (1985) identified a number of deficiencies associated with them:

1 Difficulties can occur in finding a total division of products using the classification and coding method.

2 Products of similar shape and size, but with different tolerances, tend to

be grouped together. The variety in tolerances means that the products may have to be processed on different machines.

3 Products with varying production requirements are likely to be grouped together, which may cause problems with capacity requirements within a cell.

4 Classification and coding systems are of little benefit in grouping resources into cells.

5 Classification and coding systems can be relatively expensive and time consuming to develop and implement in a factory.

6 Product families formed using the classification and coding method tend to contain parts of different sizes and varying material requirements, which could require a variety of manufacturing processes.

Classification and coding systems are not essential for the implementation of product based manufacturing, but they enable the implementation of this mode of manufacturing to be completed in a systematic manner. When deciding on the feasibility of using a classification and coding system, management must balance the deficiencies outlined above against the benefits of Group Technology codes in design and production. The main design benefit resulting from the use of Group Technology codes is the standardization of the component and product ranges within a factory (Gunn, 1982). Some of the production benefits associated with Group Technology are listed in Table 7.1 above. From a process planning perspective, Bucher, (1979) found that codes can help to standardize process plans for products from the same family, and develop plans for new products coming into production.

We will now conclude this discussion on classification and coding systems by offering a short description of some well known classification and coding systems:

- Brisch (Gombinski, 1969)
- Code (Bobrowicz, 1975)
- Multiclass (Houtzeel, 1975)
- Opitz (Opitz, 1964)
- D-Class system (D-Class, 1985)

The Brisch system was one of the earliest classification and coding systems to be developed in England. Each Brisch system is tailor made for a specific user, and allows the user to code all relevant types of objects (e.g. raw materials, components, sub-assemblies, products, tools and equipment). The code is developed in a step by step procedure using a series of coding charts (Gombinski, 1969). The code consists of a primary code and a secondary code (Gallagher and Knight, 1973). Design and shape characteristics are covered in the primary code, which is a monocode. The secondary code is a polycode and contains information on manufacturing related data.

Code is a coding and classification system which is based on the Brisch system. Through the Code system, the user can code not only the physical characteristics of the parts, but their functions as well (e.g. end product usage) (Bobrowicz, 1975). The coding system develops an eight digit mixedmode code containing alphanumeric characters. One of the main advantages of the Code system is that the system is supported by a software support system. With this software system, Code serves as a database system, containing information on the design and process characteristics of a range of products, in addition to being a coding and classification system. Despite this, Schaffer (1981) found that one drawback to the Code system is that manufacturing information, such as manufacturing process requirements, cannot be included in the code.

The Multiclass system is an enhancement of an older coding system known as Miclass and is designed and marketed as part of a complete Group Technology program. The user may use the database of coded parts for process planning, computer aided design, development of cells and scheduling of parts (Ingram, 1982). The Multiclass system uses one of nine different coding systems, one for each class of material to be coded, such as machined parts or tooling. The structure of the code includes a monocode section and a polycode section (Oir, 1983). The Multiclass system may include off the shelf coding systems or coding systems which are developed specially for the user.

The Opitz system was one of the first coding systems to be developed and it was originally developed to study component statistics. The complete system is summarised on ten sheets, but a manual is available to help the user interpret and develop codes (Opitz, 1970). The range of parts that the coding procedure can be applied to, includes; machined parts, non-machined parts, and purchased parts. The structure of the Opitz codes involves nine code fields divided into two sections; a five digit primary mixed mode code and a four digit secondary polycode (Opitz and Wiendahl, 1971). The primary code contains information on design and shape characteristics. Information relevant to manufacturing, such as tolerances and raw materials, is held in the secondary code.

A new approach in the area of coding and classification systems is known as the D-Class system. The D-Class system is a computerized generic approach to coding, which is licensed and developed by Brigham Young University in Utah USA. D-Class is a computerized system, which can accommodate user defined logic and/or an existing coding system which is not computerized such as the Opitz system. D-Class uses a tree structured decision making system, with each branch representing a condition. A specific code value is assigned at the junction of each branch and a complete code is developed by completing multiple passes of the decision tree. According to users, the D-Class system has good coding and retrieval facilities, but the task of entering user defined logic is time consuming (Hyer and Wemmerlov, 1985).

Resource Name	List of Parts which use each Resource
Machine 1.	A, B, C, D, E
Machine 2.	B, D, F

Jaccard Coefficient for Machines 1 and 2 =
(Number of Common Parts ÷ Total number of parts) =
(2 ÷ 6) = 0.33

Table 7.2. *Calculation of Jaccard Coefficient*

This brief description of a number of existing classification and coding systems is not exhaustive, but gives some indication of the range of systems available. We now turn to discuss the use of classification algorithms.

Classification Algorithms Classification algorithms have mainly been developed through operational research techniques. The main difference between classification algorithms and classification and coding systems is that in the former, algorithms rather than codes are used to determine product families and cells. We will examine the following range of techniques, which can be used to group components and associated resources:

- Similarity coefficients;
- graph theory;
- matrix representation.

Similarity Coefficients McAuley (1972) suggested the use of similarity coefficients which can measure the 'similarity' between a pair of resources and the use of the similarity measurements as the basis for gathering the resources into cells. The similarity measurement used is known as the Coefficient of Jaccard. The Coefficient of Jaccard is calculated for any pair of resources as the number of parts which use both resources divided by the number of parts which use at least one of the resources (Sneath and Sokal, 1973).

Another approach to the cell formation problem proposed by Gunasingh and Lashkari (1989) involved using a similarity index to group machines, followed by an allocation of products to the resulting machine groups. Their approach differs from other approaches involving a similarity coefficient in that an operation on a product may be completed on alternate resources, rather than on a particular resource.

After calculating the similarity coefficients for each pair of machines, the values of the coefficients are compared to a threshold value. A threshold value is used as a lower limit, and any similarity coefficients less than the threshold value are ignored. The use of a threshold value can cause problems, because it is an arbitrary value. Rajagopalan and Batra (1975) attempted to overcome this problem by developing a method for choosing a suitable threshold value based on the particular formation problem being solved. If the similarity coefficient compares favourably with the threshold value, then the resources are gathered together using a clustering algorithm. McAuley (1972) uses the single linkage cluster analysis method for grouping resources, but found that even though two clusters of resources are linked by this technique, various resources of either cluster may be quite different from each other. Additional information concerning clustering algorithms can be found in Anderberg (1973).

Graph Theory The use of graph theory to identify relationships between resources has been suggested by Rajagopalan and Batra (1975). They define a graph as a set of points and a set of lines connecting the points. The set of points in each graph is assumed to be a set of resources and the set of lines represents the relationship between resources. Graph theory can then be used to analyse the graph to discover suitable groups of resources (Haray 1971).

In the graphical technique devised by Rajagopalan and Batra (1975), there are three phases. In phase one, data entered by the user is used to develop a graph with the points representing the resources and the connecting lines represent the relationships between resources. In phase two, graph partitioning approaches are used to form groups of machines. The third phase involves allocating components to each cells and calculating the loading of each cell, so as to determine the number of each type of resource required in each cell. Within the graphical analysis technique, a threshold value in association with a similarity coefficient is used to establish the strength of the relationship between resources. Again, as with similarity coefficients, the threshold value is chosen arbitrarily. This technique should only be used to provide guidelines for making decisions concerning the grouping of resources into cells because of the complexity of the problem and the inability to incorporate non-quantifiable variables, such as higher level of production control and social considerations, into the calculation.

Matrix Representation Matrix representation can be used to describe the processing requirements of products on a series of resources, by listing all products along one vertice of the matrix and listing all resources along the other. By rearranging the rows and columns to form a series of block diagonals, resources can be grouped to form cells. Masking techniques can then be used to form block diagonals within a matrix (Fig. 7.3). A simple example of a masking technique was suggested by Iri (1968). Starting from any row, all columns which have an entry in that row are masked. Then all rows

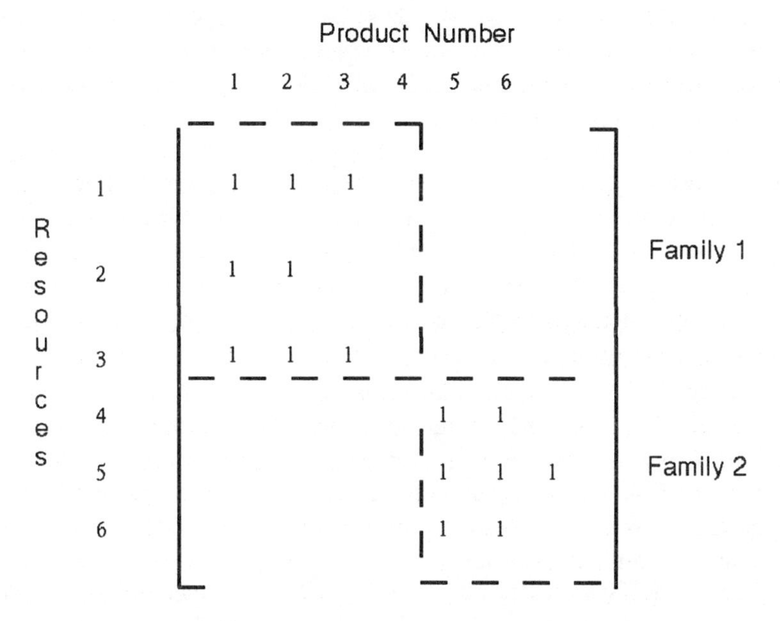

Fig. 7.3. The formation of block diagonals in a matrix.

which have entries in these columns are masked. The procedure continues until all possible rows and columns are masked. These rows and columns then constitute a block. If no blocks are found during the procedure, the whole matrix is considered to be one block. One drawback to the procedure is that it cannot be modified to take account of non-conforming elements. Kusiak and Chow (1987) present an augmented masking technique, which includes a standard cost associated with processing each part on each resource and sets a maximum value for the number of resources in each cell. This enhanced technique does not assume that the matrix decomposes into mutually separable submatrices. There are two situations where the augmented technique appears applicable; the subcontracting of parts which require processing in a number of cells in order to minimize costs, and the production of parts in a functional manufacturing layout in order to minimize production costs. Matrix representation and masking are used within production flow analysis, which is discussed in detail later in this chapter.

The main drawback to these techniques is their inability to deal with typical manufacturing complexities. This inability to solve real-life problems is often hindered by the assumptions built into these techniques. King and Nakornchai (1982) argued that typical manufacturing problems require a lengthy and complicated objective function. The objective function used by classification algorithms is only able to incorporate the quantifiable aspects of the problem.

Non-quantifiable aspects such as job satisfaction, operator preferences and operator morale cannot be included in a mathematical formulation of the problem. Despite this deficiency, some of these techniques can provide useful solutions in particular situations.

Production Flow Analysis The main objective of production flow analysis as stated by Burbidge (1971) is 'finding the families of components and associated groups of machines for group layout . . . by a progressive analysis of the information contained in route cards.' Burbidge (1977) defined the following procedure which helps to develop a complete division of products and resources going down to the cells on the shop floor:

1 Company flow analysis, plans the division of a total company into factories based on the product organization. The main objective of this step is to plan a simple inter-factory material flow system.
2 Factory flow analysis, changes the factory into a series of product producing departments. These departments cater for the manufacture of complete families of products.
3 Group analysis, divides each of the product producing departments into a series of product families and cells. Each cell is equipped with the required resources to complete the manufacture of its product family.

Each step takes data output from the previous step and develops more detailed product families and groups of resources. For example, the factory flow analysis step takes each of the factories as determined by the company flow analysis step and breaks them into more detailed product oriented departments containing all necessary resources. There is an analogy between production flow analysis and the structured analysis and design technique which is described in Chapter 3, because each technique performs an orderly break down of a subject into its constituent components. A similar procedure is used within each step which basically consists of studying the process routes both from factory to factory and within each factory and using either computer and/or clustering techniques to decide on suitable product families and groups of resources. This step by step approach, as described by Burbidge illustrates the importance of dividing the problem into manageable sizes.

El-Essawy and Torrance (1972) developed a technique called component flow analysis, which is similar in principle to production flow analysis. Component flow analysis uses three stages to sort the components in terms of their resource requirements, and develops groups of resources based on the component requirements. Finally, any necessary changes based on detailed analysis of cell loadings and product flows are made. Component flow analysis differs from production flow analysis in that it does not partition the formation problem into manageable sections and in the method of designing cells that is used (King and Nakornchai, 1982).

De Beer and De Witte, (1978) developed a technique called production flow synthesis which addresses the issues of resource duplication and different characteristics of the resources. This technique leads to a relatively large number of components requiring more than one cell, as compared to component flow analysis and production flow analysis (King and Nakornchai, 1982). Dekleva *et al.* (1988) developed an augmented version of production flow analysis which manages the problems of sub-optimal process routes by using design data gained from the process route sheets. This enhanced version of production flow analysis deals specifically with the group analysis step as described above in Burbidge's production flow analysis procedure. This design data includes information on shape, size and surface finish amongst other characteristics of each product. In addition to the matrix defining the process routings of each part, another matrix based on the design information taken from the process routings is developed. With the matrix developed from the design information, product families are identified and allocated to cells using clustering algorithms. Production flow analysis and the other related techniques are evaluative methods which involve the systematic tabulation of the components in various ways, in the hope that groups of machines and products may be found by careful analysis.

The Composite Product Mitrofanov (1959) first suggested the use of a composite product as a means of forming families of similar products and grouping the required resources into cells. The composite product technique involves conceiving a hypothetical product which incorporates all of the design and manufacturing characteristics of all the products in a particular product family. An example of a composite product is illustrated in Fig. 7.4. By arranging for all the required equipment, tooling and people to manufacture the composite product to be grouped together, it is possible to manufacture any product in the family by cancelling unwanted operations (Anon, 1960). Using this approach, it is possible to standardize process planning procedures by creating a plan for the composite product. From the composite product's process plan, a plan for any of the family's products can be developed by deleting any unnecessary process steps.

In this section we looked at the five main methods of product family and cell formation in a product based production environment. In order to outline each method, we looked at specific examples of each method so as to illustrate the general principles involved. We do not attempt to fully discuss every possible example, but rather to convey an overall picture of the approach involved in each of the formation methods. Regardless of which approach is used within an organization for the task of product family and cell formation, potential organizational problems can occur. We shall discuss these problems and possible solutions to them in section 7.2.4.

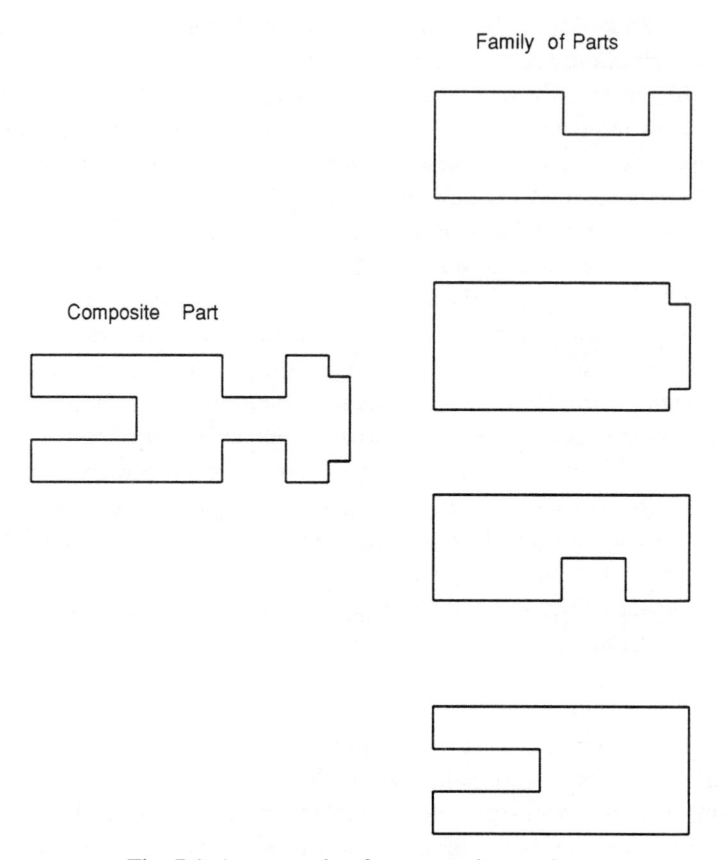

Fig. 7.4. An example of a composite product.

7.2.4 Social issues of product based manufacturing

By helping to form products into families and resources into cells to cater for the production of these product families, Group Technology principles facilitate the development of work groups. Greene and Sadowski (1984) found that the introduction of work groups can lead to an improvement in employee relations and employee skills. These benefits mainly occur because the operators are concentrating on the production of a family of products, and have responsibility for the production of specific products and the management of all aspects of production associated the product family. This helps to increase the skills of the operators and they develop a sense of ownership and pride in these products. The increased skill levels and intrinsic benefits can ensure a high commitment and performance from the team of operators. A thorough and detailed implementation and education program regarding the concepts behind Group Technology is required to achieve this level of commitment.

Fazakerley (1974) found that an unorganized attempt to introduce Group Technology can lead to employee uncertainty and insecurity and a resistance to change. The hesitation by the employees can arise over a lack of understanding of the ideas involved in Group Technology and over a fear that they may appear incompetent in coping with the new changes. Rathmill and Leonard (1977) suggested other reasons for a resistance to the introduction of Group Technology, such as the reluctance of highly skilled workers to work in a team, where they may have to do semi-skilled tasks.

Handy (1985) recognized that these problems with the formation of groups resulted from the fact that when groups are being formed, individuals are generally selected on the basis of their function in the organization (e.g. personnel from assembly, personnel from drilling etc.). He maintained that other characteristics such as a persons attitudes and preferred roles, are as important as a person's job skills. In order to help a group develop into a successful unit, each group should undergo an evolution process, which involves the following steps as suggested by Handy (1985):

- forming;
- storming;
- norming;
- performing.

In the forming stage the group is regarded as set of individuals, where issues such as leadership, the purposes of the group, and the group life span are discussed. Within this stage, individuals are attempting to establish their personal identity within the group. The preliminary agreements on the relevant issues are challenged in the storming stage, where personal hostility is evident. However if this stage is properly handled, realistic objectives can be agreed on all relevant issues. In the norming stage, the group organizes procedures and rules on how it should work, how it should make decisions and the appropriate degree of trust and openness for the group. After the successful completion of these three stages of forming, storming and norming, the group should be able to perform effectively. Periodically, discussions will still take place on issues relevant to the group such as objectives, life span and leadership. A group can mature and achieve high performance with the use of such a structured design procedure. The use of such a procedure is important in order to avoid the pitfalls that can result from a lack of understanding of Group Technology principles.

With the formation of work groups based on the Group Technology ideas, an organization can progress towards using a proper balance of global and case controls, which ensure the development of an effective overall control system (Mitchell, 1979). Global control involves deciding on the management guidelines for each work group. Only when these guidelines are in danger of being broken, is help requested from outside the group. Case control occurs within each work group, where the operators review each decision regarding the management of production within the cell, on issues such as scheduling, raw material stocks, or quality levels. This combination of case and global control ensures a distribution of responsibility throughout the shop floor which helps to solve any problems efficiently. By facilitating the formation of work groups, Group Technology helps to define the scope of responsibility for case control and provides a framework of work groups under the guidance of global control.

Group Technology is a methodology for creating product families and cells to cater for the production requirements of these families. The introduction of Group Technology helps an organization move towards product based manufacturing. In order for the benefits of product based manufacturing to be fully realized, it is necessary for the correct organizational environment to be in place.

Group Technology can also help standardize and reduce the complexity of the process planning task through the development of product families and cells. We shall describe some of the current developments in the area of process planning in Section 7.3.

7.3 Process planning

Lange (1980) describes process planning as 'being exclusively concerned with the selection of suitable processes and tools to transform raw materials into a finished product according to the design drawing.' Nau and Chang (1985) define process planning from a machining perspective as 'the task of determining what machining processes and parameters are to be used in manufacturing a part'. Current developments in the area of process planning are influenced by current market demands for greater product variety and faster introduction of new products. Therefore the emphasis in new planning advances is on the need to reduce the time taken to develop process plans and to standardize the planning procedure. The process planning step is one of the planning stages in the design and manufacturing life cycle of a product. Weill *et al.* (1982) argued that the planning stages are becoming a bottleneck within this life cycle because of the volume of manual tasks involved in the planning procedures. Traditionally, planning and production stages were of equal importance, but because of the advances in production technology, the main delays are increasingly occurring within the planning stage. When discussing the planning stage, it is important to differentiate between production planning and process planning. Production planning differs from process planning, because it involves scheduling the production requirements within a time period on all the resources as specified in each product's process plan.

Traditionally the process planning task was done manually, with an individual planner examining the drawings of products and then deciding on the most suitable sequence of operational steps and the required range of resources to complete the sequence of operations. Mill and Spraggett (1984) found that the success of the process plan depended to a large extent on the skill and experience of the planner. According to Adiga and Li (1987), this reliance on an individual planner's expertise lead to long development times, inconsistent routings and tooling requirements. The developments in Computer Aided Process Planning (CAPP) are overcoming these deficiencies and facilitating a high degree of standardization. Referring to Lockheed's CAP system, Tulkoff (1981) found that computer aided process planning systems can save up to 40% of a process planner's time, by helping to reduce the required clerical work.

However computer aided process planning of assembly operations is more difficult. Kempf (1985) indicated that the assembly planning problem is often ill-defined, with no known standard solution process apart from 'rules of thumb' and educated guesses, and there is rarely an optimal solution. Tierney *et al.* (1987) pointed out further problems:

- Once certain precedence relationships are satisfied in assembly planning, there are a number of equally feasible solutions.
- Process planning is generally the responsibility of an experienced planner, much of whose knowledge is of a diffuse and heuristic nature, which makes it difficult to represent in a computer based decision support system.

Kempf (1988) discusses some interesting observations on the methods that process planners use:

- planning by successive refinement;
- knowledge based planning;
- model based planning;
- planning by selection and sequencing.

When experienced process planners are developing process plans, they plan by successive refinement. This involves taking an initial basic plan for a product and refining it by adding more detail to each process step. The development procedure used by a process planner is also knowledge based, because the planner is using a large amount of knowledge based on past experiences. This knowledge can be in the form of:

1. Plan fragments, such as a basic machining plan involving for example the following steps; rough machine, heat treat,rough grind and fine grind.

These plan fragments combine many different pieces of knowledge into one overall framework.

2 A series of condition-action pairs involving specific knowledge. For example, a rule such as if a turning operation is required and the part is more between thirty and forty centimetres in length, then lathe no. 541 should be used.

When developing a process plan, a planner also uses model based planning techniques. These techniques involve using an abstract model of a particular process, such as a machining process to develop suitable solutions for any problems that may occur. In an abstract model of a machining process, a machine is using a cutting tool on a piece of clamped material. Two important factors in this abstract machining model are the amount of force and the accuracy of the relative position of the tool. These factors influence the accuracy of the cut and are dependent on the rigidity of the clamps, the material, the tool and the power of the machine, all of which the planner must consider in the part's process plan. A process planner, when considering all aspects of a part's process plan, uses various selection and sequencing techniques. Initially a planner selects from a number of feasible basic plans. As one basic plan is then expanded, resources must be selected and sequenced to form the process route for a part. The later stages of developing a plan involve choosing and sequencing the actual operations to be assigned to each resource. Even though a group of resources has been assigned to produce a particular part, there is a wide variety of ways of allocating the required series of operations. In deciding on the correct sequence of operations, the planner takes such issues as resource accuracy, costs and part tolerances into account.

Earlier, we referred to the increasing use of computer aided methods for process planning, which rises a number of interesting issues. Requicha (1980) discusses the importance of representing data describing each part either through geometric modelling or computational geometry. The description of a part must be complete and detailed, especially when being used by a computer aided planning system, because unlike a human planner, it cannot infer any accurate knowledge from incomplete data. Computer based methods, initially took the form of standard numerically based techniques such as database management systems and decision table processors. Current computer based methods use symbolic processing which differs from the previous numerically based techniques, in the way the knowledge is stored. In symbolic processing, knowledge is represented in a declarative fashion, and there is no control structure to define how the knowledge can be used. This increases the flexibility of the planning system because the knowledge can be applied in number of different ways.

Looking specifically at the area of assembly process planning, we find that assembly operations are not as well defined and categorized as manufacturing operations such as machining and milling operations. Tempelhof (1980) argued that the scientific definition of manufacturing tasks, as exemplified by Taylor's equation for tool life, can lead to the development of planning systems for manufacturing operations, which can automate a planner's task (Amstead *et al.* 1977). Assembly operations, in contrast are almost an 'art form', and assembly process planning is highly subjective and relies on individual expertise and experience. Bowden and Browne (1987) adopted an approach which recognized this characteristic of assembly operations, when developing a process planning system for assembly operations. Their approach towards the development of a process planning system involved three concepts:

1. The process planning system should be regarded as a decision support aid, rather than as a tool which automates the planner's task.
2. A planning system can provide a means of standardizing process plans for similar products by using a common knowledge base.
3. A process/manufacturing engineer can develop feasible sequences of operations if the assembly of a product is understood in terms of standard operations such as those defined by Nevins and Whitney (1978).

Eshel *et al.* (1988) have identified two other issues relating to the effectiveness of the process planning task, namely an efficient search method and optimality. Typically the planning task involves trading off interrelated goals, evaluating a large number of alternatives and using an incomplete range of knowledge. These characteristics can lead to the development of a large solution space. Therefore an efficient search technique is important in order to analyse all possible solutions. Optimal process plans are only developed after itemising all possible alternatives. This may be impractical in some instances, because of the large number of possible options. In this situation, the feasibility and quality of a plan become important criteria as measurements of acceptance. A feasible process plan is one where all the process steps are viable. The quality of each feasible plan is evaluated in relation to some local criteria, which provides an efficient measurement of the quality of a plan, but it does not guarantee any global optimality. With the increase in the use of artificial intelligence techniques for process planning, another important consideration is the type of planning strategy to be used. As we shall see in section 7.3.1, the current selection of strategies can be divided into two general types:

- Exploration based strategies;
- evaluation based strategies.

7.3.1 Process planning strategies

A process planning strategy is a collection of tactics that use similar guidelines to develop process plans. Process planning strategies can be divided into two general types; exploration and evaluation based strategies (Eshel *et al.*, 1988). Exploration based planning strategies generate plans relatively cheaply, but the plans have a high probability of either being altered or rejected. Plans developed using an evaluation type strategy, are less susceptible to change, but involve a lot of time and effort. Eshel *et al.* (1988) have identified the following planning strategies which we will now describe:

- Hierarchical planning;
- skeletal planning;
- opportunistic planning;
- phased planning;
- improvement planning.

Hierarchical planning develops process plans through a series of intermediate stages, where at each stage a new version of a process plan is generated. This procedure ensures that after each level, the plan is being improved into a less abstract one, until at the final level, an executable plan is devised (Sacerdoti, 1974). The use of intermediate stages within hierarchical planning is to prevent the planning procedure being impeded by unimportant details. Skeletal planning develops process plans based on a skeleton plan which consists of a series of generalized process steps (Davies and Darbyshire, 1984). An example of a skeletal planning strategy is a Group Technology based process planning system, where a generalized plan for a product family is used as a basis for the process plan of a product belonging to that family. An opportunistic planning strategy involves developing a process plan in separate pieces and putting the pieces together, when the opportunity occurs (Vere, 1983). The phased planning strategy otherwise known as least commitment planning strategy produces portions of a skeleton plan. These portions are very detailed process steps, which are used until there is a sufficient amount of evidence against them. The improvement planning strategy involves using a skeleton plan as a basis for constant improvement, which is very similar to the practise of many process planners (Sacerdoti, 1977). Planning systems which uses the plan improvement strategy do not strictly define the scope of improvement, so the volume of alterations can be difficult to quantify.

As we mentioned earlier in Chapter 2, all computer aided process planning systems use either one or more of these planning strategies within their procedures for generating process plans. The approaches to computer aided planning systems can be divided into two types: variant and generative. These two approaches will now be described.

7.3.2 *Variant and generative process planning*

There are two approaches to complete the task of process planning: a variant approach and a generative approach. The variant approach involves taking a plan from a library of existing process plans and making any adjustments to it in order to accommodate the specific manufacturing requirements of a new product. The chosen process plan is selected because of the similarity between it and the product. This similarity can be assessed using a Group Technology code for the new product and checking through all existing products for one with a similar code. Examples of variant process planning systems include MIPLAN (Tno, 1981) and CAPP (Link, 1976).

Truly generative process planning systems develop new process plans, without any reference to existing plans. Chang and Wysk (1985) found that because of the complexity involved in generative planning, current generative planning systems have a number of limitations in relation to the type of products for which they can produce process plans. These limitations can be overcome by using artificial intelligence based data representation and reasoning techniques (Nau and Chang, 1985). In the development of artificial intelligence based process planning systems two different types of data are being used. The first group of systems uses features of the product to represent geometric data (see Descotte and Latombe (1981)). In the second group manufacturing process data is the dominant data type as illustrated by Barkocy and Edelblick (1984).

Adiga and Li (1987) propose a generative process planning system consisting of four databases and five planning functions. The databases are as follows:

1. Part definition database. This database contains information on each product either in the form of a Group Technology code or as detailed information.
2. Manufacturing decision logic database. This database contains the logic required to develop a process plan. This decision logic contains a series of rules for manufacturing products, using one of four approaches; decision trees, decision tables, process decision models or artificial intelligence models (Chang and Wysk, 1985). Decision trees are hierarchical structures which contain classification systems and/or user defined logic, as illustrated in the D-Class system (D-Class, 1985). Decision tables are helpful in reducing a problem to its simplest form using conditions, rules, actions and mappings. 'If-Then' statements are one type of representation used within decision tables. Process decision models contain rules for manufacturing products in families in the form of decision logic or procedures. Artificial intelligence models seek to encode the expertise of experienced process planners.
3. Machine tool database. This database contains information on all machine tool requirements and associated operational parameters, required by all products. In a production environment which uses resources other than machine tools, the relevant operational data involving these resources is stored in this database.

4 Organizational economy database. Within this database, information on the related costs and times for processing parameters is stored.

The five process planning functions are associated with machining operations, but similar functions could be derived from other types of operations. The functions are as follows:

- Generate a sequence of operations;
- select dimensions and clamping surfaces;
- optimize the sequence of cuts for each operational step;
- complete an analysis on dimensions and tolerances;
- produce an output of the process plan.

Most process planning systems include only steps one and five and few systems contain all five functions. The generative process planning technique is more complex, but Bao (1984) suggested that it facilitates the linkage between Computer Aided Design (CAD) and Computer Aided Manufacturing (CAM). This linkage is necessary because of the data requirements of the process planning task, which includes data on design characteristics such as surface finishes and dimensions coming from a CAD system and manufacturing data such as process times and set-up requirements, coming from a CAM system.

Variant process planning systems consist of a retrieval and editing interface which connects with a database containing a range of process plans. Generative process planning systems consist of a manufacturing database and decision logic to help a planner develop new process plans. The database in a generative planning system contains design data relating to all products and manufacturing data relating to all resources within a manufacturing system.

Greater emphasis is being placed on process planning, in order to cope with reduced product life cycles and the reduced time to market for new products. New tools are needed which allow new products to be designed and manufactured in shorter lead times, with minimal cost (Gunn, 1982). Whale and Mills (1988) indicated that with this range of new tools, design and manufacturing could occur as partially parallel activities. Walsh et al., (1989) argued that with design and manufacturing occurring partially in parallel, process planning systems should have an expanded role, which incorporates an assessment of the appropriateness of a product's design for manufacturing or assembly. In the current approach to process planning, this assessment may be carried out by completing a product/process analysis, while determining a suitable process plan (Fig. 7.5). The product/process analysis incorporates various Design For Manufacturing and Assembly principles (DFM/DFA).

With this new approach towards process planning, there may be an expansion of the process planning task to include an analysis on the suitability of a product for manufacture. The use of process planning within the production environment design task of FC differs from this new approach because

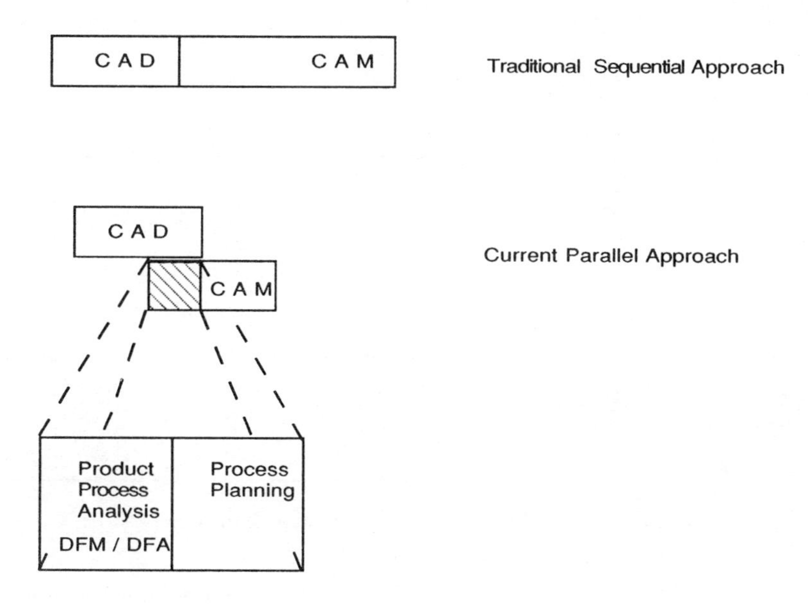

Fig. 7.5. Parallel approach of design and manufacturing.

it is concerned with allocating actual resources to carry out the operational requirements. In contrast, the new process planning approach operates at a higher level, where it is concerned with determining the correct operational descriptions for the manufacture of a product.

7.4 Manufacturing system analysis

The main purpose of the manufacturing analysis procedure within the production environment design task is to provide a diagnosis for production problems, with a view to improving the efficiency of the production system. Production problems such as large set-up times, large work in progress levels, and bad quality products clearly reduce some of the benefits of product based manufacturing.

New approaches to production management such as the Just In Time (JIT) approach are more comprehensive than the traditional approaches, because JIT involves an analysis of the product range and the processes within a manufacturing system with a view towards continuously improving existing technology, along with the option of using new technology. An analysis of

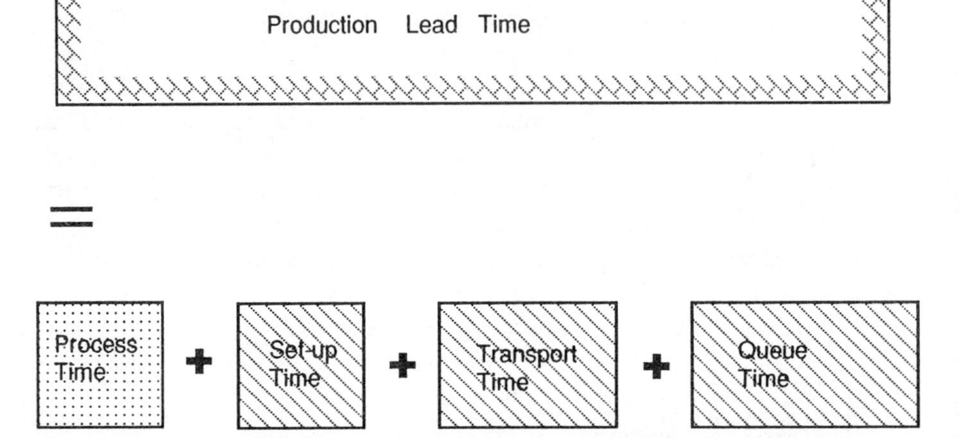

Fig. 7.6. Breakdown of the production lead time.

the production lead times of a range of products serves as a good indicator of the performance of a manufacturing environment. A breakdown of all the various elements of production lead time is illustrated in Fig. 7.6. As we have seen from Chapter 1, the elements of actual processing and set-up times within the production lead time only make up about 5% of the total time. This measurement indicates that in many cases, value adding operations only account for 5% of the total lead time. The emphasis on continuous improvement in the efficiency of the manufacturing process involves examining the various elements of production lead time and attempting to eliminate any wastage in any of the time elements.

Heard and Plossl (1984) recommended various techniques for reducing each of the elements of the total lead time. In this section we shall briefly examine some of these ideas which include the following:

1 Layout of the production process to facilitate product flow;
2 product design for ease of manufacture;
3 the use of flexible resources;
4 the use of line balancing;
5 the definition of operations standards;
6 set-up reduction.

Layout of the Production Process As we discussed earlier in section 7.2, the layout of the production process should where possible be product based. Product based manufacturing involves all products being formed into families and all resources being grouped into cells, which cater for the production requirements of each family. Group Technology is a methodology which can be used to define product families and cells. The end goal is to have a series of U-shaped cells, which allow the possible assignment of a multi-skilled operator to more than one machine, because of the close proximity of the machines.

The introduction of product based manufacturing into a manufacturing environment, should not be delayed because it may not be possible initially to have a perfectly designed product based layout with distinct product families being totally produced in distinct cells in one step. A manufacturing environment may evolve towards a product based mode, while in the interim provision for shared cells or product families having to use more than more cell to complete their process requirements may be necessary.

Product Design Intelligent product design techniques involve consideration of process issues, at the product design stage in order to stabilise or reduce the process variety while increasing the product variety. Browne and O'Gorman (1985) discussed three possible principles which are used within product design; modular design, design for simplification and design for ease of automation. Product variety can be managed efficiently by introducing modular design across the product range. Modular design ensures that component and assemblies are common across a range of products. With the use of standardized components and assemblies, inventory levels and the variety in manufacturing are reduced (Browne *et al.*, 1988). Design for simplification involves designing products which are relatively easy to manufacture. This is achieved by including as many standard components as possible in a product's structure, which reduces the need for special tooling. Product tolerances, surface finish requirements and other design details are determined with due consideration to their effect on production process. Design for ease of automation is concerned with the design of components, in order to reduce the complexity of automatic parts feeding, orienting, and assembling processes. Good practise includes designing a product and its components for unidirectional assembly and if possible, avoiding assembly operations from the side or the bottom (Owen 1985). By using these design procedures to rationalize the structure of the product range, it is possible to increase the commonalty of components in each product's structure. With this increased commonalty a greater variety of products can be processed by a reduced variety of manufacturing processes.

The Use of Flexible Resources Simplifying the production process implies an analysis of the production lead time for each product, to see if any elements can be eliminated or reduced. Possible resultant benefits from the simplification exercise include greater flexibility in handling product changeovers and fluctuations in product demand. A manufacturing system should have the flexibility to adjust the production output to satisfy the requirements of customers. Flexible multiskilled operators and flexible equipment can increase the flexibility of a manufacturing system by enabling the system to minimize order lead times and to cope with requirements for different product models (e.g. changes in colour, size etc). Hall (1987) describes flexible operators as 'people willing and able to do whatever is within their capabilities'. The use of flexible operators ties in with the concept of autonomous work groups where the variety of tasks requires the operators to be multiskilled. Flexible equipment can produce different product models with a minimal changeover. With this flexibility a manufacturing system can plan more efficiently for fluctuating product demands.

The Use of Line Balancing and Operations Standards Line balancing techniques seeks to ensure that production is the same at all processes in terms of quantities and timing. This reduces the waiting time caused by unbalanced production times and operations standards. Operations standards involves the determination of the cycle time for each operation, a detailed description of each operation and the specification of a minimum quantity of work in progress to allow smooth production. The operations standards ensure that all operators are trained in the most efficient methods for doing a particular set of tasks. Using the operations standards, line balancing techniques then seek to reduce any remaining imbalances between different processes. This approach differs from the use of line balancing in Western manufacturing, where the use of heuristics or algorithms is recommended to solve any imbalances between processes, and the operational procedures and constraints are assumed to be fixed. The main result of the use of line balancing and operations standards is a reduction in the queuing time segment of a product's lead time.

Set-up Reduction Set-up reduction is very important since it impacts throughput times, work in progress levels, and gives a greater ability to produce many different product variants simultaneously. In Chapter 1, we discussed the four rules for set-up reduction offered in the Just in Time philosophy (Monden, 1983). As the set-up procedure is minimized, batch sizes can be reduced. Smaller batch sizes result in correspondingly lower inventory costs, but involve a larger number of set-ups. However with the reduced set-up procedures, the set-up costs are minimized.

In Fig. 7.6, each of the elements of production lead time are illustrated, with processing time being the only value adding element of the production lead time. To ensure that the best manufacturing techniques are being used and that the processing time is used to its best advantage, the use of operations standards and a product based layout is recommended. Efficient product designs and a set-up reduction program can help to decrease the set-up element of production lead time. The transport time can be reduced with the use of a product based layout, because a product has a smaller distance to travel. The queuing time is reduced through the use of flexible resources, a product based layout and a set-up reduction program, all of which help to minimize the number of products waiting to be processed.

7.5 Conclusions

In this chapter we presented a review of the important ideas from product based manufacturing, process planning, and manufacturing systems analysis. Product based manufacturing is an approach towards manufacturing which involves the formation of product families and associated cells. Group Technology is one suitable methodology for developing product families and organizing groups of resources into cells. The changeover to product based manufacturing produces both production and organizational benefits. The production benefits include shorter production lead times, lower work in progress levels, and better quality levels. The organizational advantages include more flexible control procedures, increased operator morale, and increased operator commitment. The organizational benefits occur largely through the distribution of management responsibility for each cell to each group of operators.

Process planning is concerned with the development of a suitable sequence of process steps and the correct selection of resources for the manufacturing of a product. Recent developments in this area include the use of computer aided decision support systems which seek to standardize the planning process and decrease the required time for developing a plan. Computer aided planning systems use one of two approaches for developing process plans: a variant approach or a generative approach. These new developments also help to prevent a proliferation of process plans.

The techniques described in the manufacturing systems analysis section can help to solve some of the production problems, which are indicated by the analysis. The main thrust of these techniques is to decrease the amount of wasted time in the overall throughput time. This wasted time occurs either in queuing for an operation or in a resource being set-up for an operation or in handling a product from operation to operation. The techniques described attempt to reduce set-ups, organize the production layout more efficiently, reduce queue times through smaller batch sizes and organize the operational steps more efficiently through line balancing and operational standards. By reducing the production lead time for each product and simplifying the manufacturing system, a manufacturing organization can react more efficiently and quickly to market fluctuations.

In the next chapter, we turn to a different theme, that of implementing factory coordination and production activity control systems. The implementation of new technology is important because if the implementation is not handled correctly, then the potential benefits of the system may not be realized.

Part Five
THE IMPLEMENTATION OF SHOP FLOOR CONTROL SYSTEMS

Overview

Many production management specialists rightly complain that researchers in the field of production management systems frequently ignore the problems of the implementation of advanced systems in industry. In fact as we observed in Chapter 6 very few of the scheduling algorithms developed in university laboratories have found application in practice. One of the key requirements of the COSIMA project , and in fact of the ESPRIT CIM activity in general, was the requirement to develop systems which are relevant to the needs of industry. Further, in our view, the continuous interaction between the industrial and university partners within COSIMA was tremendous in terms of ensuring that the ideas and the software tools developed were continuously tested in terms of their industrial relevance.

In this part of the book we consider the problems of implementing Factory Coordination and Production Activity Control systems in manufacturing plants. Our views are based on two major implementations, one at a Digital Equipment Corporation manufacturing facility in Clonmel, Republic of Ireland, and a second at an FMS installation of Combo in Turin, Italy. In fact in Part Four we will outline the Clonmel implementation experience.

Our approach to implementation is strongly influenced by the ideas from sociotechnical design. In Chapter 8, we consider the sociotechnical design approach and try to relate the thinking behind it to the shop floor control implementation problem. The proponents of sociotechnical design argue strongly for an approach to design and implementation which considers equally the technical and social sub-systems of a proposed solution and seeks to achieve a 'best fit' of the two sub-systems. We try to describe the environment of shop floor control, in terms of the technical and social sub-systems and to present ideas for achieving a 'best fit' of the two.

In Chapter 9 we present a outline design and a prototype of a design tool which can be used to help create appropriate PAC systems and control sub-systems in FC. This design tool, which was developed within the COSIMA project, allows a manufacturing systems analyst to describe the manufacturing system under consideration, select an appropriate combination of scheduler, dispatcher and monitor, and then test this proposed solution on a detailed emulation model of the shop floor, prior to implementation.

8

An approach to the implementation of factory coordination and production activity control systems

8.1 Introduction

In the preceding chapters, we described the necessary functional and software requirements of a shop floor control system, which consists of a Factory Coordination (FC) system controlling a number of Production Activity Control (PAC) systems. In this chapter, we are concerned with the design and implementation of such a shop floor control system. The structure of the chapter is as follows:

1. We shall discuss some principles which are relevant in terms of ensuring a successful implementation of the functional and software requirements of FC and PAC systems. In particular, we stress the importance of preparing a suitable environment both socially and technically for FC and PAC systems to function. We will argue that this best fit of the social and technical subsystems, achieved through the sociotechnical design approach, helps to create a suitable manufacturing environment.
2. We shall discuss the contribution that sociotechnical design can make towards the implementation of production management systems, by using the experiences from the implementation of Materials Requirement Planning (MRP) systems and Flexible Manufacturing Systems (FMS) to outline the argument in favour of sociotechnical design.
3. Finally, we shall examine the characteristics of the most suitable social and technical sub-systems within an organization for the operation of FC and PAC.

We shall start with a description of sociotechnical design, which ensures that both social and technical considerations are incorporated in the implementation of new technology in an organization.

8.2 Sociotechnical design

In our opinion, sociotechnical design is a methodology that enables the members of an organization to improve the performance of their own organization, through their own ideas and initiative. Davis (1982) describes organizations as 'social inventions, that is, created units of society where people are brought together and provided with a technology and structure for doing work.' The introduction of new technology and new strategies into an organization may affect the aspirations and attitudes of the people in the organization. Traditionally the emphasis on the introduction of innovative technology has been on the range of technical and financial benefits that accrue from its usage. In formal hierarchical organizations, this narrow emphasis may cause the status attached to work to be tacitly ignored, which may lead to a lack of consideration of the possible effects of new technology on the organization or its members (Browne, 1954).

Skinner (1985) traces the relationship between the introduction of new technology and the resultant impact on the quality of working life. The impact on the quality of working life comes from the effect of new technology on existing working procedures. Based on their particular attitudes and aspirations in relation to the work environment, employees react in different ways to the introduction of new technology. One of the end results of the introduction of new technology may be a perceived change in the quality of each employee's working life. Skinner (1985) concludes that the traditional attitude towards the development and introduction of new technology accepts that such an introduction may cause some personnel and organizational problems, but assumes that these problems may be addressed through using various procedures involving communication, selection, training, or financial incentives. Today this attitude is changing to one of greater awareness of the importance of a good working environment, because employees' attitudes and expectations are becoming more demanding (Kerr and Rosow, 1979). Reflecting this new awareness, Lodge (1975) has identified a new ideology known as communitarianism. Communitarianism focuses on the needs of the whole organization and the interdependence between different departments, rather than viewing each department within an organization as independent from all others.

Schonberger (1986) describes the current manufacturing environment as one where there is an emphasis on 'continual and rapid improvement' in all aspects of manufacturing, such as time to market and product quality levels. In order for organizations to survive in the current business climate, it is not only necessary for them to have the best strategies and product ranges for a correct target market, but the personnel working in the organization, must be committed and prepared to support the range of products and/or services (McEwen *et al.*, 1988). In addition to the need for new technology and new strategies, employees must possess the characteristics of competence, commitment and capacity for change (Hayes and Fonda, 1986). Competence involves people being willing to operate in teams and having the right skills and capabilities to make decisions within their scope of authority. People become committed when authority is delegated to them, and they feel more involved with decision making and problem solving within an organization. When people view change as a positive development, and adopt a cooperative approach towards new developments at work, they then possess a capacity for change. This attitude involves more than simply regarding people as an important resource within an organization. They must also have a role in the design and management of an organization. Therefore employees must also be involved in the development and introduction of new technology into their organization, since the introduction of new technology influences the management of an organization.

One approach which provides for increased involvement of employees in the development and introduction of new technology into an organization is known as the sociotechnical approach. This methodology was initially proposed by the Tavistock Institute in the late 1960s and has been subsequently enhanced over the years (Trist, 1982). The sociotechnical design approach seeks to promote human qualities and bring about required changes in an organization, using the organization's own staff. Thus a sociotechnical design procedure is concerned with implementing, rather than developing FC and PAC systems. As an introduction to sociotechnical design, we shall now describe some basic principles underlying the approach. Then we shall discuss the basic procedures for analysing both the social and technical sub-systems within an organization. This section on sociotechnical design concludes with a description of the sociotechnical analysis procedure, through which organizational personnel examine both the social and technical sub-systems within the organization and develop proposals for making changes to the various processes/procedures within both sub-systems.

8.2.1 The basic principles of sociotechnical design

In this section we shall describe the basic principles underlying organization design using the open sociotechnical approach:

1. Open systems theory;
2. the technical and social sub-systems;
3. an organization philosophy;
4. participation in design and implementation;
5. minimum critical specifications;
6. open-ended design process.

Much of this material is based on work done by Cherns (1976) and Davis (1977). The principles relate to the structure of an organization, the different sub-systems operating within an organization, and the support systems which are necessary to maintain the sociotechnical approach.

Open Systems Theory Traditionally, organizations have consisted of different levels controlled by a hierarchy and the organization was not greatly influenced by the outside environment. Managers and supervisors were required at each level to control the employees at the next lower level, and tasks were grouped according to function or location. This traditional theory is based on writings by Taylor (1911) on scientific management and Weber (1964) on bureaucracy. The principle behind this type of organization is differentiation, in which control and understanding are achieved by dividing the skilled tasks into small elements which may be performed with minimum skill at minimum cost (Smith, 1776).

However the classical theory is giving way to a new theory which considers organizations as open systems, consisting of a series of sub-systems, very similar to an organic structure (Burns and Stalker, 1961). Open system organizations are active organizations, converting a set of inputs into outputs in order to achieve certain objectives. To illustrate the applicability of this notion of a system we shall use an example of a factory, where the inputs are production personnel, raw materials, sophisticated equipment and a production schedule, and the desired outputs are high quality products with the objective of generating revenue. Within the factory, raw materials are converted into high quality products by production personnel through the use of production equipment. The production personnel use the production schedule to help plan the manufacturing of the products. The sale of the range of high quality products in turn generates revenue for the organization. This notion of an organization as a system may be extended to include the concept of feedback. With the use of feedback, a system is capable of managing itself with reference to a series of goals and without excessive supervision (Fig. 8.1) (Mitchell, 1979). Within a production environment, a production system receives feedback as well as other inputs, such as raw materials. This feedback is gained from comparing what the system actually produced with what was the desired output. For example the feedback may include information on how efficiently products are being produced and whether all order due dates are being fulfilled.

By viewing an organization as an open system, it is regarded as being part of a larger environment from which it receives its inputs, such as raw materials and customer orders and to which it gives its outputs, such as final products (Emery and Trist, 1960). The organization also receives some useful feedback from its environment (e.g. customer's reactions to new products on the market).

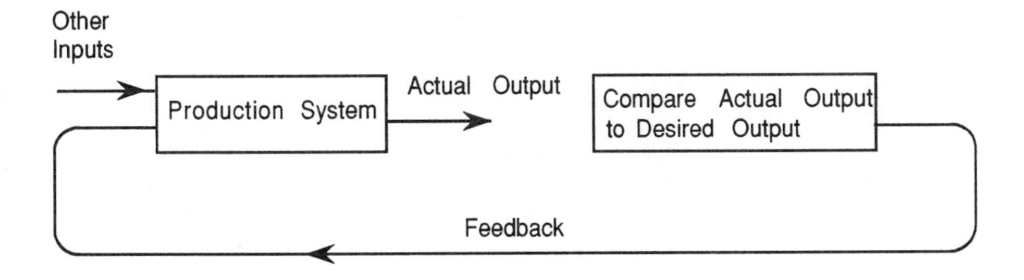

Fig. 8.1. Inclusion of feedback in a system.

Pava (1983) suggested that the environment of an organization may be divided into two type: transactional and contextual (Fig. 8.2).

A system can influence elements in its transactional environment even though it is considered to be outside of the system. If we consider a shop floor production system as an example of an open system, then we argue that a materials department is situated in the transactional environment of the production system, because it influences the development of a production schedule by giving estimates of arrival dates for raw materials which are required to satisfy production requirements. The production requirements cannot be fulfilled until the raw materials are on-hand in the production system. The production schedule is important to the production personnel because they use it to plan their production tasks and capacity requirements. Therefore the materials department influences the production department's decisions through the production schedule, but it is not under the direct control of the production department. However the production department can ask the materials department to reconsider the raw material delivery due dates, so as to make the production planning task easier. The contextual environment of a system contains external groups which affect the performance of a system, and over whom the system has no control or influence. In relation to the example of the production system, the contextual environment includes such groups as competitors, consumers and government.

The relationship between an organization and an environment is a two way one, with organizations providing either a product or service to an environment, while organizational structures are strongly influenced by the state of the environment and the demands of the environment (Emery and Trist, 1973). Vickers (1965) viewed an organization from the technological,

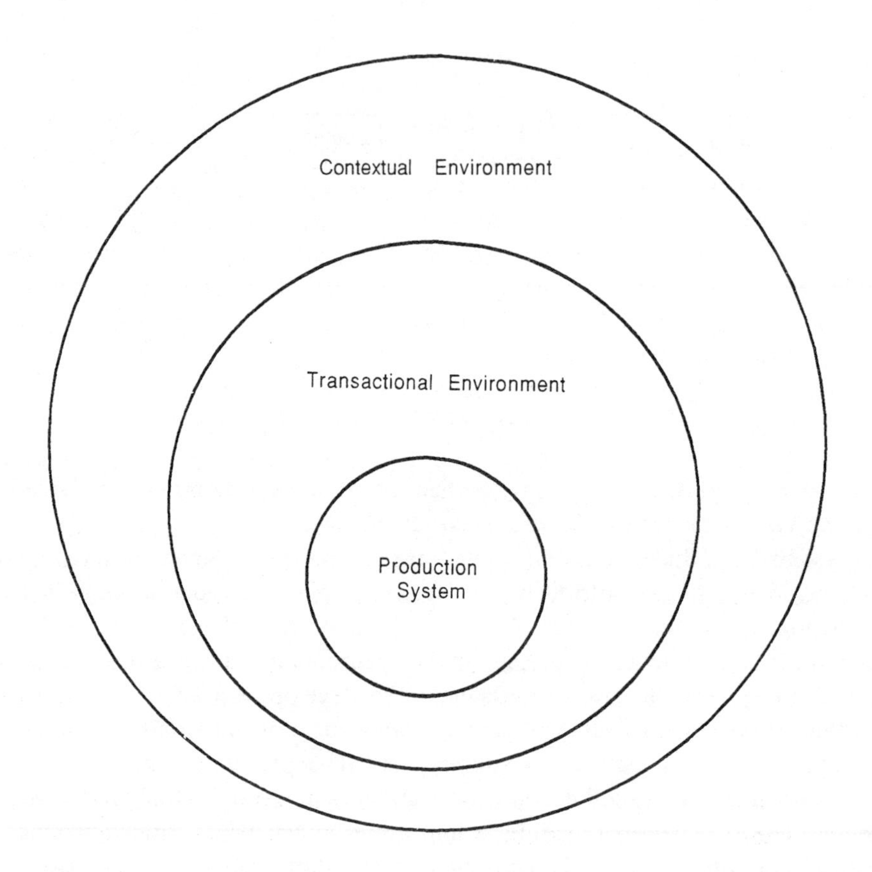

Fig. 8.2. The transactional and contextual environments (Pava, 1983).

social, economic and political perspectives, in order to understand the different states and influences that an environment exerts on an organization. The technological aspects are a major influence of the environment because they may cause a large amount of instability within a system. Two examples of technological issues are shortening product life cycles and an increasing rate of development of production processes. Technological developments are also changing the meaning of work and skills. We find that with the use of advanced technology, greater reliance is placed on the employees to operate and manage this new technology within an organization. This increased reliance on employees arises because advanced technology involves an increase in capital investment and a corresponding reduction in the numbers of employees. Thus, there are less employees responsible for a larger array of new technology The social environment may influence an organization through the values, attitudes and objectives of its employees. The social environment is the society within which an organization functions. The economic environment is becoming progressively unstable because of such factors as competition in the market and currency fluctuations. The economic environment exerts an important influence because all investments within an organization are assessed on financial criteria. Financial analysis becomes more difficult because of increasing instability in the economic environment. The political environment influences an organization through the imposition of different legislation and regulations (e.g. regulations relating to such issues as racial inequality, product liability, and the use of labour).

Thus, the technological, social, economic and political aspects of an organization's environment are changing at an increasing rate. Trist and Emery (1965) describe this phenomenon of constant rapid environmental change as environmental turbulence. In order to manage the resultant uncertainty, organizations must become more adaptable. Ashby (1956) argued that this adaptability involves having a greater range of responses to deal with all possible contingencies and having the capability to reorganise based on past experience.

Having looked at the influences of the environment on an open system, we shall now examine the two main sub-systems within an open system; the social and technical sub-systems.

The Technical and Social Sub-systems An organization while being influenced by the changes in its environment, brings together two sub-systems; a technical sub-system and a social sub-system (Davis, 1982). Both sub-systems are necessary in order for an organization to fulfil its objectives and convert all inputs into outputs. The social and technical sub-systems are part of any organization and it is within these sub-systems that new production management tools, such as FC and PAC, have to function.

The technical sub-system consists of the equipment, procedures, and rules which are the means by which the members of an organization convert its inputs into outputs. Trist *et al.* (1963) found that behind each choice of a particular technical system design lies a series of implicit assumptions. For example, a possible rule could be that each worker is responsible for his or her particular task. This rule is based on a social assumption that all operators require supervision to prevent poor performance and leads to the design of small individualised tasks and a management hierarchy to supervise them. The effective operation of the technical sub-system, which is the responsibility of the social sub-system, is an important factor in helping an organization achieve its objectives.

Lawrence and Lorsch (1967) defined the social sub-system as consisting of personnel and procedures, which are required to manage and operate the technical sub-system and to solve any problems that occur because of changes in the organization's environment. The procedures are necessary to ensure the proper coordination of work within an organization. This involves assigning tasks and using either a hierarchy, a series of rules, or an informal code of procedures to ensure that all tasks are organized towards the same set of goals.

The description of an organization as an open system recognises that all aspects of the organization are interrelated and that they should be integrated. Using this type of systems thinking, the role of each group or department is explained in terms of its function within its organization. Similarly, the role of an organization is explained in terms of its function within its environment, and may be explicitly described in an organization philosophy.

An Organization Philosophy An organization philosophy is an important element of the sociotechnical approach to organizational design, because it articulates the objectives and procedures involved in the design procedure (Hill, 1971). Based on the guidelines from this philosophy, decisions are taken regarding the technical sub-system, the social sub-system, the role of employees in the design procedure, the overall structure of the organization and the necessary support systems. These guidelines ensure that all aspects of the sociotechnical approach are completed in a thorough and efficient manner. The process of developing and documenting an organization's philosophy provides an opportunity for open discussion and for the creation of an understanding of the organization among personnel from different departments.

This understanding may lead to a deep commitment to the goals and objectives of the sociotechnical approach. Mason (1969) and Jayaran (1978) each propose a different method for the development of an organization philosophy which explores the values and assumptions of the people involved in the design process and illustrates the potential discrepancies between the existing organization structure and the demands being made by the environment. The use of another organization's philosophy may be regarded as a shortcut, but it may cause harmful results, because each organization has its own unique characteristics in relation to its environment and the range of technologies it uses. The understanding and discussion which is generated in the development phase of the philosophy may also be lost. This understanding is very important because its significance is often deeper than the actual statements used in the philosophy document.

The organization philosophy then serves as a set of guidelines throughout the sociotechnical design process, which may help to settle any conflicts or problems that may occur. Apart from the use of an organization philosophy to ensure a successful completion of the design process, it is important that all relevant members of an organization participate in the design process, in particular those who may be affected by any new proposals arising from the design process.

Participation in Design and Implementation Keen and Morton (1978) argued that the ideal approach towards the design and implementation of new measures within an organization is to involve the members of the organization. The involvement of members of an organization increases the acceptability of any proposals which are initiated, concerning the design and development of any new measures. Gunn (1987), referring to strategic plans, argued that 'without heavy involvement by the users in their creation, such plans have little chance of successful implementation'. Lack of involvement reinforces an attitude among manufacturing personnel that planning serves no useful purpose, and as a result a 'work harder not smarter' work culture may develop within an organization. Ackoff (1975) suggested to sociotechnical system planners that 'once they give up the idea of redesigning the work of others and involve the workers in redesigning their own work, there are no difficulties in bringing about changes that lead to improvements'. A further reason for involving operators is, through their continuous involvement with a particular process, they may have experience of the occurrence of particular problems and their causes. Blumberg (1969) found that this participation in the design process may enhance the operator's satisfaction with his/her work. By participating in the design process, each member of an organization may bring his/her own experiences to bear, provided that the design process does not get loaded with too much detail.

Minimum Critical Specifications Herbst (1974) points out that details of the new organizational design proposal should not be overspecified and he refers to this as the principle of minimum critical specifications. This principle relates only to organizational design, rather than areas such as machine design, where every component of a machine must be specified. By specifying the minimal amount of detail in an organizational design proposal, all relevant parties may use their own experiences and ideas to influence the development of design specifications, and the possibility of other feasible options is not closed off. In addition to using their own experiences, the members of the organization become more knowledgeable and creative and are prepared to contribute to the development of a proposal. This principle of minimum critical specifications involves a change of attitude towards the measurement of the success of a design proposal. As Cherns (1977) noted 'we measure our success and effectiveness less by the quality of the ultimate design than by the quantity of our ideas and whether our personal preferences have been incorporated into the design'. With the use of minimum critical specifications, each design proposal is not regarded as the final statement, but rather as a step in a process of continuous improvement of the design of the organization.

Open-ended Design Process The sociotechnical approach to organizational design is a continual process in that, as options and problems are discussed and solved, new issues inevitably arise. The objective is not only to produce a good design proposal but to develop and maintain the procedures and personnel who generate new design proposals on an ongoing basis. The use of sociotechnical approach as an ongoing process is essential in order to cope with the changing environment of an organization (i.e. environmental turbulence). This notion of continuous improvement of design proposals is similar in principle to the Kaizen* principle of ongoing improvements involving all members of an organization, including operators, supervisors and managers (Imai, 1986).

8.2.2 An analysis of social and technical sub-systems

In the last section, we discussed some of the principles of sociotechnical design. We shall now move on to examine some of the main characteristics of the social sub-system and the technical sub-system within an organization. Technical and social sub-systems traditionally have being organised in a functional hierarchy. Handy (1985) describes this type of organizational culture as a role culture and compares it to the structure of a Greek temple. Within the role culture, people, procedures and resources are organized into departments according to function (e.g. the production department, the finance department, the sales department etc). These departments can be regarded as the pillars of the temple which are coordinated by a small number of managers at the top of the organization. In fact, managers are considered to be the only form of coordination required within a role culture because the various procedures are supposed to take care of any potential problems. The structure of the role culture ensures that management's decision making is distanced from where the decisions take effect (i.e. the lower levels of the organization). Thus, this type of organization is unlikely to cope with an unstable environment, because it is likely to be slow to identify changes in the environment. Referring back to the metaphor of the Greek temple, once the foundations become shaky, the temple may collapse.

An emerging theory suggests that responsibility should be distributed throughout an organization, rather than remaining at the upper levels as in the role culture. Drucker (1969) recommends the integration of activities according to the principle of federal decentralization, which may lead to an

* Kaizen is a trademark of the Kaizen Institute Ltd.

increase in the simplicity and efficiency of an organization. Federal decentralization involves organizing operators into autonomous groups which are responsible for a range of tasks and for managing their own affairs. By giving the authority and capability to each group, problems may be solved more quickly, because the required skills for solving any problem are located closer to where the problem occurs (Emery, 1967). The type of organization which brings together the appropriate people and resources at the right level to complete a set of tasks is known as a task culture (Handy, 1985). An example of a task culture within a manufacturing environment is a group of operators concerned with the production of a family of products.

This distribution of responsibility helps the social sub-system engage and develop a high commitment in operators and management towards the goals of the organization. Walton (1980) found that this commitment is important because carelessness, boredom and indifference on the part of operators and management may weaken the impact of any new technology. Emery and Thorsrud (1976) developed a list of task characteristics relating to the distribution of responsibility which help to maintain an operators commitment to a task. By incorporating these characteristics within a task or series of tasks, the tasks may become more satisfying:

1. There should be a large amount of management responsibility supported by a set of clear goals or guidelines associated with each group of tasks. This helps to bring the decision making process closer to where any problems might occur and the set of goals or guidelines help to coordinate the decision making with the overall goals of the organization.
2. Each operator should have the opportunity of continuous learning so as to help him/her develop a range of skills. The greater the range of skills an operator acquires, the more beneficial he/she is to the organization.
3. Reasonable challenges must be presented to an operator and suitable feedback must be given on his/her performance, in order to help an operator to learn. Challenges may be provided by offering each operator a variety of tasks.
4. Within each group, operators should have the opportunity to exchange information and use their experiences to help each other.
5. Operators should feel they are making a significant contribution to the overall organization. For example, within a Group Technology based group of operators, each operator can sense he/she is making a meaningful contribution, because the group is concerned with the complete production of a family of products.
6. Each operator should feel that his/her future holds the same characteristics of autonomy, new skills, challenges and job security.

An individual's psychological contract is an important consideration, in relation to the examination of the above criteria. Schein (1980) describes a psychological contract as a set of expectations and results that a member expects from an organization. Problems may occur if an organization and the member do not view a psychological contract in similar terms. The organization may feel that the member is not cooperating and the member may feel he/she is being exploited. If a member's expectations are concerned mainly with financial rewards, then he/she may not be satisfied with greater autonomy and a variety of skills, unless there is an explicit financial support involved. A person's psychological contract is very important, when determining how satisfying a set of tasks are for that person.

The range of responsibilities and tasks available to a group of members depends on the boundaries drawn up by the group. The boundaries define the scope of each group's responsibilities in relation to other groups within an organization. Cherns (1977) noted that with the development of groups, the function of the supervisor or manager of the group becomes that of boundary management, rather than controlling the activities of the members of the group, who should be managing their own activities. Susman (1976) defines the supervisor's boundary management task as involving coordination with other groups and with the environment of the group. For example in a manufacturing system, a group supervisor is concerned with ensuring that high quality raw materials are received on time in order to facilitate his/her group's production requirements. This task involves communicating with personnel in the warehouse or with the vendors, all of which are located in the group's environment. Hackman and Oldham (1980) recommend that the group supervisor should be concerned with the development of a consensus through open discussion and a facilitative style of leadership. With this style of leadership a work group may become more flexible and perform more effectively.

Zander (1983) suggested that an important characteristic in making a group effective is the perceptions by the members themselves that they are a group. This self-perception is supported by a high degree of commonalty of objectives among all group members, predetermined hierarchies, defined membership criteria, ongoing training and selection procedures and a financial incentive system which supports group work (Whyte, 1969).

The design of the social and technical sub-systems within an organization, involving principles such as group work and distributed responsibility, should ensure a best fit between the two sub-systems. With a suitable overall combination of the technical and the social sub-systems, an organization is better equipped to achieve its goals despite the uncertainties in its environment (Davis, 1982).

We will now examine one particular procedure, which was initially developed by the Tavistock Institute in England in the late 1960s and which incorporates the sociotechnical principles in designing an organization (Trist, 1982).

8.2.3 A sociotechnical analysis procedure

We shall now discuss a sociotechnical analysis procedure devised by the Tavistock Institute, which allows the principles we have just described to be put into practice, by providing a methodology for design teams to use when analysing the social and technical sub-systems within an organization. Some minor changes have been made to the original Tavistock method, but the basic procedure has not being altered in any major way. In Fig. 8.3, we illustrate the overall procedure and the various outputs from each stage. The procedure consists of five steps:

1. Initial scan;
2. analysis of the technical sub-system;
3. analysis of the social sub-system;
4. proposals for the design of the system;
5. approval and enactment of design proposals.

During the first step initial scan the design team members view the organization as a system and compile information on:

- The influential factors in the environment which affect the performance of the organization;
- the major inputs and outputs of the organization;
- the organization's long term and business objectives;
- the organization's philosophy for personnel management.

With this wide ranging information, design team members become more familiar with the organization. The information gathered during the initial scan allows the team to start the analysis of the social and technical sub-systems with a better understanding of the organization. Some of this information may be concise, especially in relation to the mission, business objectives and management philosophy of the organization. This conciseness may be deceptive, because many trade-offs may be made in deciding on the philosophy and the list of objectives. The initial scan also gives a team a new understanding of the various organizational problems that may appear unrelated and unpredictable (Pava, 1983). Cherns (1977) describes the initial scan phase as 'a structured, but flexible approach to learning as quickly as possible the organization's political, social, economic, technological and physical environments'. When the initial scan has concluded, the design team may then proceed to examine both the technical and social sub-systems in detail.

One of the key objectives of the second step – analysis of the technical sub-system – is to examine any potential problems associated with the transformation process, so as to enable the establishment of an efficient system for solving any problems. The initial step in this technical analysis is to identify and specify each step in the transformation process and the methods used to

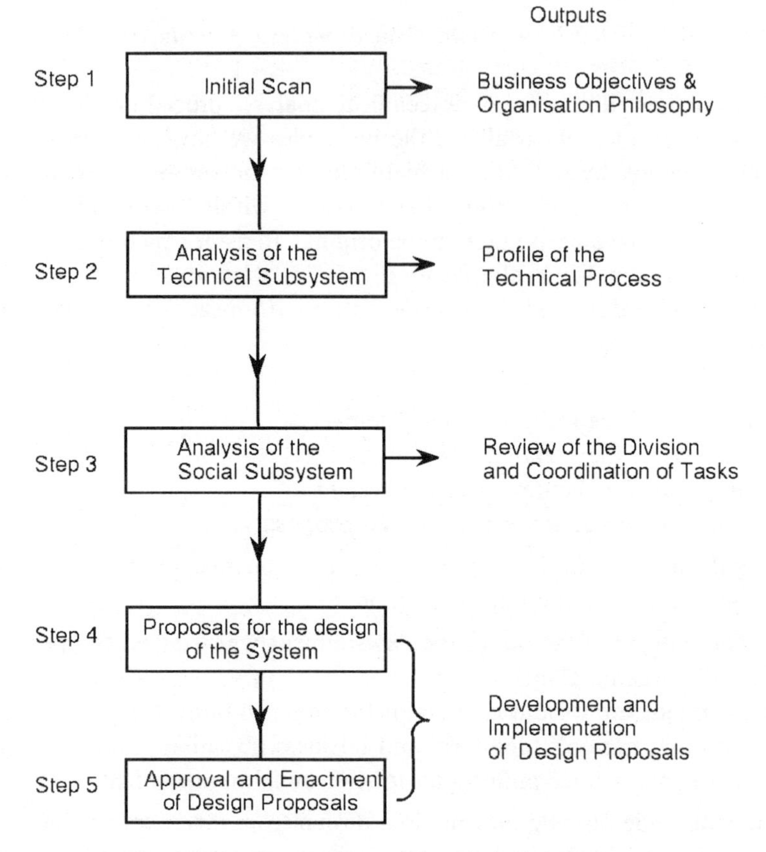

Fig. 8.3. An approach to sociotechnical design.

convert the inputs to the outputs within an organization. These steps are then grouped together into units. For example in a manufacturing system, each process step involved in the production of a series of products is described, together with the relevant methods and technology. When the process steps have been established, the design team identifies any potential problems associated with each process step. The purpose of the problem identification is to identify and correct the conditions which may be causing problems, rather assigning any blame to individual personnel. When these conditions have been identified, the design team determines the relevant information, procedures and responsibilities for controlling the problems and improving the process are determined. For example if a production process has a consistent problem with material shortages which affects its performance, the design team would examine the production process, the current procedure for receiving raw materials and any other relevant information. Based on their analysis, procedures would be devised which ensure regular supplies of raw materials. Ideally in this example, the design team would include members from the production and materials departments. For example, these procedures could involve the use of a signal to alert the materials personnel to prepare a particular batch of raw materials, such as the use of Kanbans in a pull type production system.

Through the analysis of the technical sub-system, the design team seeks to understand the problems in terms of the characteristics of the conversion process, rather than in terms of individual error. For example, an error in data transfer, could be explained by a software transfer procedure which confuses the user (i.e. characteristic of the conversion process) rather than by a clerical error (i.e. individual error). The technical sub-system analysis may be summarized as a scrutiny of the work system within an organization, which involves questioning all existing arrangements for task assignment, with the intention of improving the work system, rather than criticizing individuals for prior mistakes (Pava, 1983).

The third step of the sociotechnical design procedure is the analysis of the social sub-system, which involves reviewing the division of labour, the mechanisms for coordinating the different tasks, and the degree of fulfilment of psychological job requirements. The objective of the design team in this analysis is to study the interdependencies between different tasks and how these tasks interact to control problems (Ackoff and Emery, 1972). The social sub-system analysis also attempts to regularize all tasks using the psychological job criteria as a standard (Emery, 1977). This aspect of the social sub-system analysis involves a comparison of all tasks within an organization on the basis of how well the tasks evoke and sustain commitment and involvement on the part of the personnel carrying out the tasks. As we explained earlier in the chapter, the psychological job criteria, such as the degree of self management within a task or the opportunity to continue learning on the job, provide an accurate assessment of how well each operator is enjoying his/her range of tasks.

The fourth step – proposals for design of the system – involves taking the information compiled during the first three phases and developing proposals for adapting procedures and structures within the organization. One of the aims of the design team is to develop proposals which provide for a social and a technical sub-system which enhance each other and the overall organization. This proposal phase may be the most difficult of the entire design procedure, because the design team are attempting to develop a best overall combination of the social and technical sub-systems by careful analysis and judgement. When developing the proposals, the team must be in agreement about the business objectives and philosophy of the organization, so as to ensure that any proposals do not contradict any of the objectives or the philosophy. Each proposal contains information on work boundaries, the responsibilities of each work group, and the necessary support schemes for the new organizational structure.

The implementation of any proposals which takes place in the final step – approval and enactment – should not be considered as occurring sequentially to the proposals for systems design step. Ideally these two steps may occur in parallel, because with participation in the implementation process, the design team members gain useful experience in the development of proposals (Davis, 1982). Since these two steps may occur partially in parallel, the design members also receive contributions from personnel who are not directly associated with the design procedure. Keen and Morton (1978) stressed the importance of a successful implementation, when they argued that with a successful application of the design procedure, a series of improvements occur throughout the organization and the attitude of its members may be more supportive towards the use of sociotechnical design. This may help to deflect any opposition or antagonism from similar units who are organized on a traditional basis. Otherwise these units may become defensive about their structure and methods of operation, and the resultant antagonism may hinder the success of the design proposal.

Change within an organization may occur in one of two ways: reactive or anticipatory (Albrecht, 1978). In a reactive organization, change is perceived as threatening, and pressure may be required for any change to occur. The members are familiar with their existing norms and resistance to change occurs because there is a mistrust of management and new technology (Pugh *et al.*, 1971). White (1980) writes that in an anticipatory organization, change occurs because the members feel there is a genuine need for change in anticipation of future conditions, rather than because of a current temporary problem. We believe the use of the sociotechnical design process helps to develop an anticipatory organization.

In this section we have provided a description of the basic ideas and procedures used within sociotechnical design. In the next section, we shall examine the contribution sociotechnical design can make to the implementation of new production management ideas, such as Factory Coordination (FC) and Production Activity Control (PAC) systems.

8.3 The contribution of sociotechnical design to the implementation of PMS

In relation to the introduction of new production management concepts, we believe that the sociotechnical approach has a great deal to offer. This arises from the methods and concepts used within sociotechnical design which allow people to analyse the effect of the new ideas on the existing production management arrangements being used (i.e. technical sub-system), and on the existing culture within the organization (i.e. social sub-system). The significance of the use of this analysis can be illustrated by briefly looking at the lessons learned from the implementation of materials requirement planning systems and flexible manufacturing systems.

8.3.1 Materials Requirement Planning systems

The discussion relating to the implementation of MRP concepts, includes both Materials Requirements Planning (MRP) and its successor Manufacturing Resource Planning (MRP II). MRP is probably the most widely implemented large scale production management system since the early 1970s. Latham (1981) found that many of these implementations have not been completely successful in attaining their objectives of reduced inventories, improved customer service and increased operating efficiency. In Chapter 1 we took a brief look at the some of the reasons for the many disappointing MRP installations. We referred to a study by Lawrence (1986), in which he lists various reasons for the failure of a range of MRP implementations, which are clearly relevant to our discussion:

- Lack of top management commitment to the idea;
- lack of education in the concepts behind MRP for the potential users of the system;
- inaccurate data and impractical master production schedules.

The implementation of a MRP system requires top management commitment. Because of the requirement for accurate data and discipline to maintain the system, it may affect a variety of different departments within an organization (Kukla, 1984). For example, purchasing personnel must provide accurate purchasing lead times, detailed bills of materials are required from engineering personnel and materials and production personnel must have the discipline to support the system by following schedules and maintaining accurate inventory data. However within an organization there may already exist an informal system for completing these different tasks and with the introduction of an MRP system, a change in the organization's culture may occur (Safizadeh and Raafat, 1986). The introduction of a formal system for production management such as an MRP system, may encroach on the existing informal system. Grzyb (1981) suggested that this infringement by the formal system may lead to resistance to the new system because the evolution of the informal system has resulted in individuals acquiring certain power, influence and decision making responsibilities. Employees may view the informal system as an outlet to gain job satisfaction and the formal system may therefore be regarded as a threat.

Latham (1981) states that great emphasis has been placed on the technical aspects of MRP systems, while the skills necessary for dealing with the social sub-system have been largely ignored by the personnel responsible for the implementation of these systems. Additional social skills may enable these personnel to answer any doubts that the employees may have concerning MRP type systems. With this understanding, resistance to the introduction of MRP type systems may be minimized. The implementation of MRP systems should involve a best fit between the social and technical sub-systems, in order to ensure a successful application. This balanced perspective between the social and technical requirements of an implementation of an MRP system may also help encourage meaningful communication between the users of such systems and the personnel responsible for implementing them. Sociotechnical design can provide a means for communication, since it involves the sharing of information and the development of an understanding through a shared design procedure using a common language. Elaborating on this point, Hinds (1982) states that 'it is during . . . . the education process . . . that the success of MRP is often determined . . . . Education is the first key to successful implementation'. In the case of MRP systems, proper education in the internal functions of MRP systems and the operation of such systems is important. Within sociotechnical design, the technical and social sub-systems analysis steps help to identify the necessary education and training required in the use of new concepts and technology.

8.3.2 Flexible manufacturing systems

Browne *et al.* (1984) describe a flexible manufacturing system as 'an integrated computer controlled complex of automated material handling devices and numerically controlled machine tools that can simultaneously process medium sized volumes of a variety of part types'. Flexible manufacturing systems are designed to achieve the efficiency of well-balanced transfer lines, while taking advantage of the flexibility of job shops to produce a wide range of products simultaneously. Willenborg and Krabbendam (1987) have examined the link between organizational change and the introduction of new technology using Flexible Manufacturing Systems as an example. We shall now outline some of their conclusions.

The FMS must have a perfectly controlled manufacturing process because of its high capital investment, high level of automation and a low process flexibility (Willenborg, 1987). The FMS must be heavily utilized and all precautions taken to prevent any faults occurring with machine breakdowns or problems with raw materials or tooling, in order to recoup the large capital investment. The need to prevent breakdowns emphasizes the importance of planning the operation of an FMS.

Most FMS operators are highly skilled, because they must ensure that the FMS is organized so that the instructions given by the planning department are completed. This organization task ensures that the typical range of tasks for an operator is very diverse, because the operator is responsible for coordinating, controlling and organizing several machines in the FMS. Apart from coordinating several machines, the organization task also involves ensuring that all the necessary ingredients are available at the prescribed time including materials, part programs, tools and fixtures. Responsibility for organizing the operation of an FMS is distributed throughout the FMS workforce.

Within an FMS, changes in product design can only occur without violating the specifications of the system. This constraint requires design engineers to be aware of the manufacturability of their products and production engineers to give feedback to the design engineers on any potential production problems resulting from the design of the product. Design engineers also provide information of future products, so as to assist production engineers in long term capacity planning. This exchange of information requires the formation of interdisciplinary teams from the design and production departments.

The use of a sociotechnical approach in the introduction of FMS may facilitate the distribution of responsibility among operators in organizing the operation of an FMS and the formation of interdisciplinary teams of design and production engineers. However, as we have already indicated, management's attitude and commitment is important, particularly in relation to the organization of labour. Jones and Scott (1986) argued that 'the implementation of flexible manufacturing systems into small batch production has renewed the conflict of interpretation between 'sociotechnical' and a more critical 'labour process' perspective. The 'sociotechnical' management perspective stresses the potential of FMS for task variety, and relatively autonomous work groups. The 'labour process' management perspective associates computerized control with 'further subdivision, deskilling and increasingly hierarchical organization', which is not supportive of the use of FMS. Table 8.1 summarizes a comparison between the traditional and alternative work organization within an FMS which illustrates the depth of conflict between the two different management perspectives. The 'labour process' perspective places an emphasis on the individual operator learning all of his/her specialized skills by trial and error. This may lead to a lack of motivation on the part of the operator. The alternative 'sociotechnical' approach recommends the use of comprehensive training methods for each operator and encouraging operators to work in groups to solve problems within their scope of responsibility. With this different approach towards the operators training and work practises, each group of operators may be highly motivated.

	Traditional	Alternative
Division of Labour	job specialization	job rotation
Training	on the job trial and error low training costs	comprehensive theory high training costs
Relations	individual	group
Culture	lack of motivation	high motivation
Software	standard	interactive
Implementation	smooth	radical restructuring
Problem solving	low capabilities	high capabilities

Table 8.1. *Comparison between traditional and alternative work organizations within FMS (Willenborg and Krabbendam, 1987)*

8.3.3 Overview of the contribution of sociotechnical design

With the introduction of a new technology, the sociotechnical design approach may achieve the following:

Top management commitment Management realize that sociotechnical design may ensure the successful introduction of technology. With this new technology, an organization may adapt more efficiently to future potential problems.

Employee support Fears or doubts shared by employees can be answered by determining a best fit between the social and technical sub-systems, which is one of objectives of sociotechnical design.

According to Dumaine (1989), the successful organization of the future will be a learning organization, where members are free to identify and pursue problems and opportunities. We believe that sociotechnical design can help an organization develop into a learning organization. Skinner (1985) suggests that the current manufacturing characteristics of shortening product life cycles and increased product diversity give rise to structural problems within an organization and that the traditional focus on productivity and technology is no longer sufficient. Structural and strategic changes are required to help to cope and they can be facilitated by sociotechnical design with its emphasis on developing the most suitable combination of social and technical sub-systems. The importance of the sociotechnical approach is highlighted by Pava

(1983) who observed that sociotechnically designed systems tend to be more robust than systems which have been developed with little reference to the organizational environment.

We shall now discuss how to apply some of the principles of sociotechnical design to create a suitable production environment for implementing FC and PAC systems.

8.4 The environment for Factory Coordination and Production Activity Control

In this section we shall discuss some of the characteristics of the social and technical sub-systems which support the use of FC and PAC systems within an organization. The social sub-system involves the coordination and division of work within the production environment and includes the distribution of responsibility using autonomous work groups. The technical sub-system includes the procedures and processes used within the production environment to manufacture a range of products, such as the layout of the manufacturing process, the matching of a range of products with the demands of the market and the development of efficient manufacturing procedures and processes. In addition to looking at how the social and technical sub-systems can support the use of FC and PAC systems, we shall also examine how our approach to shop floor control compliments the development of suitable social and technical sub-systems.

8.4.1 The social sub-system.

The social sub-system should encourage distributed responsibility through the use of autonomous work groups, where each group is allocated responsibility for the manufacture of a product family. This responsibility involves managing the production, quality, maintenance and any administration tasks associated with the manufacture of the product family. The inherent variety of tasks ensures that each operator within a group has a range of skills at his/her disposal. If this is not the case, each operator should be encouraged to expand his/her range of skills. One of the objectives of the allocation of responsibility for a variety of tasks, is to encourage a high degree of job satisfaction.

Fazakerley (1974) found that one of the benefits of autonomous work groups is that operators are given the opportunity to get on with their tasks and can make productive use of their resources without any frustration and unnecessary delay. This management responsibility is similar to the sociotechnical design principle of participation in design and implementation as outlined earlier in this chapter. The responsibility for managing its own affairs, and having the capacity to complete a variety of tasks, instils in each work group a commitment to higher performance and to the achievement of organizational goals. Our approach to shop floor control compliments this type of social sub-system in a number of ways:

1 FC supports this concept of autonomous work groups and product families by providing a facility for developing and maintaining product families based on Group Technology principles (see Chapter 2).

2 FC and PAC together form a decentralized approach to controlling production, which facilitates the use of distributed responsibility among the work groups on the shop floor.

One potential problem with the use of autonomous work groups is the coordination of all groups in relation to overall organizational goals. Stincombe (1959) proposed two methods for ensuring that operators consistently make appropriate decisions: the use of centralized planning and supervised control or the use of goals and guidelines. Goals and guidelines may ensure that the local perspective within a work group corresponds closely to the global perspective of an organization. We feel the use of supervised control and centralized planning may reduce the autonomy of a work group and thus the organization may not have the flexibility to cope with the changes brought about by the manufacturing environment. FC adopts the guidelines approach by using a factory level planning and control system to provide coordination guidelines. These guidelines are used by each work group to coordinate its activities in relation to other work groups. The factory level systems do not specify every detail on each process step for each product. Rather they provide guidelines to each work group relating to the completion of a series of operations for the manufacture of each product order, which are within its scope of responsibility. The thinking behind the use of factory level planning and control systems is similar to that proposed by Campbell *et al.* (1970), namely that by specifying targets rather than specific rules, and allowing operators to choose their own behaviour appropriate to the tasks, a more efficient production performance may be achieved.

If any of the guidelines relating to a particular cell given by the factory level planning system, have to be broken, the cell supervisor is responsible for any adjustments that have to be made. In some situations, where other products or other cells production activities are affected by any changes, authorization may have to be sought from the production manager. For example if a machine breaks down, it may result in the alteration of a product order's due date because of the consequent delays. The supervisor would alert the production manager about the change in the order's due date and ensure that the change does not cause problems for other orders. However if the particular order is important, some contingency measures may have to be taken, such as the use of overtime or the alteration of another order's due date. These tasks are all part of the supervisor's responsibility in managing the boundary between his/her work group and its environment.

The work group's environment consists of the remainder of the manufacturing organization (e.g. other work groups, the sales department, the finance department etc). Similar to an organization, a work group may be considered as an open system which is open to the influences from its environment. For example, if the marketing department of a company wish to launch a new product on the market, it alerts the work group concerned with the manufacture of the new product about the launch date. The work group will then have to organize the necessary resources so that the production of the new product may commence in order to meet its launch date. These requirements may include new set-up procedures, new quality criteria, operator training and the purchase of new equipment and tools, amongst others. If a situation arises where the work group is not able to start production of the new product in time for its launch, then the supervisor in his/her capacity as boundary manager, liaises with the marketing department to develop a compromise.

One of the reasons that work groups function effectively as open systems and can adapt to changes in their environment is that they have the responsibility for managing their own activities and a focus on a particular product family. Schonberger (1985) uses the term 'responsibility centres' for such work groups and argues that the centres help generate an attitude which seeks continuous improvement in all areas related to their responsibility. Skinner (1985) illustrates the importance of having a proper focus by using the concept of a 'Plant Within a Plant' (PWP), where each PWP has its own facilities to complete a manufacturing task. The focus on a particular set of products by a work group may lead to continuous improvements in quality characteristics, set-up procedures, and materials handling procedures etc. Knight (1974) lists the various economic benefits of using work groups and product families, especially in the areas of scrap and rework, work in progress levels, set-up procedures, and direct labour. One of the reasons advanced for some of these savings, especially in the area of scrap and rework is the familiarity of the operators with the products and their components. Quality circles are a practical example of the concept of responsibility centres, where operators are given responsibility for quality related issues and encourages them to suggest ways of making improvements in activities related to product quality (Schonberger, 1982). This concept of continuous improvement should form the basis of all the activities within the sphere of responsibility of a work group and is similar to the sociotechnical principle discussed under the label of an open-ended design process, which we discussed earlier in this chapter.

The use of work groups and product families facilitates the development of a product based layout (see Chapter 7). A product based layout supports the concept of product focus and the awareness of continuous improvement because all resources and procedures related to the manufacture of a product family are organized or situated within close proximity of each other. Holstein and Berry (1970) found that in contrast to the product based layout, it is more difficult to focus on a group of products in a process based layout. One reason for this difficulty is the process routings for each product in a process based layout are more complicated and relatively longer than in a product based layout.

When an organization decides to introduce autonomous work groups, which involve increased distributed responsibility and greater variety of tasks of each operator, it is important that the social sub-system is ready for such a change. Susman and Chase (1976) suggest four important properties that must be present in the social sub-system to support the concept of work autonomous groups:

1 Values
2 goals
3 roles
4 incentive schemes.

An organization should appreciate its human resources and actively encourage their skills and creativity. The values of such an organization include the recognition of the achievements and loyalty of the employees. With the use of product families, operators can pursue goals which are identifiable with a specific range of products. These goals may involve quality or production output levels among others. The scope of responsibility given to each work group may involve a variety of tasks, which allows each operator to assume different roles. In order to assume these different roles, each operator must have a range of skills at his/her disposal, which may help to generate increased job satisfaction. Management should back up their commitment to the use of work groups and the distribution of responsibility by having a proper support system in place in the organization. This support system should promote the concept of group work as illustrated with the use of group incentive schemes. Susman and Chase (1976) suggest that the incentive scheme should be 'paying for knowledge rather than for the position held'. This encourages group members to pursue goals consistent with the goals of the organization and to expand their range of skills.

The above characteristics of the social sub-system relates to the methods of organizing and coordinating a range of tasks associated with the manufacturing of a selection of products. We will now move on to discuss the necessary characteristics and procedures which can be used within the technical sub-system to help the implementation of our shop floor planning and control systems.

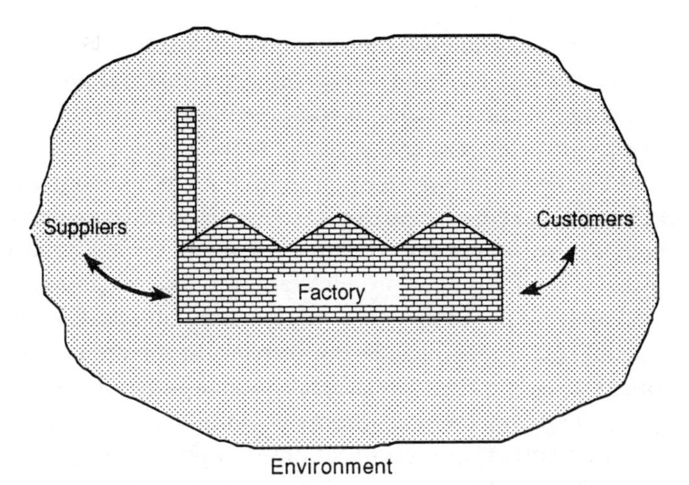

Fig. 8.4. View of an organization within its environment (Browne *et al.*, 1988).

8.4.2 *The technical sub-system*

The technical sub-system consists of the production environment in which the concepts of FC, PAC and the work group based social sub-system are implemented. Traditionally, the emphasis within the production environment has been on productivity and this has diverted attention from the potential performance improvements which may be gained by making structural and strategic changes to the production environment (Skinner, 1985). The analysis of the technical sub-system within the sociotechnical design procedure focuses on the required structural and strategic changes which support the following three key elements:

1. The provision of a product range to match market demand;
2. the development of product families and a product based layout;
3. the establishment of relationships with suppliers to achieve efficient deliveries of high quality raw materials.

These three elements are part of an approach within sociotechnical design which regards an organization as situated within an environment as illustrated in Fig. 8.4.

Earlier, we alluded to the changing manufacturing environment, with particular attention being given to the decreasing product life cycles and the need for product diversity. These two features have a major influence on the selection of a suitable product range to satisfy market demands. Browne *et al.* (1988) suggested that in order to satisfy market demand, an organization must design 'a range of products which anticipate the market requirement and include sufficient variety to meet consumers' expectations and which can be manufactured and delivered to the market at a price which the market is willing and able to pay'. One of the methods of achieving this objective is through modular design. Modular design involves rationalising the product range by examining the commonalty of components and sub-assemblies across the product range (Browne and O'Gorman, 1985).

Once a suitable product range is agreed upon, the production process should be structured so as to enable efficient manufacture of products under fluctuating market demand. As we discussed in Chapter 7, Group Technology is a common approach to the task of forming product families and developing a product based layout. Lewis (1986) stated that the resultant structural benefits from the use of Group Technology include the increase in the predictability of the manufacturing system, and the standardization and integration of production processes. As was noted in Chapter 2 some environments may not support an ideal product based layout. In such situations, the principles of sociotechnical design are still relevant, despite any compromises concerning the sharing of resources that may have to be made.

When an organization has developed a product based layout and is producing a range of products suitable for the market requirements, attention must be given towards ensuring a regular and stable supply of raw materials. This demands close cooperation with all the organization's suppliers and should only start when the organization's manufacturing system is in an efficient state, because there is little benefit in having material delivered on time and then allowing it to remain in the manufacturing system for an abnormally long time. Hall (1983) describes this cooperation as 'concentrating on linking the supplier's process to the customer's process, whereas inventory buffers these two processes'. According to Lubben (1988), by building a long term working relationship with a limited number of suppliers, an organization seeks to:

1 Maintain a steady flow of quality components;
2 reduce the component order lead time;
3 reduce the component inventory levels;
4 reduce purchasing costs.

In return for this commitment to frequent deliveries of high quality components, the supplier is assured of his customer's continued business. In order to assist the supplier achieve regular deliveries, an organization must exchange information on relevant production topics, such as quality standards and modifications to the master production schedule. Hall (1987) suggests that an organization and its suppliers view the cooperative relationship as one involving continuous improvement in component quality, delivery methods and frequency of deliveries.

We may view a manufacturing system as a series of interrelated sub-systems and associated material and data flows (Fig. 8.5). The product design flow contains data relating to the design and process plans for each product. The design information is provided by Computer Aided Engineering (CAE) and

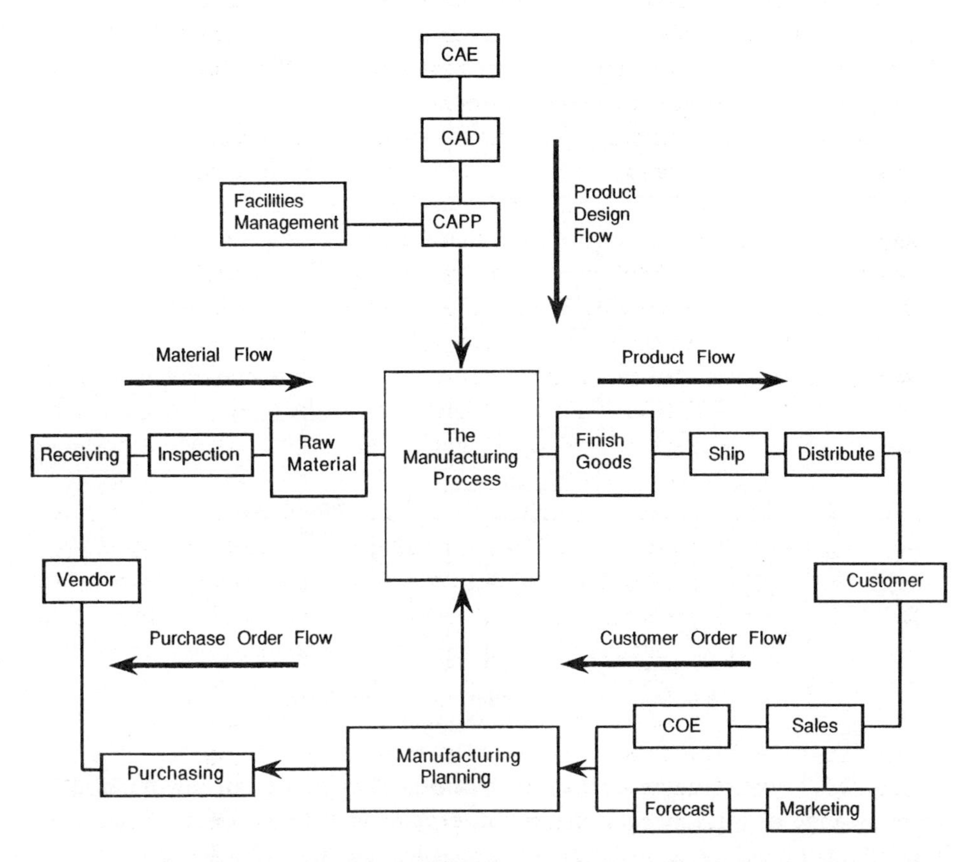

Fig. 8.5. The different flows in a technical sub-system.

Computer Aided Design (CAD) systems, while the outline process plans are developed by a Computer Aided Process Planning system (CAPP). As we have already indicated, the detailed process plans, where the allocation of specific resources are specified, are developed by the process planning function within FC. The customer order flow identifies the production requirements for each customer which is passed by the sales/marketing functions to the manufacturing planning function. During the manufacturing planning stage, various schedules are developed for the manufacture of products to fulfil the customer's requirements. These schedules may be used to organize the different stages of manufacture, for example the purchasing of raw materials and the actual production of products. The schedule for organizing the purchase of raw materials is a timetable which ensures that the raw materials are available in the correct quantities and at the right time. The purchase order flow generates a flow of raw materials into a factory according to this timetable.

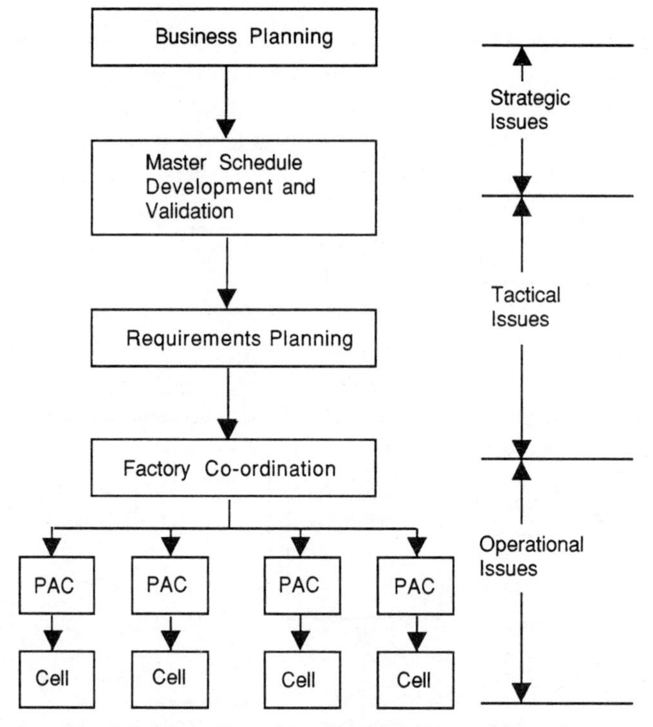

Fig. 8.6. Manufacturing controls systems hierarchy.

The material flow commences with the receipt and inspection of raw materials from a range of vendors. The inspection process may be necessary to ensure that all incoming raw materials are of the highest quality. After inspection, the raw materials are then stored either in a warehouse or on the shop floor, until they are required within the manufacturing process. The raw materials are processed in the manufacturing process according to the production schedule developed in the manufacturing planning stage.

In our approach to shop floor control, the tasks of developing a production schedule and controlling the flow of products within a factory are dealt with by a FC system and a number of PAC systems. The FC and PAC systems form part of a hybrid model for manufacturing planning, which we described in Chapter 1 (Fig 8.6). This model consists of three hierarchical levels; strategic, tactical and operational. The FC and PAC tasks are situated on the operational level, and are concerned with integrating the requirements planning task with activities on the shop floor. As we explained earlier in Chapter 2, FC is responsible for the flow of products between cells and within the overall factory, while each PAC system is responsible for organising the flow within each cell. It is perhaps worth noting that in the context of the simple model illustrated in Fig. 8.5, FC and PAC are situated at the interaction between the various product and data flows. FC and PAC receive the outline process plans from the CAE function. Within the constraints imposed by the material flow and the customer order flow, they coordinate the flow of products throughout a manufacturing system to enable the products to the customers as required.

Once the manufacturing process is completed, there is a range of finished products ready for distribution to their required customers. This distribution of finished products is labelled product flow. One of the key features of the product flow is a fast efficient delivery of finished products to each customer.

We have just described some of the individual requirements of the social and technical sub-systems in a manufacturing environment, which are necessary to support the use of FC and PAC systems. We would like to conclude this topic by examining the principle of a 'best fit' between the social and technical sub-systems in the next section.

8.4.3 A best fit between the technical and social sub-systems

The principle of a 'best fit' is very important because traditionally the emphasis of new technology has been solely on optimising the technical sub-system, with little consideration for the social sub-system. As we have already indicated, sociotechnical design adopts a different approach, which stresses the importance of analysing both sub-systems and their interrelation, so that they may be designed to enhance each other. We shall now use the scheduling task as an example to demonstrate how FC and PAC systems incorporate this 'best fit' principle.

From our discussion on scheduling in Chapter 6, it is clear that the scheduling problem is difficult because of the large number of schedule permutations that are possible. Thus, optimal schedules for manufacturing problems are often unattainable. One possible way to overcome this problem is the use of heuristics for developing schedules. However because of the assumptions made in developing scheduling heuristics, the resulting schedules often lack the sophistication necessary to deal with typical problems on the shop floor. Events such as material shortages or equipment breakdowns may render invalid the assumptions and make the schedule obsolete. To develop a more realistic schedule, we recommend the use of a two stage scheduling procedure which uses a combination of heuristics and the human experience and knowledge available on the shop floor (Fig. 8.7).

The first of the two stages involves choosing a suitable heuristic from a selection of scheduling heuristics, with which to develop a first pass schedule. In the second stage, the first pass schedule is manipulated, using the experience and knowledge of a production scheduler, to provide an acceptable schedule. We term the second stage interactive scheduling. By encouraging human interaction, the interactive scheduling task produces a schedule which reflects

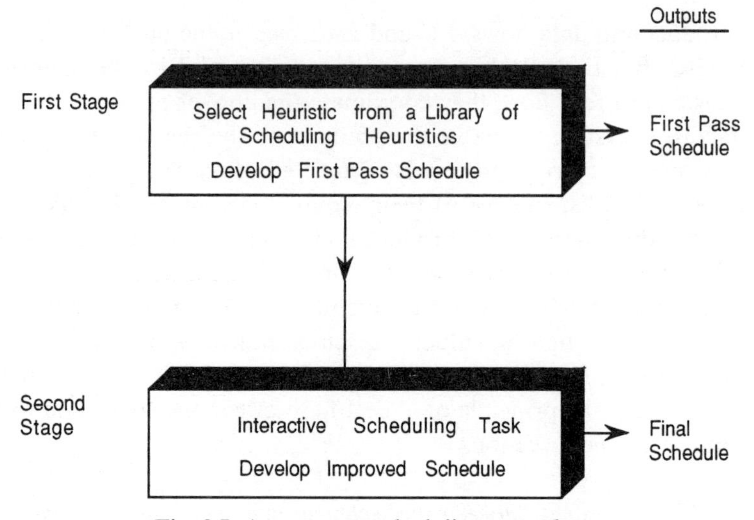

Fig. 8.7. A two stage scheduling procedure.

a more balanced view of the shop floor requirements and capabilities. One suitable method of representing the output of a first pass schedule involves using a computer based Gantt chart with a list of resources displayed along the vertical axis and a time horizon presented along the horizontal axis (Fig. 8.8). The user is then provided with a graphics editor to manipulate the Gantt chart.

When viewing the scheduling task as a work system, we see that it has various structured and unstructured inputs and its output is a solution in the form of an acceptable production schedule (Fig. 8.9). Jackson and Browne (1989) differentiate between structured and unstructured inputs.

Structured inputs represent the technical sub-system, which include the various scheduling heuristics and algorithms, the graphical representation provided by the computer system and the information provided by a database.

Unstructured inputs represent the social sub-system, which include the past experience, knowledge and intelligence of the person using the interactive scheduler.

The improved schedule is more effective and practical because the interactive scheduling task allows the unstructured inputs to be incorporated into the development of a schedule. By manipulating the first pass schedule through an interactive scheduler, a user is giving feedback about his/her intuitive feelings of what is 'wrong' or 'right' about the first pass schedule. The improved schedule can then be used in the PAC system to coordinate the production activities within the cell. This approach promotes the idea of distributed responsibility by allowing each work group with

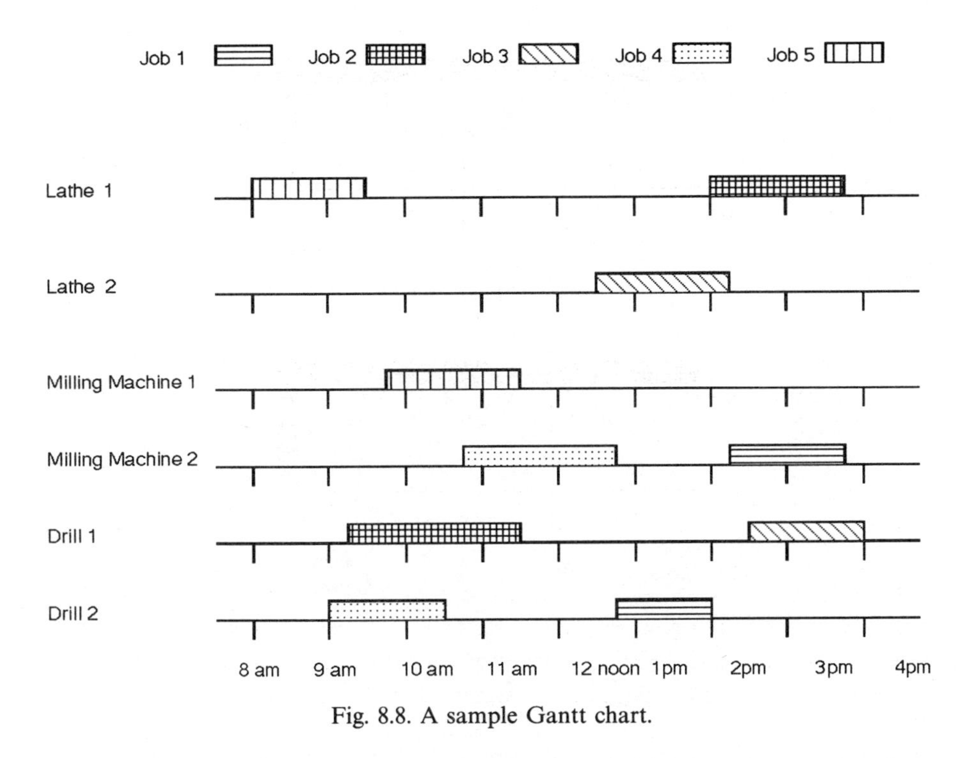

Fig. 8.8. A sample Gantt chart.

the support of a PAC system to plan the specific details for the cell's production according to the guidelines specified in the factory level schedule.

8.4.4 Overview of the environment of FC and PAC.

FC and PAC systems compliment many of the ideas in sociotechnical design. In relation to the social sub-system, the use of distributed responsibility is encouraged with the production environment design element of FC supporting autonomous work groups with the development and maintenance of an organized product based layout. The combined use of FC and PAC to control the flow of products within a factory is a decentralized approach which also supports distributed responsibility. We have seen from Fig. 8.5, where the FC and PAC tasks fit in relation to the different data and material flows within a technical sub-system. Using the scheduling task as an example, we demonstrated how the FC and PAC tasks incorporate the principle of best fit of the social and technical sub-systems.

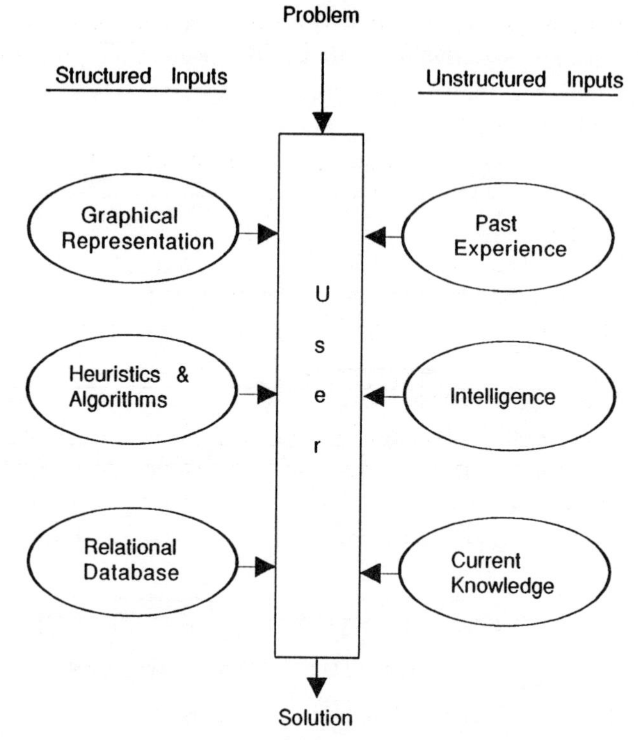

Fig. 8.9. The two types of inputs used in the interactive scheduling task.

8.5 Conclusions

Toffler (1970) remarked, when referring to the turbulence in society and the rapid change in organizational structures, that just as 'the new nomads migrate from place to place, man increasingly migrates from organizational structure to organizational structure'. The nomad is a vivid metaphor for depicting the changes occurring within organizations. Hence the importance of the sociotechnical design principles to help organizations cope with change in an effective and systematic manner. Before any changes are made to an organization, Weinberg (1975) stressed the importance of agreeing about the attributes and characteristics of the particular system under observation. By viewing an organization as an open system within this larger turbulent environment, a framework for performance improvement may be set up. This framework involves developing both the technical and social sub-systems of an organization to ensure a better overall combination. A best fit between the social and technical sub-systems helps to develop increased flexibility and more efficient problem solving methods throughout an organization by distributing responsibility and encouraging participation of members of the organization in any new technology developments within the organization. Cherns (1977) argues that this promotion of human versatility and capacity for discretion, increases the probability of an organizations adapting to the environmental instabilities in modern markets.

We feel that with this type of organizational structure, which operates on a basis of continuous improvement, the implementation of new production management strategies such as FC and PAC systems may be achieved more easily. FC and PAC systems may then operate within the social and technical sub-systems of a manufacturing organization to provide more efficient techniques for the organization, planning and control of the manufacturing system.

In Chapter 9, we shall discuss the specific features of a prototype design tool for the development of FC and PAC systems. Then we shall show how the sociotechnical design principles were applied to implement FC and PAC systems in an electronics assembly environment in Chapters 10 and 11.

9

A design tool for shop floor control systems

9.1 Introduction

This chapter describes a design tool which can be used to support the design, development and implementation of shop floor control systems within a particular manufacturing environment. This chapter illustrates the use of the design tool to support the development of PAC systems at the cell level of manufacturing. It is important to emphazise that the tool may also be used to support the development of suitable control structures for FC, although it is not suitable to experiment with production environment design techniques. This design tool, the Application Generator (AG), is collection of software modules which interact together in a distributed computing environment. The interesting aspects of the AG are twofold: firstly it uses an artificial intelligence approach to aid in the selection of suitable control strategies (scheduling and dispatching); secondly, its distributed simulation approach identifies a clear migration path from the experimental world of simulation modelling to the reality of real-time manufacturing. Within this chapter the overall objectives of the AG will be discussed, and each of the software modules which constitute the AG will be described in detail.

The work described in this chapter was designed and developed by the participants of an ESPRIT (European Strategic Program for Research and Development in Information Technology) project in the Computer Integrated Manufacturing Area, entitled COSIMA (Control Systems for Integrated Manufacturing), and the software system has been demonstrated at research and industrial exhibitions in different European community countries as well as the United States of America.

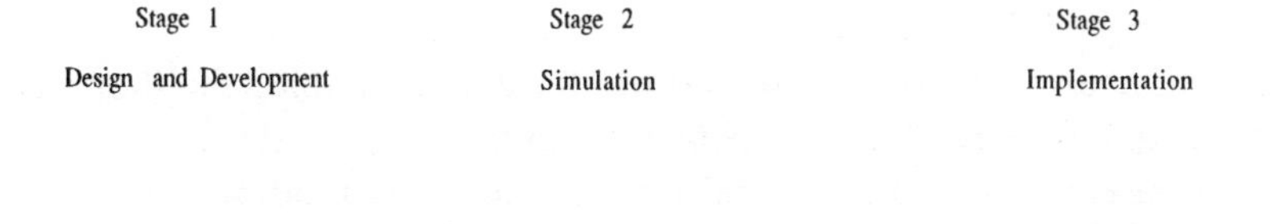

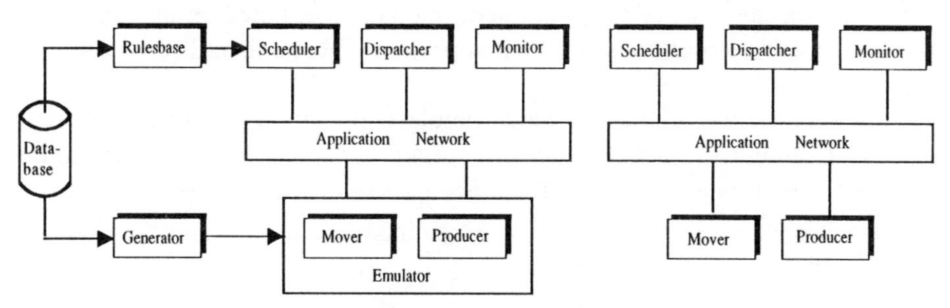

Fig. 9.1. The PAC life cycle model.

9.2 The application generator

Having developed an architecture for shop floor control (described in detail in Chapters 2 and 3), the task of developing implementable systems to support this design is the next logical step in the realization of operational PAC systems. Achieving a workable PAC application requires that each of the five distinct building blocks (scheduler, dispatcher, monitor, mover and producer) work together in a distributed computing environment, planning and controlling the day to day activities on the shop floor. Before this implementation can take place, it is essential that a model of PAC be constructed, to allow the concepts and the architecture to be tested in a controllable simulated environment.

The PAC life cycle model, illustrated in Fig. 9.1, facilitates the migration path from systems design, through simulation, and onto implementation. This life cycle consists of three distinct stages:

1. Design and generation of an accurate model of the manufacturing system;
2. simulation of different control strategies with a view to selecting the most appropriate for the particular needs of the manufacturing system under study;
3. an implementation of these strategies in a test environment.

The first two stages are supported by the application generator, which allows a designer to experiment with possible PAC solutions before implementing these solutions on the shop floor. This chapter is solely concerned with the first two stages of the PAC development life cycle, and the final stage of implementation is dealt with in Chapters 10 and 11.

The AG is a software application designed to allow a user to experiment with

PAC methodologies for a particular manufacturing environment. It consists of three integrated modules:

1 A manufacturing database, which stores the data describing the shop floor and the different attributes associated with the manufacturing cycle.
2 A rulesbase, which is an artificial intelligence based module which analyses the contents of the manufacturing database, based on its knowledge base, and recommends the most appropriate planning and control strategies.
3 A PAC simulation model, which simulates the interaction of the different PAC modules in a distributed environment. The results of the simulation are used to assess the potential effectiveness of the various modules.

We will now briefly outline each of these modules in turn, and then discuss each in detail in sections 8.3, 8.4 and 8.5.

1 **Manufacturing Database**
   A vital component of any PAC system is a manufacturing database, as this provides the data foundation upon which the PAC system carries out the planning and control tasks. This database is structured to reflect the needs of the PAC system, and it allows the relevant attributes of the manufacturing environment to be described in detail. Through an user-friendly interface, a designer or analyst may incrementally build up a model of a manufacturing system by describing:
   (a) The physical model of the system (workstations, buffers, conveyors etc.).
   (b) The raw materials used (material name, reorder point, etc.).
   (c) The process and product model, which details all of the parts and respective operational steps (operation number, process time, etc.).
   (d) The customer orders which provide the PAC system with the stimulus to carry out its functions.
2 **Rulesbase**
   The rulesbase analyses the data in the manufacturing database, and predicts the potential effects of customer orders on the manufacturing system. Partly based on the Optimized Production Technology philosophy (Goldratt, 1986), it searches through the structure of the manufacturing system and identifies potential bottlenecks. A bottleneck is a point in the manufacturing process that constrains the amount of product that a system can produce. It is essential from a control viewpoint to manage bottlenecks effectively. The rulesbase uses a system of rating the potential of a resource to be a bottleneck, based on its position in the process (the earlier in the process a bottleneck occurs, the greater the effect will be on the overall system), the number of products which use the resource, and predicted utilization of the resources.

As a result of this analysis, the rulesbase recommends a certain scheduling strategy and an appropriate schedule is developed. This schedule is then made available to the PAC simulation model, which can model its execution, taking unexpected events (such as machine breakdown and operator unavailability) into account. The results of the simulation may be analysed to determine the potential effectiveness of the proposed scheduling solution.

3 **PAC Simulation Model**
This simulation model is driven by the data specified in the manufacturing database. It consists of four distinct software modules which communicate with each other through a distributed messaging facility known as the Application Network (AN). These modules, a direct mapping of the PAC architecture, are: a scheduler, dispatcher, monitor and emulator. The emulator models the actions of the movers and producers on the shop floor.

The advantage of having a distributed simulation model is that the scheduler, dispatcher and monitor, used with the emulator in the PAC simulation model, can be used later in the live environment. Thus, the application software modules readily enable the migration from PAC simulation to PAC implementation. Ideally, the user tests and experiments with different control strategies using the rulesbase and the PAC simulation model. When satisfied, the shop floor emulator is decoupled from the simulation model, and the real system of producers and movers links to the other modules of the PAC simulation model to form a coherent and implementable PAC system.

Prior to our discussion of the individual elements of the AG, it is important to state that we see it having two possible modes of operation, namely:

1 Mode 1, where it can support the selection of the most appropriate control strategies for a manufacturing system. We envisage the utilization of this mode on an infrequent basis, perhaps when major changes are enforced on the manufacturing system. Therefore, if new products are being introduced, or existing process details changed, this design tool can be used in this mode re-evaluate the control strategies.

2 Mode 2, where it is used as an off-line simulation support tool for manufacturing personnel. This mode could be used on a frequent basis, perhaps prior to each production shift. The primary advantage of the design tool in this mode, is that the user has an interactive what–if analysis capability for analysing potential disruptions to the production plan. Typical production shift scenarios, such as preventative maintenence, operator unavailability and shortages of raw materials, can be simulated with relative ease, thus illuminating possible production shortcomings.

9.3 The manufacturing database

The manufacturing database is essential to the realization of the Application Generator and indeed any PAC system, because it provides a consistent source of manufacturing data which can be used by all software applications within the AG. This consistency is vital to the design and development tasks of the AG, because the use of a common database means that all the applications have access to the latest versions of manufacturing data. The use of the database, along with a user-friendly interface, allows an engineer to describe the manufacturing system in a logical structured manner using the day-to-day terminology which describes the manufacturing environment. One advantage of this is to place a tool like the AG within easy reach of manufacturing personnel who might not have a great deal of software experience, thereby facilitating the implementation PAC systems in a more effective manner.

The manufacturing database for the AG is a relational database structured using the Digital Equipment Corporation Rdb product. The main advantage of a relational database is the ease with which information may be retrieved and there are standard relational operations available in Rdb which allows the user to access data from different relations. The editing, deletion and creation of data may also be carried out using Rdb, and this is a necessary requirement for the manufacturing database, since the manufacturing data is unlikely to remain constant over a period of time (i.e. new product routings, process times, planned orders etc).

The manufacturing database therefore contains a data model for the manufacturing system, and stores sufficient relevant data for the PAC functions to perform their respective tasks. This data model is constructed using a layered approach to model building. This is achieved by firstly describing the basic features of the manufacturing system, which are the physical devices such as workstations, buffers and transportation systems. These are viewed as the fundamental resources of the manufacturing system, since no production can occur unless these devices, which together transform raw materials into finished products, are available for use.

When the physical model has been defined, the next layer of the overall data model to be described are the raw materials. The raw materials combine together at different steps of the process to form a finished product, and the raw materials model consists of the relevant raw materials data.

After the raw materials and physical resources have been described, the next step is to describe the list of products which are manufactured. Data entered here includes process information such as the number of operations, details of the operations, set-up times and process times. This product and process model describes how each of the resources are used to complete customer orders.

Finally, an order requirements model is described, containing the customer requirements for the particular PAC system, and these requirements act as a starting point for the PAC process of planning and control. The requirements, along with the other three categories (physical, raw materials and product/process), are then used by the different modules of the AG to design, develop and simulate PAC solutions in a controlled environment.

Thus, we have identified four distinct data categories from which a model of the manufacturing system may be constructed and used for the purposes of design, development and implementation of PAC systems. A more detailed analysis of this database model is now presented under the separate headings of:

1. Physical model;
2. raw materials model;
3. product and process model;
4. order requirements model.

The Physical Model Table 9.1 shows the main data relations which describe the physical model of the system within the Application Generator. The first relation to be completed is that describing each of the buffer locations within the modelled system, and these buffer locations may contain either Work In Progress (WIP) or raw materials. The final relation describes the structure of the cell, and details here include the list of stations which belong to the cell and the types of parts which are processed by the cell.

Raw Materials Model The raw materials model of the manufacturing database contains the relevant information which describes the range of raw materials which are used within the manufacturing cell under consideration. Table 9.2 gives a description of the attributes of this relation, and includes important characteristics such as re-order point and lead time to receipt of the raw material from the source. This information is used by the scheduler, dispatcher and monitor in performing the PAC task.

Product and Process Model The product and process model defines details of the operational steps for each product in the manufacturing cycle. It includes part data, process data and materials needed data. This data, described in detail in Table 9.3 is used by the PAC modules to estimate start and finish times of jobs, and also to route each job to its correct location.

Depending on the complexity of the manufacturing model, additional information (Table 9.4) may be added to describe different types of manufacturing operations. This relations are made available so that, theoretically, any type of manufacturing environment may be described and simulated. Nof (1985) suggested that all manufacturing operations follow four basic patterns and the AG is designed to cater for these types of operations. The four patterns of manufacturing operation identified by Nof are:

Relation Name	Fields
Buffer Data	Buffer Name Buffer Type Buffer Capacity
Station Data	Station Name Station Capacity Failure Rate Repair Time
Station Buffer Data	Station Name Buffer Name Buffer Type Buffer Class
Cell Data	Cell Name Part Families Average Capacity Station name

Table 9.1. *Physical data model relations*

Relation Name	Fields
Raw Materials	Material Name Buffer Name Maximum Quantity Reorder Point Lead Time to Receipt Source of Material

Table 9.2. *Raw material data model relation with example*

1 Combinative. With this operation separate parts are assembled together at one operation to form a new part. Typically, each of the input parts are components or sub-assemblies which combine to realise a parent part on the next level in the hierarchy of the bill of materials. An example of a combinative operation is the assembly of components to a Printed Circuit Board (PCB). This type of operation implies a 'many to one' relationship.

2 Sequential. With this type of operation one piece of material is progressively modified by successive operations. Machining operations, such as

Relation Name	Fields
Part Data	Part Number BOM Level Lead Time Final Destination
Process Data	Part Number Station Name Operation Number Operation Type Set-up Time Process time Input Buffer Output Buffer
Materials Needed	Part Number Station Name Operation Number Material Name Buffer Name Quantity

Table 9.3. *Part and process data model relations*

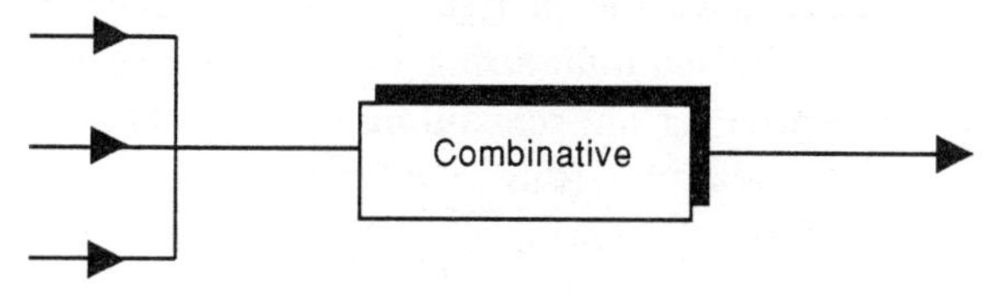

Fig. 9.2. Combinative operation.

turning, boring, milling, drilling etc. are typical 'one to one' sequential operations.

3 Inspection. With this type of operation material is inspected and a certain percentage possibly rejected and re-routed. In some cases rejected material may have no further use and may be sent to scrap. However, in other cases the part may be rendered useful by performing a repair operation.

4 Disjunctive. With this type of operation material is disassembled into components. Thus, this is a 'one-to-many' relation, an example of which is the parting off of bar stock into components in a machine shop.

Thus, the manufacturing database has been designed to cater for these four categories of operations, and in doing so, the possibility of representing

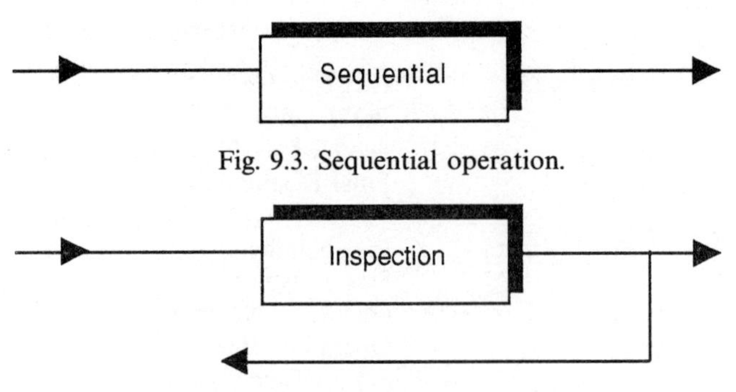

Fig. 9.3. Sequential operation.

Fig. 9.4. Inspection operation.

discrete manufacturing scenarios is realized. This database was populated for each of three manufacturing case studies, namely: electronics assembly (Digital Equipment Corporation), car engine assembly (Renault Automobiles) and flexible manufacturing systems (COMAU).

Order Requirements Model Finally, it is necessary to list the order requirements for the PAC system in the manufacturing database. These requirements (Table 9.5) enable the PAC process to start, and the ultimate goal of PAC is to meet the requirements within the specified time frame in the best possible way. Therefore, the rulesbase takes the planned requirements in order to analyse the potential effect on the manufacturing system, and, where possible, this rulesbase will recommend a certain scheduling strategy. This strategy develops a particular plan for production (the schedule), and this is released to the other modules of the PAC simulation model for the purposes of a detailed simulation run.

9.4 The rulesbase

Much of the research on scheduling to date has concentrated on the development of specific techniques for particular manufacturing systems using operational research and expert systems approaches (Copas and Browne, 1990). However, not a great deal of insight has been gained into the actual selection of schedulers for use in varying environments – hence the development of the rulesbase within the AG. The function of the rulesbase is to aid in the process of selecting an appropriate scheduling strategy for a given manufacturing system. This is achieved by analysing the potential effects of the order requirements on the resources of the manufacturing system. There are three checks on the different system resources, namely:

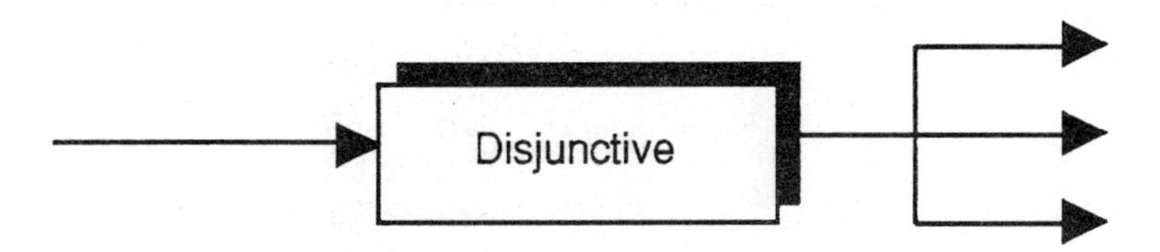

Fig. 9.5. Disjunctive operation.

Relation Name	Fields
Combinative Data	Part Number Station Name Operation Number Sub Part Number Quantity Buffer Name
Disjunctive Data	Part Number Station Name Operation Number Split Part Number Split Class Quantity Buffer Name
Inspection Data	Part Number Station Name Operation Number Scrap Rate Scrap Destination Repair Operation Repair Station

Table 9.4. *Additional relations for part and process data model*

1 To evaluate the number of workstations in the system;
2 to check for materials availability;
3 to calculate a capacity check of system resources.

1 **Evaluate workstation number**

The purpose of calculating the number of workstations in the system is to test for trivial scheduling solutions. For instance, if there is only one machine in the system, then rules such as SPT, EDD, etc. will produce optimal schedules. If there are two machines in the system then Johnson's algorithm can be used to generate an optimal schedule with respect to

Relation Name	Fields
Order Requirements	Part Number Release Date Release Time Job Number Due Date Batch Size

Table 9.5. *Order requirements data model relation with example*

the criterion of throughput time. However, it is most likely that the systems under analysis will be more complex than the simple scenarios presented here, hence the need for the other two system checks.

2 **Check raw materials**
The raw materials availability check of the material stocks is then performed to ensure that sufficient raw materials will be available to fulfil the production orders from a higher planning system. This check on raw materials availability consists of taking the daily orders and locating the corresponding bill of materials for each of these orders. If the materials for a certain order are not available in sufficient quantities to complete the order, the order is delayed until the stocks of materials are available. If stocks are available, then the order is made available for the scheduler.

3 **Check capacity**
Finally, a capacity check of the system resources is performed by the rulesbase. This calculates the pre-production utilisation levels of the workstations in the system and these utilisations are then checked to evaluate potential bottlenecks. If a potential bottleneck is found, the jobs going through the bottleneck are given a higher priority. If no bottlenecks are found then, depending on the users criteria, some other heuristic is chosen to generate the schedule. This heuristic may be earliest due date, shortest processing time or least slack time.

Where it is possible, the rulesbase selects what it considers the most appropriate scheduling strategy. This scheduling strategy (five different strategies are described in the following section) is then used to develop a detailed schedule for the PAC simulation model, and this schedule is then simulated passing through each of the operational steps.

9.5 The PAC simulation model

The role of the PAC simulation model is to simulate PAC solutions in a distributed environment in order to test the likely effectiveness of the PAC system. This is an important feature of the AG, since it facilitates the achievement of an implementable PAC solution. The first aim of the PAC simulation model is to provide the user with a realistic simulation environment within which different planning and control strategies may be tested, the overall aim being to find the most suitable strategies for the particular environment. These strategies may be recommended by the rulesbase, or alternatively, the user may select them from a pre-defined software library.

Secondly, because of the unique PAC simulation architecture that accurately mirrors the PAC architecture, the PAC simulation model provides a clear and well defined migration path from simulation to full implementation. The five different software applications which run together in a distributed environment to simulate PAC solutions are now described, and consist of:

1 Scheduler;
2 dispatcher;
3 emulator;
4 monitor;
5 application network.

- The Scheduler consists of a number of different heuristics and algorithms from which the user selects. In addition, the scheduler contains the interactive scheduler, which is a graphical software planning tool based on the Gantt chart method, that allows an experienced user to manually alter a given schedule if it is considered necessary.
- The Dispatcher takes the plan from the scheduler and proceeds to implement it, based on the conditions of the emulated producer and mover systems. In effect, the dispatcher controls what happens within the emulator by issuing instructions to the various workstations.
- The Emulator simulates all of the events that occur on the shop floor, including machine breakdown, workstation usage and the creation and depletion of work in progress (WIP). This emulator was initially developed using artificial intelligence based tools which simulated the events of the shop floor using Petri nets structures.
- The Monitor gathers all of the information in real time from the emulator and then displays this for the user in terms of workstation availability, work in progress and material status. It also issues reorder commands if material stocks are too low. When the simulation is finished, the user is provided with the opportunity to analyse the results using a historical reporting module.

- The Application Network is a distributed software bus that allows each of the above applications to communicate with one another during the simulation run.

The Scheduler The scheduler contains a library of a number of different rules which are selected the rulesbase, or perhaps directly by the user. The function of these algorithms is to take the guidelines from a higher planning function, and select a sequence of operations to be performed so that the job is completed to satisfaction. The scheduler specifies the resources needed to perform each operation, and also assigns recommended start and finish times for each of those operations. The advantage of the scheduling approach developed for the AG was the development of a library of schedulers, which may be referenced when selecting the most appropriate scheduler. This approach ensured that further scheduling developments may be easily adopted by building up this library with new approaches, techniques and heuristics. The rules currently available within the scheduling library include:

1 A bottleneck scheduler, based on the Optimized Production Technology philosophy. OPT revolves to a great extent on the identification of bottlenecks and the generation of a schedule which ensures that the bottleneck is never idle. The bottleneck scheduler is only used if a bottleneck is identified in the system (using the rulesbase to evaluate whether or not a bottleneck exists). It works by assigning a higher priority to jobs that are processed on the bottleneck workstation. This is designed to keep jobs flowing up to and through the bottleneck to utilize fully this heavily loaded resource.
2 Shortest processing time, which schedules jobs in order of ascending processing times, i.e. the job with the shortest processing time is processed first.
3 Earliest due date, which sequences jobs in the order in which they are required, i.e. to sequence the jobs such that the first processed has the earliest due date, the second processed the next earliest due date, and so on.
4 Least slack time, which is a due date based rule i.e. it is based on due dates rather than processing times. Slack time is the time remaining before a jobs due date.
5 Johnson's Algorithm, the widely known algorithm for the two machine problem, discussed earlier in Chapter 6.

Each of these five rules produces a schedule which is considered to be a first pass effort in trying to plan for production. Because of the complex nature of the manufacturing environment the task of designing an optimal scheduler is practically impossible, hence the approach taken for developing this scheduling system, i.e. to build a library of scheduling heuristics from

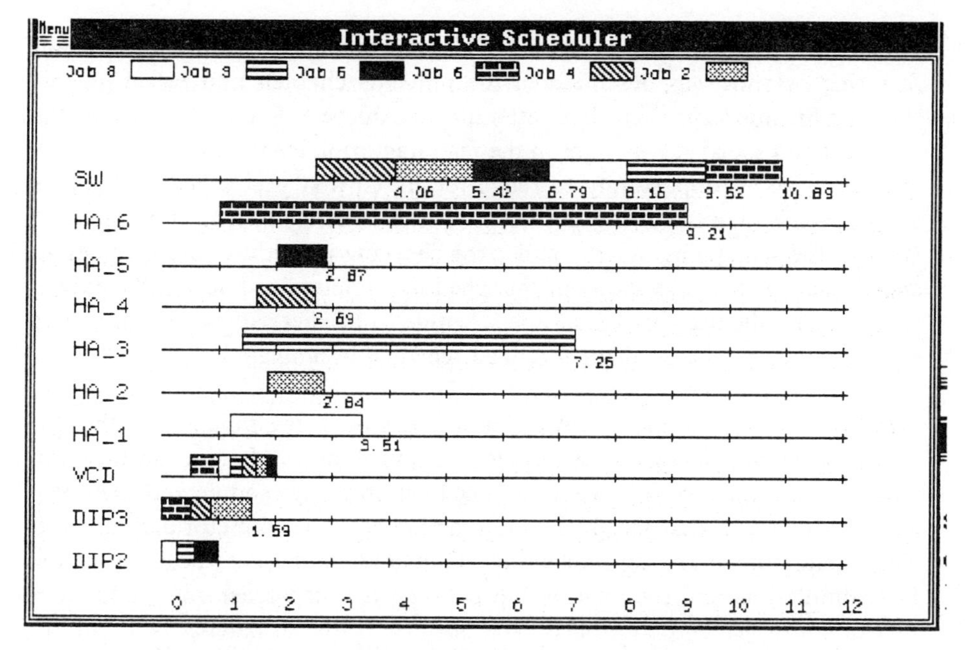

Fig. 9.6. Sample output for the interactive scheduler.

which the best one for any manufacturing system can be used. However, heuristics are only sophisticated 'rules of thumb' which are designed to give good first pass schedules, and in some cases this is not good enough. Thus, an interactive scheduler was developed as part of the Application Generator, and the purpose of this software module is to allow an experienced user modify the first-pass schedule using their intelligence and experience.

The Interactive Scheduler

Jackson and Browne (1989) describe the interactive scheduler as:

> 'A decision support system that attempts to achieve a balance between available computing power and human intuition and experience, the former to alleviate the more mundane and tedious scheduling tasks and the latter to provide the intelligent input necessary to generate a solution.'

Essentially, the interactive scheduler is a computerized Gantt chart, and provides the user with a graphical display composed of workstations on a vertical axis, and the time scale on the horizontal axis. Individual jobs are denoted by different coloured 'blocks' which represent busy time on various workstations (Fig. 9.6). The display may be changed interactively by the user through the use of a range of in-built editing features.

The sociotechnical approach to the design and implementation of manufacturing systems was described in the previous chapter, and based on this thinking the interactive scheduler attempts to achieve a best match between the technical and social sub-systems in the manufacturing environment. It provides the user with an opportunity to refine the initial first pass schedule which is stored in the manufacturing database. The interactive scheduler is the final stage in the scheduling system, and error checking procedures are included to ensure that the logical validity of the schedule is maintained during the editing phase. When the user has edited the schedule to his/her satisfaction, it is then sent to the dispatcher and the modified schedule is simulated.

The Dispatcher Within the PAC simulation model, the function of the dispatcher is similar to that of a dispatcher controlling production on the shop floor. It takes the schedule from the scheduler software module and then proceed to implement that schedule by interacting with the emulator and monitor. The interaction of the dispatcher, emulator and monitor is the basis of the PAC simulation, and the relationship between the dispatcher and emulator is that of controller and controlled. The purpose of the dispatcher is to control the emulator and it does this by issuing schedule specific instructions telling the emulator to produce and move jobs through the manufacturing system. In turn, the dispatcher receives relevant information regarding the schedule status from the monitor, and this information enables it to calculate whether or not the schedule is in control.

In controlling the flow of work through the emulated manufacturing system, the dispatcher reacts to information from the emulator and monitor ensuring that the schedule is implemented within the specified guidelines. Within the dispatcher, there are three main elements which together achieve the dispatching goals. The first of these is the static data module, which holds all of the relevant static manufacturing data to be used during the simulation run. The main information detailed here is the process data for all of the manufactured products and the schedule details which have been received from the scheduler.

The second part of the dispatcher is the dynamic data module, which contains information that is continually updated by the events of the emulator, which essentially means that the current status of the shop floor is held within this module. It is evident that an accurate picture of the current state of the shop floor is necessary for informed and accurate decision making. Typical data stored here includes current work in progress levels, machine status and contents of queues.

Finally, there is the dispatching rulesbase, which carries out the task of managing the work in progress so that the schedule is met. This rulesbase contains a set of rules designed to react to any predictable event within the manufacturing system, and make a decision which will result in the best possible course of action.

Within the PAC simulation model, the dispatcher cannot be tested in the absence of the emulator, as the emulator simulates the producer and mover activities. The emulator is based on the simulation of a Petri nets (discussed in Chapter 4), and overview of the simulation of Petri nets is now presented.

Petri Nets and Simulation Part of the process of simulation performed by the PAC simulation model (i.e. the emulator) is based on a simple Petri net model of the manufacturing system which simulates the flow of work throughout the manufacturing cycle. In this section we show how the idea for a shop floor emulator based on simple Petri nets evolved.

While Petri nets provide an excellent opportunity to detail the dynamics of a particular production system, their applicability may be further enhanced if the execution of a Petri net can be simulated using computer based methods. A direct relationship between the execution strategy of Petri nets and the rules based AI language OPS-5 was identified, and a software system ESPNET (Duggan and Browne, 1988a) was developed which takes as its input the logical structure of a Petri net, and subsequently develops a program which simulates the dynamics of the Petri net, yielding useful information such as resource utilizations and data flow through the model.

AI Based Petri Net Simulation OPS-5 (Brownstone *et al.*, 1985) is a rulebased language and is a member of a class of programming languages known as production systems. A production system is a program composed entirely of conditional statements called production rules. These productions operate on expressions stored in a global data base called working memory. The productions are similar to the if–then statements of conventional programming languages. The condition part of a production is called the Left Hand Side (LHS) and the action part called the Right Hand Side (RHS). A production which contains n conditions C_1 through to C_n and m actions A_1 to A_m means:

When working memory is such that C_1 through C_n are true simultaneously then actions A_1 to A_m are executed.

The production system interpreter executes a production system by performing a sequence of operations called the recognise–act cycle. The recognize function involves two steps, that of match and conflict resolution, while the act is a one step function.

1 **Step 1. Recognize**

- Match
  Evaluate the LHS of the production to determine which conditions are satisfied given the current contents of the working memory.

- Conflict Resolution
  Select one production with a satisfied LHS. If none of the productions have a satisfied LHS then halt the interpreter.

2 **Step 2. Act:** Perform the action specified in the RHS of the selected production.

3 **Step 3. Go to Step 1**

The firing of transitions in a Petri net is similar to the recognize–act cycle of OPS5. This was the main factor in deciding to simulate Petri nets using the rulebased language. The logic of a Petri net is captured by the production rules of OPS5. Each transition in a Petri net is modelled by a rule in which the LHS contains the conditions which are necessary for the transition to fire. When the rule fires, a token is removed from each of the input places and added to the token values of each of the output places.

The input of the data describes the basic logic of the Petri net to be modelled. It includes the number of transitions, their preconditions and postconditions and also the timing of the transitions with respect to each other. The OPS5 program is then written by means of a series of routines, and this simulates the execution of the Petri net. Typical output from a model generated using ESPNET gives a listing of all the transitions which occurs during the execution of the net, and provides information on the utilization of the various resources and the total throughput time for each of the jobs.

This work done on Petri nets and AI simulation provided the foundation to the design, development and implementation of the emulator.

The Emulator The emulator simulates the functions of the movers and producers within the PAC simulation model. The messages produced by the emulator are similar to those which would emanate from an active shop floor, and the existence of the emulator as a separate building block within the PAC simulation model enables the migration path from simulation to live implementation to be realized. Conventional simulation techniques do not have a clear distinction between the controller and the controlled within a particular manufacturing model, and hence these models cannot provide the same migration path as the approach presented here.

The emulator is based upon the use of Petri nets to develop models of manufacturing systems. The main drawback of this work was that while Petri nets are a very powerful tool for modelling work flow, they have drawbacks when they are used to model complex decision making activities. Trying to model the complex real-time decision making necessary to control movers and producers may not be appropriate for a single Petri net model, and as a result the Petri net based emulator performs simple activities based on instructions from the more complex dispatcher software module. This limitation of Petri

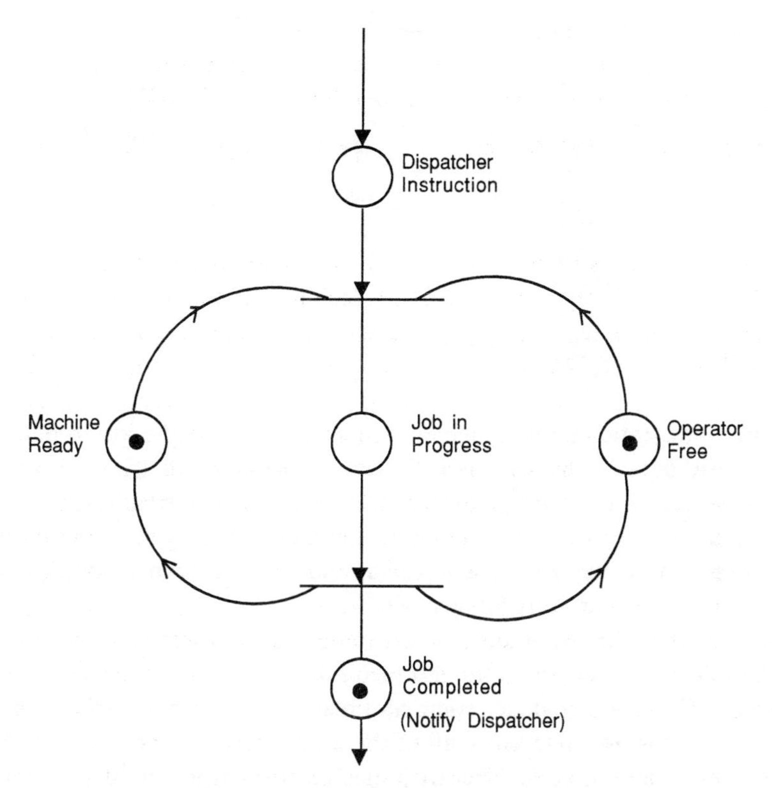

Fig. 9.7. Petri net model of a completed producer task.

nets was used to the advantage of the PAC simulation model, where the Petri nets developed were used to simply to simulate mover and producer events on the shop floor.

In order to understand how the emulator works with the dispatcher, consider the following simple illustration which shows how a simple Petri net model can be used to simulate the actions of the producer which is controlled by the dispatcher. Fig. 9.7 shows a Petri net model of a producer (machine 1) which has just completed an operation on a job. This is indicated by the presence of tokens on three of the places of the Petri net, that is, the places representing the states of machine idle, operator free and job completed. Notice that none of the transitions of the Petri net are enabled, in other words it cannot fire another transition unless a token is entered into the dispatcher command place. Because the Petri net has just modelled the completion of a task, this information is sent across the application network (the communications system) to the dispatcher.

Thus, the dispatcher receives the job completion message, and a chain of events commence, the end result of which is to instruct the producer (machine 1) to commence work on a subsequent job. The dispatcher must then access

Job	Batch Size	Priority	Decision
Job 1	25	12	*** Selected
Job 2	20	10	_______
Job 3	14	9	_______
Job 4	24	3	_______

Table 9.6. *The dispatcher selects the highest priority job*

the relevant static and dynamic data in order to select the job to be processed. The static data in this case is a list of job priorities that have been received from the scheduler, the dynamic data is a list of jobs which are contained in the input queue to machine 1. Table 9.6 shows that in this case the job with the highest priority (job 1) is selected, and this selection is then passed back to the emulator (i.e. start job 1 now).

Within the Petri net model of the emulator, this information (received via the application network) is transformed into an attribute carrying token, which is placed into the dispatcher command place. This is illustrated in Fig. 9.8, and the Petri net is now live, since all of the input places to the first transition (job starts) now have tokens. The attributes received from the dispatcher include the job number, the part type, job priority and the batch size. The emulator can now access the processing time of the part from the manufacturing database, and this time is then used to schedule the following transitions in the Petri net.

Figs. 9.9 and 9.10 show how the Petri net simulates the start and completion of Job 1 on Machine 1, and in doing so, starts off the cycle of dispatcher selection of a new job. When transition 1 fires, each of the tokens are removed from the three input places and a token appears in the job in progress place. The timed Petri net, using the time of Job 1 on Machine 1, schedules transition 2 to fire at the appropriate time. The firing of this transition removes the token from the job in progress place and returns one to each of the three original input places and the job completed place. The dispatcher is then notified of the completion of Job 1 on Machine 1.

Thus, the interaction of the dispatcher and emulator enable a simulation to take place. However, the usefulness of a simulation depends on the relevant information which can be extracted from thousands of events which may happen throughout the simulation run. One of the main functions of the monitor is to collate all of the data which it receives from the emulator, and organise it so that the modules of the PAC simulation model, and of course the user, use it to good effect.

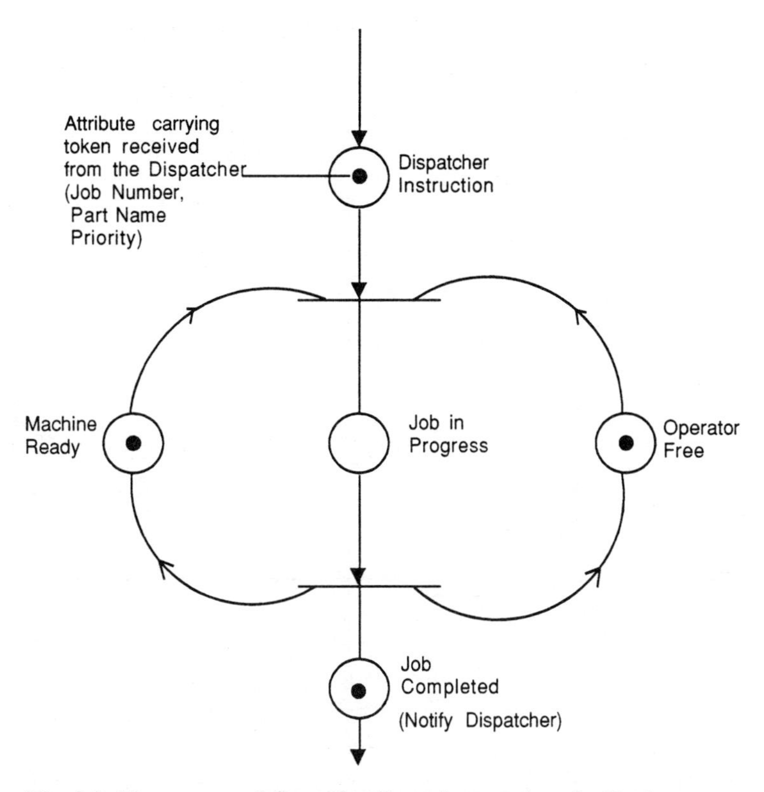

Fig. 9.8. The command from the dispatcher reaches the Petri net.

The Monitor It is clear that effective Production Activity Control depends on accurate, timely recording of data from the shop floor. The monitoring function is concerned mainly with activities which have already taken place. Data collected as a result of these activities are translated into information which may be the basis for real–time control or later management decisions. A necessary condition for monitoring is a data collection system which operates reliably, quickly and accurately. The monitoring function analyses the data and provides real–time feedback to the other functional elements or produces reports which may be used later.

Thus, the monitoring function can be seen as comprising of three different elements, namely:

- Data collection;
- data analysis;
- decision support.

These activities have been been described in detail earlier in Chapters 2 and 3. Essentially, the data collection system in the monitor collects all the relevant information from the shop floor and this data is then 'massaged'

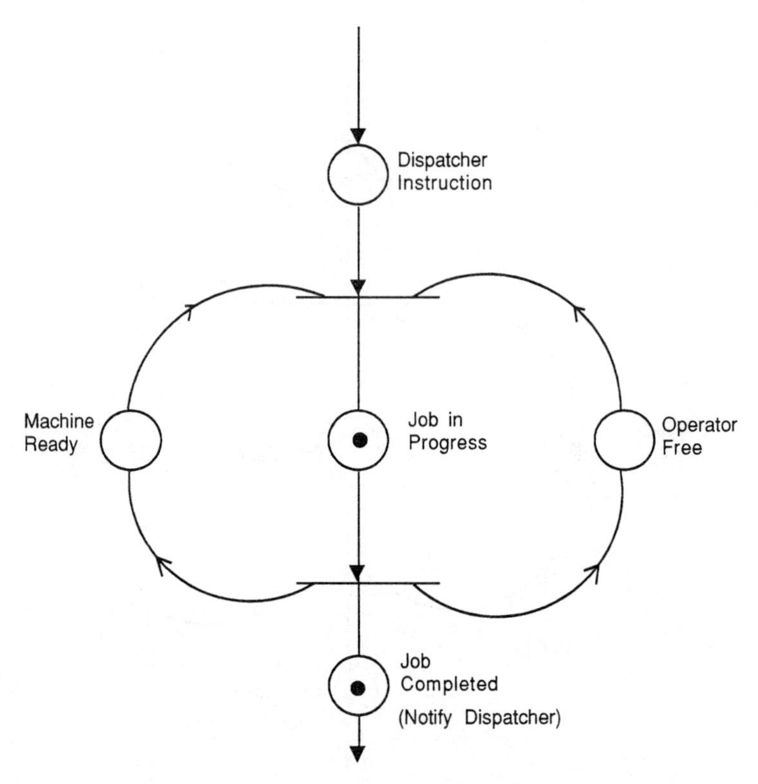

Fig. 9.9. Petri net fires transition 1.

(data analysis via different layers of software to produce both real–time and historical reports (decision support). Examples of real–time reporting include the generation of information on current utilisation levels, inventory levels etc. Historical reporting involves the production of graphs and reports on a variety of items of interest to manufacturing personnel. As well as acting as a reporting mechanism, the monitor also has a decision support capability. For example, if the level of materials on the shop floor falls below the desired level at a certain point in time, then the monitor signals this fact to the dispatcher software module. A further important feature of the monitor, is that is informs the scheduler when a new schedule is needed, and this occurs when it calculates that the schedule has gone out of control.

The monitor for the AG consists of a number of separate modules which combine to support decision making activities. The monitor receives notification of events from the emulator, each of these messages having a distinct identifying code. The relevant content of the messages is analysed, and data records holding useful monitoring information are updated. This process is performed by the following three modules:

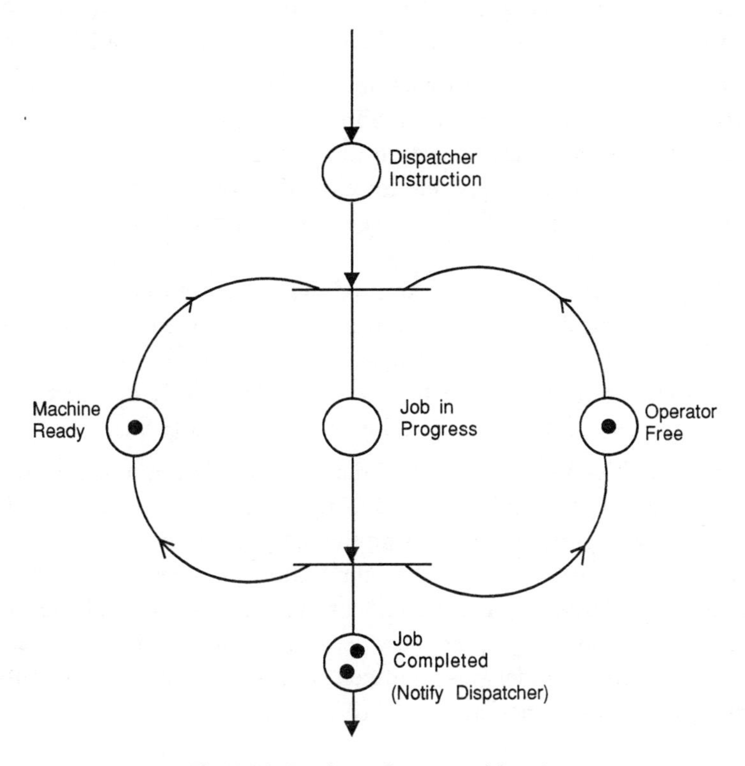

Fig. 9.10. Petri net fires transition 2

1 **Overall Monitor Program**
This is the program which is used to link all the other modules together. It is the controlling software for the monitor, and contains all the necessary routines to communicate with the Application Network. This routine ensures that incoming messages are routed to the appropriate data analysis routines, and that outgoing information is sent to its correct ultimate destination.

2 **Jobs, Workstations and Materials Monitor Modules**
These modules receive the messages from the overall monitor program and depending on the content of the message update all dynamic data in the system. They contain all the calculation and screen routines for the collection and real–time reporting of data. They also contain routines which store the data in the database at the end of a simulation run.

3 **Historical Reporting Module**
The historical reporting facilities provided by the monitor software are threefold. They include screen, graphical and hard copy reports. Items of interest here include data on late jobs left in the system after a

simulation run, the average time of a job in a resource queue, total time in manufacturing system etc. The options presently available are:

(a) Throughput time for each job in the system.
(b) Time in queue for each job in the system.
(c) Tardiness or slack for each job in the system.
(d) The number of reorders for each material in the system.
(e) The average level of a particular material at each location on the shop floor.
(f) The overall utilization level of each workstation.
(g) The times down, idle and time in set-up for each workstation.
(h) The exact state of the jobs that did not get through the system. This indicates where they were at the end of day and how much remaining work needs to be carried out on them.

Monitoring is an important part of the PAC simulation model as it provides information on system performance during the simulation, and also allows the user to identify important trends using the historical reporting facility. Because the monitor is separate to the other simulation modules, it can also be used within a real manufacturing environment. This adaptability of the monitor is highlighted later on in Chapter 11, where the monitor initially developed for the PAC simulation model was modified to link in with a live manufacturing system in a pilot implementation of Production Activity Control.

The Application Network The Application Network is a distributed software bus which allows the building blocks to communicate with each other in a controlled environment. The Application Network (AN), together with the Manufacturing Data Dictionary (MDD), provides a comfortable application environment for the building blocks of a PAC system. This includes features necessary to define the external interfaces of the building blocks in the PAC system, and to enable communication between the building blocks through the so defined interfaces. The AN communication facility includes intelligent data passing and data access mechanisms.

Each building block in such a system consists of one or more processes, which are called Application Network User Processes (ANUP). Within the PAC simulation model there are four ANUPs (scheduler, dispatcher, monitor and emulator). The MDD contains the definitions of the information entities that are handled at the interfaces between the ANUPs and the AN. These information entities are called Data Transfer Units (DTU). A typical DTU structure is the following:

1 DTU Name: New Schedule.
2 Description: A message which contains the latest schedule.
3 Source: Scheduler.

DTU Name	Source	Destination	Parameters
New Schedule	Scheduler	Dispatcher	Job, Part, Release Time, Due date, Priority

Table 9.7. *Scheduler to dispatcher information flow*

4 Destination: Dispatcher.

The AN contains the information about the software configuration of an actual PAC system. The only external interface the ANUP has to handle is the AN interface. The addressing information is stored in the AN and in the MDD. This allows the development of single ANUPs without needing to know the software configuration of the whole PAC system or the structure of every other ANUP in the PAC system. Another advantage is the possibility of exchanging parts of the PAC system without affecting the implementation of the concerned ANUPs.

Information Flow in the PAC Simulation Model To further understand how the internals of the PAC simulation model function, it is useful to view the different messages that flow to and from the four software modules through the AN. Each of these messages is defined within the MDD, so that the AN can match the DTU name with the relevant source and destination. Each DTU corresponds to a certain message in the PAC simulation system, and it is likely that in reality of a fully operational PAC system, the categories of messages would be similar. The following messages are outlined (this is by no means a complete message defintion for PAC):

1 Scheduler to dispatcher;
2 dispatcher to emulator;
3 monitor to scheduler;
4 monitor to dispatcher;
5 emulator to dispatcher;
6 emulator to monitor.

1 Scheduler to Dispatcher
The scheduler passes the daily schedule to the dispatcher so that it may be simulated passing through each of its operational steps, and the typical structure of the information passed is shown in Table 9.7.

2 Dispatcher to Emulator
Because the dispatcher is the real time controller of the PAC system, it sends control commands to the emulator. This control commands are to mover and producer devices whose actions are then simulated within the emulator. The move commands instruct the emulator to simulate the moving of work in progress to a specific location, while the

DTU Name	Source	Destination	Parameters
Move Command	Dispatcher	Emulator	Location, Destination Device, Job, Time
Select Job	Dispatcher	Emulator	Queue, Station Job, Time
Shutdown	Dispatcher	Emulator	Station
Maintenance	Dispatcher	Emulator	Station, Device

Table 9.8. *Dispatcher to emulator information flow*

DTU Name	Source	Destination	Parameters
Job Arrived	Emulator	Dispatcher	Location, Device, Job, Time
Job Completed	Emulator	Dispatcher	Location, Quantity Device, Job, Time
Breakdown	Emulator	Dispatcher	Station, Time Cause of Breakdown
Repair	Emulator	Dispatcher	Station, Time
Materials Arrival	Emulator	Dispatcher	Material Name, Time Location, Quantity

Table 9.9. *Emulator to dispatcher information flow*

produce commands orders the emulator to start work on a particular job (Table 9.8).

3 Emulator to Dispatcher
Since the shop floor emulator emulates all of the physical activities on the shop floor, it must inform the dispatcher of any relevant events which have an effect on the realisation of the production plan. Table 9.9 shows the messages which are passed to the dispatcher, and these include notification of job arrivals, machine breakdown, job completion and repairing of machines.

4 Monitor to Scheduler
Within the design tool, an important function of the monitor is to keep track of the schedule performance. We believe that the Monitor should simply highlight the fact that the schedule is drifting out of control. The

DTU Name	Source	Destination	Parameters
Request Reschedule	Monitor	Scheduler	Reason for request

Table 9.10. *Monitor to scheduler information flow*

DTU Name	Source	Destination	Parameters
Reorder Materials	Monitor	Dispatcher	Material Name, Location, Quantity.

Table 9.11. *Monitor to dispatcher information flow*

user should then determine if it is necessary to generate a new schedule. Table 9.10 illustrates the message sent to the scheduler.

5 Monitor to Dispatcher
Within the design tool, an important function of the monitor is to act as a decision support mechanism to the dispatching module. The monitor keeps track of the raw materials (e.g. components) as their levels are depleted during the simulation run, and when these levels fall below a certain predefined level (i.e. the reorder point, a message (Table 9.11) is sent to the dispatcher so that an order release for raw materials is issued.

6 Emulator to Monitor
Within the PAC simulation model, this information route is the one most widely used. These messages form the data capture activity of the monitor, and all data produced by the emulator is sent to the monitor so that it may be transformed into understandable information which gives a clear and concise picture of the shop floor activities. Table 9.12 shows the different messages which are passed.

The list of messages outlined in this section was developed for a particular manufacturing model within the electronics assembly industry. It is not a comprehensive list which would cover most discrete part manufacturing industries. However it does indicate some basic requirements for a messaging system which provide the basis of information flow within a PAC system, and because of the modular design of the simulation model, new messages may be added to the MDD to cater for more complex manufacturing systems.

9.6 Using the AG in the electronics industry

This section illustrates the potential application for the AG in an electronics assembly environment, namely, that of Digital Equipment Corporation, Clonmel, Ireland. This plant, described in detail in Chapter 10, factory produces

DTU Name	Source	Destination	Parameters
Started Processing	Emulator	Monitor	Job, Part, Station Time, Quantity.
Finished Processing	Emulator	Monitor	Job, Part, Station Time, Quantity.

Table 9.12. *Emulator to monitor information flow*

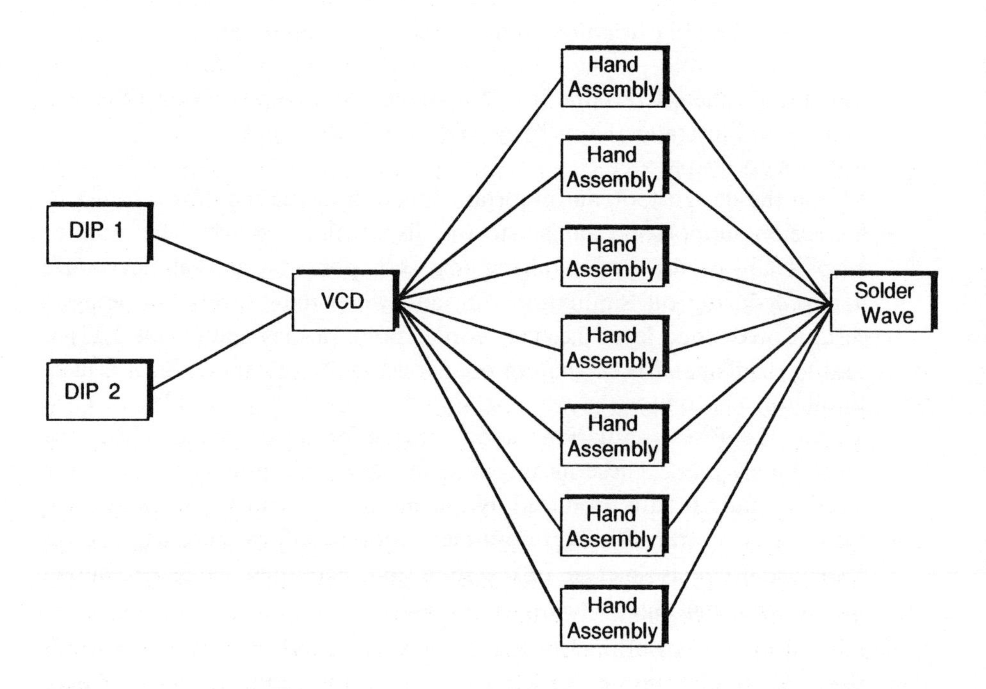

Fig. 9.11. Test area layout for the application generator.

power and communications modules which are subsequently used in the companies wide range of computers. One of the requirements of the COSIMA project was to develop a simulation test bed to test the AG concepts on the real manufacturing data of each of the project partners. Therefore, a section of the production process was selected (Fig. 9.11), the appropriate manufacturing data collected, and finally, the AG software was tested. The overall end result of this experiment was that the PAC architecture was validated in a controlled and simulated environment.

Because initially a limited product range was selected as it was not necessary to model all of the workstations in the factory. However later project work took a more global view of the factory, and a simulation model was then developed for all of the products (Duggan, 1989 and Bowden, 1989). Four different categories of workstations were modelled:

- **DIP (dual inline package)**
  Computer controlled insertion machines are supplied by many vendors for the insertion of DIP components. Equipment of this nature requires pattern programs which instruct the machine to an X and Y coordinate on the PCB and simultaneously instructs the machine to pick a component from an operator loaded component station on the machine. The pattern programs are interpreted for the machine by an operating system. The operation of the machine requires that the operator must load and maintain the material on the machine, load and unload boards from the machine and load the correct pattern programs. These patterns may be loaded using paper tape, floppy disk or via the use of a host computer.
- **VCD (variable centred distance)**
  Material must be prepared at a preceeding step on a machine called a sequencer in order to allow a mix of the components to be inserted in one pass of the VCD. The sequencer is set up with specific part numbers dedicated to an input station. The normal range varies from twenty to one hundred and twenty stations on a sequencer. Material is automatically picked from these input stations to form an output reel which is sequenced to the requirement of the VCD. The objective of running the sequencer is to minimize the set-up changes from product to product through proper machine scheduling and/or extending the number of input stations. Sequenced material can be inserted by the ability of the machine to adjust the board insertion centres within the control of the board pattern program. The pattern in this case governs the X,Y coordinates of the PCB as well as the centre distance of each component being inserted.
- **Hand assembly**
  This process involves the hand insertion of components which cannot be physically inserted by either VCD or DIP machines. Examples of such components are transistors, connectors, crystals etc. With the drive towards automation and the use of productivity analysis there will be a marked reduction in hand assembly operations over the next couple of years. A large percentage of the operation time in hand assembly is taken up with de–kitting and job set-up.
- **Solder wave**
  Solder wave is the process that is required to ensure reliable electrical contact between the circuits and components. It consists of a number of

> stages and is performed on one machine only. The board is placed on a conveyor which grips the board by its edges and carries it over a fluxing station. Flux is required to deoxidize the board and assist in the flow of solder over the joint to be soldered. As it continues its journey through the machine the board is heated to a specific temperature through the use of a series of pre-heaters under the conveyor. The purpose of this stage is two-fold. The first is to remove any excess moisture that might be trapped inside the board and the second is to dry the flux off the board. The board then passes over a bath of solder which is set up in such a way for a particular board type that it will deposit a precise amount of solder where required. As flux acts as a contaminant which can affect the electrical performance of the board, the board must be cleaned of flux deposits. Typically a water soluble flux is used which can be removed by water and detergent.

Having selected the test area, the process of developing a model for the AG commenced. The first logical step was to collect all manufacturing data needed to perform the tasks of scheduling, dispatching and simulating the manufacturing system. This data included details on the physical description, a list of raw materials used, the product and process data and finally, the order requirements. All of this information was then entered into the manufacturing database through the manufacturing database interface.

The manufacturing database interface is an essential feature of the AG and it allows the user to populate the database with the appropriate manufacturing data. This interface consists of a series of menu driven applications which allow the user to incrementally build up a data model of the manufacturing system through the systematic addition, deletion and modification of data relations. There is an individual menu for each of the relations in the database, and these may be accessed with ease by the user. The sample screen shown in the diagram illustrates the type of manufacturing data which describes a typical process step for the Digital Clonmel model. When an accurate picture of the manufacturing system has been developed, the rulesbase may then be used as the initial step for selecting the most appropriate PAC building blocks.

The rulesbase uses the manufacturing data specified through the user interface to perform analysis on the potential effects of the customer requirements through the system. For the particular Clonmel model, this involves analysing the effect of the customer requirements on each of the eleven workstations which perform the four categories of process operations. The first check is to assess whether or not enough raw materials are available to meet the perform each of the operations. The procedure for carrying out this check is to calculate the raw material needs for the day and compare this figure to the total number of raw materials on hand (specified in the raw materials data relation of the manufacturing database). Any shortage in raw materials will result in a modification of the customer requirements to meet the unforeseen circumstances.

The next step for the rulesbase is to calculate utilization figures for each of the workstations based on the customer requirements and process details (set-up times, process times etc.). Sample output from this analysis of the Clonmel model is shown in Fig. 9.12. This information is then collated with two other measures to form an index on the likelihood of a station to be a bottleneck during the duration of the day. The other two measures are its position in the process and the number of different products that pass through the station, and as a result of this analysis a list of possible bottlenecks is developed. Following this, an appropriate schedule is generated.

The scheduler may select from a number of different algorithms and heuristics stored in the scheduling library. Whichever method is used, the results are always of the same format, that is, the recommended start and finish times of each job on the eleven individual workstations. The dispatching function then takes the schedule and proceeds to implement it by issuing instructions to the emulator to move and produce each of the customer orders at the planned time intervals. The emulator and dispatcher run as separate processes on the operating system, and the messages are passed using the application network. The messages which emanate from the emulator then form the basis for the analysis and decision support functions of the monitor.

The monitor makes sense of all of the PAC simulation model events by transforming the many messages from the emulator into usable information which describes the current status of the shop floor. This information is divided into categories describing raw materials, workstations and work in progress. The monitor also has a historical reporting facility which allows the user to reflect on the results of different simulations, and trends may be established on the basis of these results.

Thus, the feasibility of the AG has been tested and PAC architecture has been validated within a specific simulated manufacturing environment. As the PAC life cycle diagram showed at the start of this chapter, the AG accounts for the first two stages in the development of implementable PAC systems. The next logical step in this cycle is to decouple the emulator and plug in the shop floor. In theory this seems a straightforward proposition, but in reality it is a complicated task from a technical and social point of view. These issues will be discussed in the concluding chapters of this book.

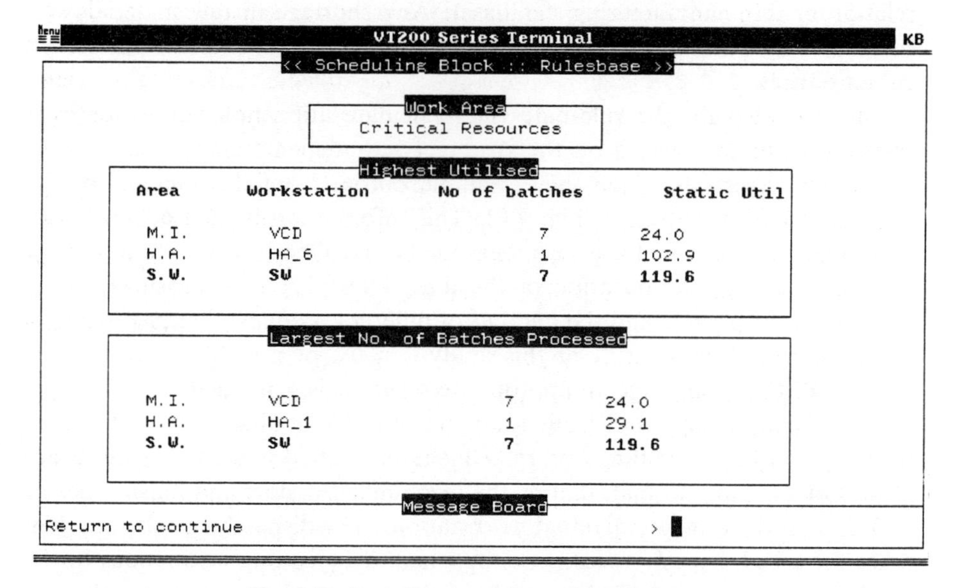

Fig. 9.12. Sample output from the rulesbase.

VT200 Series Terminal KB

<< TEST BED SIMULATOR ::: MONITOR ANUP >>

Clock Time: 152 — Time of Day: 8 : 45

<< JOB PROGRESS ENQUIRY >>

Job Information	Part Number	Number Completed	Time in System	Time in Queue	Lateness	Slack/ Time left
3 WIP-DIP2	H7186	0	15.4	14.4	-944.6	2.1
5 Not rel.	H7061	0	0.0	0.0	-2384.6	0.0
8 Not rel.	H7250	0	0.0	0.0	-3824.6	0.0
2 WIP-DIP3	DHU11	0	15.4	15.1	-1424.8	7.7
4 Not rel.	DF124	0	0.0	0.0	-1904.8	0.0
7 Not rel.	DMB32	0	0.0	0.0	-2384.8	0.0
6 WIP-VCD	H7253	0	15.2	0.2	-1904.8	3.3

Messages

<< Type r/R to return >>>

Fig. 9.13. Monitor screen during the simulation run.

9.7 Conclusions

In this chapter we have been concerned mainly with the development of a design tool to support the creation of PAC systems and the control elements of

FC systems. In effect, this implies the development of schedulers, dispatchers and monitors. Of the three building blocks, the tasks of the scheduler is the most difficult, and earlier in Chapter 6, we identified five essential elements of our overall approach to the development of viable scheduling solutions in manufacturing, namely:

- We identified the need for a well defined control architecture as an essential prerequisite to any implementable scheduling solution. This architecture should be incoporate both flexibility and decentralization, and decision making should be made at the lowest possible level of the hierarchy. We believe the architecture for FC and PAC can ensure both flexibility and decentralization of shop floor control related activities.
- We recognized the important role of information technology in the implemention such a scheduling solution.
- We recommended that a sound strategy is required to develop detailed schedules for any particular shop floor. No one scheduling heuristic can be viewed as the most appropriate in all cases, Hence, we recognize the need to construct, what we term a scheduling library, which contains a wide range of possible scheduling solutions. This can enable the selection of the most appropriate scheduling strategy to developed a plan for the shop floor. This important concept has been illustrated in this chapter.
- We make an important distinction between finite and infinite scheduling approaches. As stated previously, finite scheduling should be used at the PAC level of the production management hierarchy. For FC, finite scheduling is the preferable option. However, this may not be feasible if the level of manufacturing detail is too high, and infinite scheduling may be necessary.
- Finally, we identified that in most scenarios, scheduling occurs in a human environment, and thus the scheduling system should take full cognisance of the important role that humans have to play in intelligent decision making. We believe that scheduling systems must be designed to achieve a worthwhile balance between available scheduling heuristics and the experience of humans. Thus, the interactive scheduler, proposed in this chapter, allows humans to use their experience to enhance possible scheduling solutions.

We view our design tool for shop floor control systems as an important mechanism which can be used to support the implementation of FC and PAC. In fact, the design tool has been used to directly support the implementation of PAC in a modern electronics assembly facility, The remaining chapters of this book deal this pilot implementation of PAC.

Part Six
AN IMPLEMENTATION OF A PAC SYSTEM

Overview

The interface between architect and engineer is often difficult. Conflict between aesthetic design and the rigours of the built environment must be resolved in order to realize a durable, functional, and yet elegant, structure. Similarly, in the development of manufacturing systems, the emergence of implementable control mechanisms from a global architectural model poses a formidable challenge. In earlier sections, we proposed a functional and information architecture for operational control in the modern production management environment of discrete parts manufacture. In Chapter 8, we reviewed the social and technical organizations within which operational control must function. We also described a design tool which can be used to develop and evaluate site-specific operational control systems in accordance with our proposed architecture.

In this Part, we will use a case study to illustrate how the principles of PAC and FC can be applied in practice to shop floor planning and control. Here we will focus on the implementation of such systems, based on the architectures described in Part Two. This is the final stage of the PAC life cycle, and charts the progress from architecture, through design and simulation, to final implementation in the manufacturing environment.

The case study is drawn from the electronics industry and is based on our experience of production management at the computer peripherals manufacturing facility of Digital Equipment Corporation, in Clonmel Ireland. Digital Equipment Corporation is the world's second largest producer of computer systems and is a leader in the field of distributed systems and networking. The Clonmel plant is one of eight Digital Equipment Corporation manufacturing facilities in Europe and 35 worldwide. Many of our insights on the practical application of PAC systems have been gained during an implementation of PAC at this plant, in 1989.

We have argued, in Chapter 8, that the success of any operational control method relies on the design and organization of the implementation environment. We will thus begin our case study by examining the influences of the current manufacturing environment, at the study site. While many of these organizational and technical features are specific to the Clonmel plant, we believe the environmental and implementation issues highlighted are common to most discrete parts manufacturing industries.

The part is divided into two chapters. Chapter 10 examines the implementation environment of the Clonmel plant in relation to the contextual, transactional and sociotechnical viewpoints, discussed in Chapter 8. This not only provides a framework for positioning the operational control architecture, but also emphasizes the interaction of technical, social and information technology (IT) sub-systems in implementing effective shop floor control methods. In Chapter 11, details of a PAC implementation at the Clonmel plant are presented. Here, the use of our application generator, manufacturing database and other functional blocks are illustrated in a real manufacturing environment.

Finally, the lessons learned from our implementation are examined, with a view to providing guidelines for future Factory Coordination and Production Activity Control implementations.

10

The environment of the case study

10.1 Introduction

During the past decade manufacturing industry has been confronted with a host of new sophisticated tools (and indeed a new vocabulary – CAD, CAM, CIM, CIB, DFM, etc.) which purport to enhance business competitiveness. By and large, investments in such technology have been viewed with caution by manufacturing managers. This has been due both to the financial commitments required and to the often intangible nature of expected benefits for the manufacturing business. However, there is little doubt that, where planned investments in manufacturing management systems are made, a substantial competitive advantage can be gained. This is certainly evident in our case study of the Digital Equipment Corporation plant located in Clonmel in the Republic of Ireland. But, even if the value of such investments can be determined in advance, there remains the challenge of implementing systems which are appropriate for the particular plant, given its business, level of automation, process design, personnel and skill levels, and commitment to information technology (IT).

We do not believe that IT and systems solutions alone can improve manufacturing performance. Indeed, automation of poor business practices or processes may give rise to nothing more than automated bad management. For instance, the varying degrees of success of MRP II implementations across a range of industries, bears evidence to the unrealistic expectations of managers towards computerized production management systems. This illustrates that MRP II is neither a panacea for manufacturing management, or an off-the-shelf implementation package. Such systems typically require customization to meet the specific needs of the host plant, and demand a commitment from managers towards their continued support and development, once installed.

There is a need, then, for a clear understanding of the implementation or target environment for manufacturing control systems. This understanding must view the proposed systems solution within the context of a production management architecture and strategy for the manufacturing site. It must also extend to the assessment of existing manufacturing control practices, and set about altering these where necessary.

It is our belief that the success of any manufacturing control system relies, almost entirely, on the manner in which the system is implemented. The opportunity to implement innovative shop floor control systems in new green-field facilities will occur only infrequently. For the most part, FC and PAC systems will be implemented in well-established facilities, where existing work practices, processes, personnel, technologies, planning and control techniques already combine to provide, at least some degree of, effective manufacturing management. This case study is based on one such site.

In this chapter we explore the environmental context for PAC at the Digital Equipment Corporation plant in Clonmel. It begins by reviewing the business pressures which have brought about a focus on CIM-based manufacturing. We then study the current approach towards production management; this can best be characterized as a hybrid approach, encompassing MRP II planning allied with JIT thinking and design principles. Organization of the technical and social sub-systems, which make-up the manufacturing process, is next discussed. Finally, we describe the underlying information environment, within which production planning and control systems must function.

10.2 The business environment

The competitive business environment can be viewed as the context in which modern manufacturing takes place (i.e. the contextual environment, as discussed in Chapter 8). For most discrete parts industries today, excellence in manufacturing is a prerequisite for business success. The computer and electronics industry is a classic example of the new manufacturing environment. As this industry grows to maturity and profit margins tighten, most of the larger producers have begun to focus on manufacturing performance as the key to sustained profitability. Thus, supported by their core information technologies, the computer companies have been to the forefront in the development of automated manufacturing systems.

The Digital Equipment Corporation plant in Clonmel is a development and test site for certain CIM technologies, including manufacturing control systems. However, its principal function is discrete parts manufacturing. The business environment for the plant includes both the final marketplace for its products and the Digital Equipment Corporation itself, which is also an important customer of the plant. The products manufactured at Clonmel include:

1 Network and Communication (NAC) products;
2 communication back-plane modules;
3 power supply units.

The network and communication product set covers a wide range of computer peripherals, which enable the transmission and distribution of information over Local- and Wide-Area Networks (LAN and WAN). These products are marketed either individually, through the Digital Equipment Corporation's sales subsidiaries, or incorporated in large systems orders (i.e. computer systems with supporting network and other peripherals), delivered from the computer systems manufacturing plants. The remaining products (i.e. both back-plane modules and power units) are transferred to computer systems plants for final assembly into their parent computers. Thus, the plant's principal customers are other Digital Equipment Corporations manufacturing plants. However, this does not insulate the plant from the vagaries and pressures of the final marketplace.

Digital Equipment Corporation began manufacturing at Clonmel in 1978, as a computer assembly and test site. The current, new facility was planned as a volume (that is, aimed at producing a high volume of products at low cost) manufacturing plant in the late 1970s, during a boom period for the computer industry. However, by the time this plant opened in 1983, the industry had entered a worldwide recession, and the growth prospects for volume manufacturing sites were slim. The plant management responded to this difficult external environment by:

1 Focusing on the development of manufacturing competitiveness and excellence.
2 Extending the plant's charter beyond its planned role of volume manufacturing.

These initiatives, supported by substantial investments in the development of a CIM-based approach to production management, have been rewarded by the rapid growth of the plant's business and profitability.

Today, the marketplace pressures upon the Clonmel plant's products include: price competition, product quality, demand for enhanced capabilities and functions, and timeliness. As a corporation dedicated to engineering development and continuous enhancement of its entire product range, Digital Equipment Corporation is always faced with the encroachment of low-cost competition on its market share. In marketing products (commonly referred to as clones) which emulate Digital Equipment Corporation's designs, such companies do not have to underwrite the costs of large R&D and engineering efforts, and thus compete on the basis of product price alone. The challenge, then, for the Clonmel plant (and for Digital Equipment Corporation as a whole) is to ensure that its product range, quality and functionality are significantly better than those of its competitors, thus offering greater price-performance. As competitors become more sophisticated, however, the ability to keep ahead of the marketplace relies more and more on manufacturing performance, in terms of cost, responsiveness, flexibility and reliability.

Within manufacturing, the specific challenges of marketplace competition are to:

1 Reduce the cost of manufacture;
2 guarantee very high product quality;
3 ensure process flexibility;
4 guarantee high process yields;
5 reduce manufacturing cycle-time;
6 reduce the time taken to get products to the marketplace in adequate volume which is commonly termed 'Time-to-Volume'.

In the years since the present Clonmel plant was opened in 1983, the business environment has become significantly more competitive. Product life cycles are continuing to contract. In 1983, the typical life cycle for Clonmel's products was five years. This has shortened to just two years, on average. Product diversity has also escalated, from just a small range in 1983 to over 60 active products today, some of which are manufactured externally. The rate at which new products are brought to market has also increased dramatically, and today the plant introduces (and retires) in the region of 15 new products each year.

This environment has meant that the manufacturing plant had to become competent in managing the entire life cycle for its products, from product and process design through materials sourcing, shop floor production, inventory control, and delivery to meet scheduled orders. It also needed to develop the expertise and systems that would allow it to manage the introduction of new products without disrupting the main revenue earning activity of steady-state manufacturing. Indeed, the concept of steady-state manufacturing is now becoming obsolete at the Clonmel plant, as the product set is continually changing in response to market demands. The pressures imposed by this environment on manufacturing operations are substantial, and they create a demand for effective and versatile manufacturing control systems.

Success of the Clonmel manufacturing business has depended greatly on the integration of the product design, process development and manufacturing operations. Development of closer relationships with material suppliers and customers, together with advances in business planning, have also been central to the business competitive strategy. Today, the plant views the business environment in terms of an integrated model (Fig. 10.1), supported by an information technology, that enables business integration.

We have identified the competitive business environment for the Clonmel

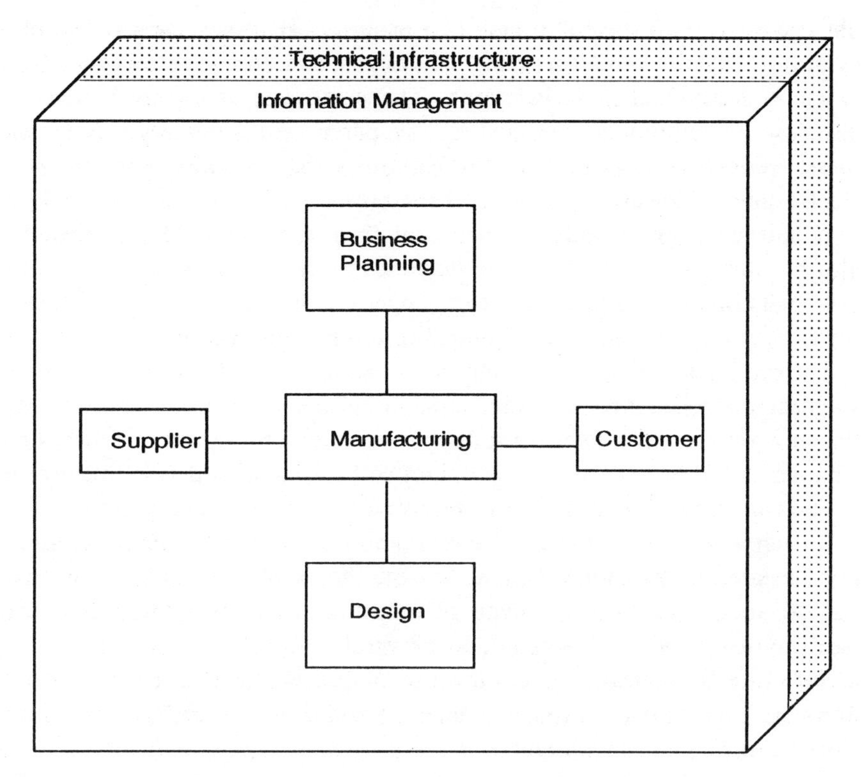

Fig. 10.1. Integrated manufacturing business model.

plant and, more specifically, the pressures imposed on manufacturing operations to support the business strategy. The plant's approach to production management will now be discussed.

10.3 The production management system environment

As a discrete parts industry, the Clonmel plant can be characterized as a batch production, make-to-stock environment. (However, this does not reflect the type of manufacturing process, which can be described as Continuous Flow Manufacturing (CFM). The process will be discussed in detail in the next section. The volume and diversity of products manufactured at Clonmel, means that the process is organized for medium production runs, with reasonably frequent set-ups being required for different products. Business planning relies almost totally on forecasted demand.

As the plant's customers are either other Digital Equipment Corporation manufacturing plants or sales subsidiaries, the Clonmel business has little direct contact with the end-users of its products. However, most of the plant's current products are built to standard designs, and do not entail configuration to the end-customer's specifications. The unit value of Clonmel products is also low, in relation to the cost of the parent computer systems to which these products are connected. Production is thus conducted in a make-to-stock context. However, pressures of the business environment are demanding that finished goods inventories are kept to a minimum. The product set is also becoming more complex, and demands more detailed specification of end customer configuration requirements. Together, these factors impose a demand for greater responsiveness and control in production operations.

The production management environment at the plant encompasses a philosophical approach and a formal systems element. We can now recognize this environment as a hybrid production management system, as discussed in Chapter 1. But, when the Clonmel plant embarked on its development of production management, it did so in response to urgent business pressures. The approach which evolved over time was (and remains) one that was considered appropriate for the Clonmel plant. As the theory of production management has advanced, we can now avail of a formalized hybrid PMS architecture (see Chapter 1) which accommodates several different PMS approaches. This architecture is applicable across a range of discrete parts industries, and also describes the production management environment at the Digital Equipment Corporation's plant in Clonmel.

The philosophy at Clonmel The underlying business philosophy at the Clonmel plant is one of continual improvement across the full spectrum of the plant's activities. This broad philosophy can be reduced to a more pragmatic and tangible strategy in the production management area: Simplify Integrate Automate. While this strategy appears to be axiomatic, it is the essence of the plant's success in production management. This approach is founded on JIT principles for process and product design, and is the hallmark of the Clonmel plant's process, in particular. (We will discuss process organization further in section 10.4).

The plant management believe that simplification of business and production processes is the key to successful manufacturing. However, simplification alone can not guarantee business competitiveness. The integration of all business process, in terms of people, information and technology, eliminates delays and unnecessary re-work in all business areas. Finally, the automation of integrated business and manufacturing processes provides a fast and resource-effective production management environment. The idealized model of the plant's business, presented above in Fig 10.1, is itself an expression of this philosophy.

What place then do sophisticated production management (or manufactur-

Plant level benefits	Corporate level benefits
Inventory reduction	Improved Time-to-Market and Time-to-Volume
Overhead reduction	Improved customer satisfaction
Reduced manufacturing lead time	Improved business and manufacturing control
Improved quality	Improved business and manufacturing predictability
	Improved flexibility
	Business integration

Table 10.1. *Benefits of CIM based production management*

ing control) systems have in such manufacturing environments? Indeed, this points to the heart of the argument presented in this text – that attention to design of the target manufacturing environment is essential to the successful implementation of PAC and FC systems. The inclusion of the production environment design component of FC is clear recognition of the dependency between manufacturing control systems and manufacturing process design (see Chapters 2 and 3).

Production Management Systems From a business systems perspective, the Clonmel plant is categorized as a MRP II Class A site. This indicates that the plant has been recognized (with respect to the APICS standard) for its achievement of a very high level in MRP II implementation. The initial implementation of MRP II at Clonmel, in 1983/84, was aimed at reducing the plant's inventory carrying cost, through better planning and information integration. Today, the plant manages over 3000 active parts and 250+ suppliers; so, the materials management aspects of MRP II are more valuable than ever. But the plant has expanded its implementation, from just materials planning (MRP), to the full scope of manufacturing resource planning and management (MRP II). Today the plant's MRP II system is the central business and production planning mechanism used at Clonmel.

Together, the underlying philosophy and planning systems have already resulted in significant business improvements for the Clonmel plant. These benefits have accrued both directly within the plant and in the wider corporate arena, as shown in Table 10.1.

Success of MRP II The plant's success with MRP II is itself a lesson in effective implementation of manufacturing systems. However, success did not occur immediately, nor without difficulties in the initial learning period. As in many other MRP II implementations, the system was primarily viewed as a tool which could aid materials planners in forecasting and managing inventory levels. Little improvement in materials management was achieved in the first year of MRP II implementation. These planners had little confidence in the new system and frequently over-ruled its projections, based on their own experience. The level of understanding among the MRP II system users was also poor at first.

An even more fundamental reason for its poor performance was that the system itself could not reduce inventory buffer stocks, as often believed. In fact, automation of inventory management merely concealed and computerized the management of inflated inventory buffers. Only when the plant environment was examined and redesigned, were these buffers identified and eliminated. In time, the coordinated development of the manufacturing process and the entire business planning cycle resulted in success for the MRP II implementation.

We highlight the following lessons from the plant's MRP II experience, as these have a direct bearing on the implementation of other manufacturing systems, including shop floor planning and control applications:

1. Attention must be given to design of the target environment, and this must be developed or altered, where necessary;
2. The system users must be involved at all stages of the implementation to impart a feeling of ownership and to guarantee support for the implementation;
3. Formal training should be provided to ensure user confidence;
4. Accuracy and integrity of the system database must be ensured to maintain user confidence;
5. Performance of the installed system should be continually monitored to maintain its effectiveness.

We will return to such practical wisdom in our discussion of the pilot implementation of PAC in Chapter 11.

The MRP II planning environment The MRP II implementation spans the entire planning cycle at the Clonmel plant, from business planning down to operational production planning and materials requirements planning, (Fig. 10.2). This can be viewed as the tactical level of the hybrid PMS architecture. It is concerned with planning alone, and not the execution of production plans at the operational level. It includes the development of a master schedule from the latest business forecast and firm orders, planning the plant's capacity and materials requirements, and development of a weekly production plan (i.e. the 'build plan'). Daily production schedules by product are currently developed manually.

The system operates on an 18-month planning horizon, with daily to quarter-yearly resolution, depending on the integrity of available forecast information.

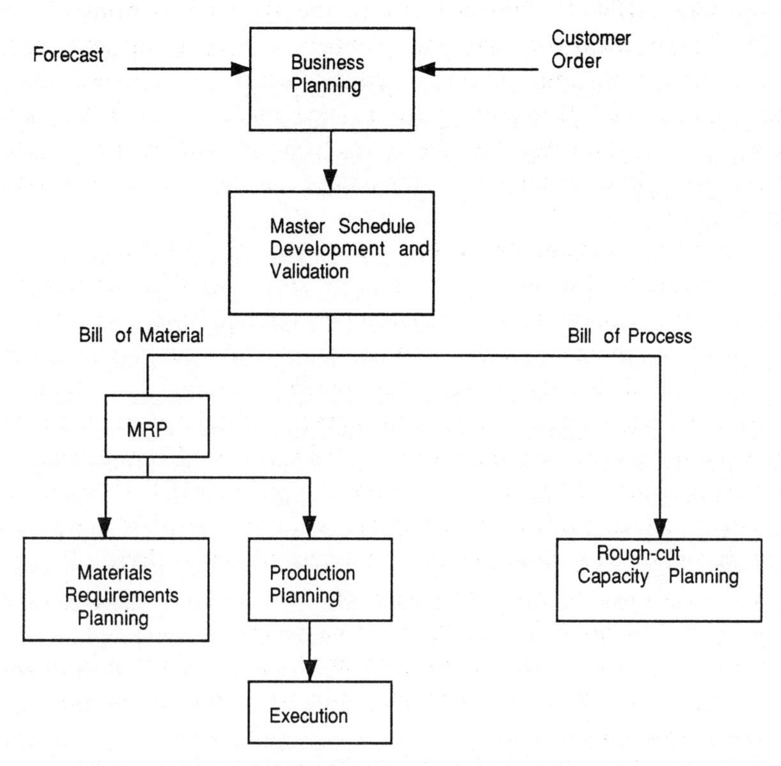

Fig. 10.2. The Planning cycle at the Digital Equipment Corporation's plant in Clonmel.

The Master Schedule is reviewed and approved on a monthly basis for most parts; but this is re-visited more frequently for parts which exhibit significant changes in forecast volumes over short periods. All products manufactured at the plant, including design prototypes, are included in the master schedule, so no informally planned production takes place. The underlying production planning cycle is one week, but the system issues daily build-plans, where week-to-week product volumes are reasonably steady.

MRP II implementation technology The plant's MRP II implementation is based on the commercial Production and Inventory Optimization Software (PIOS) package, which consists of the following eight modules:

1 Part master
2 Bill of materials
3 Inventory control
4 Purchasing/receiving
5 Costing
6 Order fulfilment

7 Master schedule
8 Shop floor control.

This system has been enhanced to meet the plant's requirements and is also supported by additional applications, developed by the plant's Management Information Systems (MIS) department, particularly in the business forecasting area. PIOS and its supporting applications are written in the traditional third generation COBOL language, and is run at Clonmel on a VAXcluster computer environment under the Digital Equipment Corporation proprietary operating system, VMS. The PIOS database is the repository for most of the plant's business data, from part and Bill of Materials (BOM) information to the master production schedule. Any new manufacturing systems must, thus, interact with PIOS. This requirement was addressed during our implementation of PAC at the plant, and is discussed further in Chapter 11.

With the exception of shop floor control, all of the PIOS modules are implemented at the plant. The fact that shop floor control has not been applied is a reflection both on the short-comings of MRP II as a production control paradigm and on the plant's philosophy of production management. (This comment is not specifically addressed to PIOS, but to the general theory and application of MRP II.) The plant management clearly recognize the value of MRP II as a planning tool, and have developed an environment where its application is effective and central to business and production management. However, its use for shop floor control is not appropriate, in an environment where continual improvement, simplicity and JIT philosophies are applied. Among its key failings in this area are (Browne *et al.*, 1988):

- Planned lead times are not conducive to continual improvement;
- too cumbersome and complex for real-time control;
- JIT-based inventory consumption not adequately supported;
- interfaces to shop floor systems are poor;
- database inaccessability during planning mode;
- emphasis on work in progress tracking inappropriate in JIT environment.

Having examined the production management (or 'transactional') environment for the Clonmel plant, we will now discuss the operational environment, for which the PAC based systems are intended.

10.4 The technical and social sub-systems

The Digital Equipment Corporation's Clonmel plant is a modern production facility covering 15,000 square metres, and houses the manufacturing process, offices, laboratories, design centre and other functions of the business. The plant employs about 350 people, of which 100 are directly engaged in the manufacturing process. As we have seen above (section 10.2), the plant manufactures in the region of 60 different products and introduces over 15 new products each year. The total weekly volume of products manufactured here is in excess of 3000.

Since the plant opened in 1983, a major re-structuring of the manufacturing operation has taken place, in order to meet the competitive demands of the marketplace, as highlighted in section 10.2. Today, production is organized as a Continuous Flow Manufacturing (CFM) system, embracing JIT thinking and process design principles. However, the Kanban card system is not used as a control mechanism. As the process is a non-repetitive one, and certainly not steady-state, application of the Kanban approach is inappropriate here. Organization of the process, in terms of people and equipment, has broadly followed the sociotechnical method, as discussed in Chapter 8. It also represents a successful application of the production environment design element of factory coordination, in simplifying the process for better operational management and control.

The Technical Sub-system The Clonmel process may be described as semi-automated. While the process is centred on computer controlled machines, most of the materials handling, equipment set-up, inspection and test relies on operator intervention.

The Process The principal stages of the process are fairly typical of those found in most modern Printed Circuit Board (PCB) assembly plants in the electronics industry. They include:

1 Surface mount: In the surface mount manufacturing cell, microprocessor and memory chips are attached to the PCBs, by surface adhesion techniques. This method of PCB assembly allows for the population of boards at a very high density, thus reducing the size of the end-product. This is a relatively new technique in PCB assembly. But already, a high percentage of the products manufactured at Clonmel use this process.

2 Machine insertion: In the Machine Insertion area, components are inserted through pre-drilled holes on the PCBs by two distinct types of computer controlled machine:

- Variable Centred Distance (VCD) machines, which insert axial components (such as resistors) at various X–Y coordinates, with varying component leg centre distances. Components are drawn from master reels which have been previously sequenced (by a multi-station sequencer machine) to minimize set-ups in the VCD operation.
- Dual In-line Package (DIP) machines are used to insert integrated circuits (of fixed leg centre distance) at various X–Y coordinates.

Components are drawn from operator loaded cartridges on the DIP machine.

The machine insertion area at the Clonmel plant consists of three DIP and two VCD machines.

3 Solder wave: All components placed in the machine insertion area are soldered to guarantee reliable electrical contact. The wave solder process involves: fluxing, heating, solder bath, post-solder washing and drying. When the PCBs are loaded onto a feed conveyor, the process continues automatically, except for routine in-line inspection.

4 Hand assembly and Final assembly: Hand assembly operators attach Light Emitting Diodes (LEDs), connectors, transformers, and other components, which do not lend themselves to machine placement or insertion, to the PCBs. Final assembly of products is also performed manually.

5 Unit assembly and final test: Unit assembly operators attach the PCBs from final assembly with other components to make up units or modules. These units and modules undergo a final test, which includes functional test, voltage and temperature cycling. There is a final inspection before the products are packaged for shipment.

Process Organization When the Clonmel plant opened in 1983, the production operation was organized as a series of discrete functions, and products moved in steps through the process. This was a multi-level process, with multi-level Bill of Materials (BOMs), and assemblies were cycled between the stockroom and production until products were finally completed. Material was pushed through the process and work orders were associated with each job lot, requiring a considerable administrative overhead in paperwork. Upon analysis, the plant management exposed highly inflated work in progress and material buffers throughout the process. Furthermore, the complexity of the process meant that, of the total time taken to manufacture products, only five per cent was spent in value adding assembly operations. Quality levels, material availability and management of the work order process were also poor.

Today, production is organized as an efficient Continuous Flow Manufacturing (CFM) operation. The CFM approach is aimed at 'flowing' material through the process on an 'as needed' basis. It is focused on simplification of the process organization in line with JIT principles. Key elements in the development of CFM at the Clonmel facility have been:

- Re-organization of the process around product groups;
- elimination of process steps and application of single-level bill of materials;
- point-of-use materials delivery and consumption;
- re-organization of operations personnel into product based workgroups.

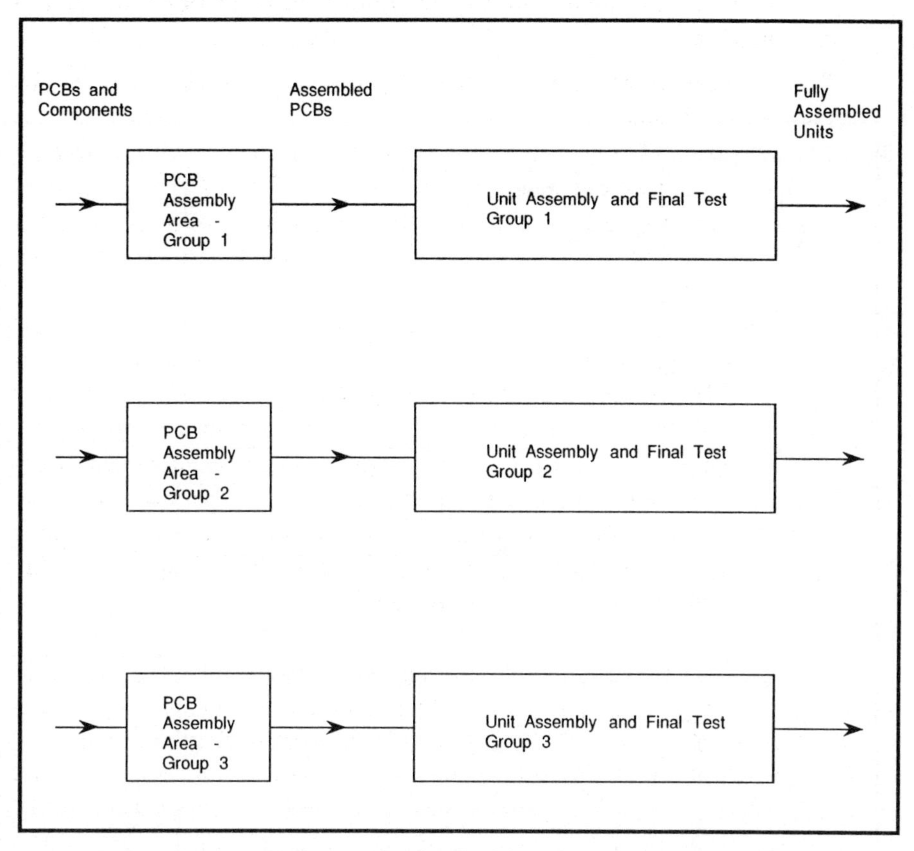

Fig. 10.3. An Ideal layout under continuous flow manufacturing.

Initial re-design of the process relied on Group Technology principles to isolate three distinct product categories, on the basis of process operations required by discrete products. This immediately resulted in simplification of the process, and today there are less than half the original number of process steps. However, management's aim is to reduce this even further to a single step process. The Bill of Materials (BOM) structure was next reduced to a single level, which requires only a single MRP II transaction to open or close jobs in the process. Delivery of material to its point-of-use accompanied the development of a JIT relationship with suppliers. This has resulted in the elimination of kitting (that is, the preparation of 'kits' of components and sub-assemblies for issue to the shop floor) and de-kitting operations and phasing-out of the materials and work in progress stockrooms.

Ideally under CFM, the shop floor layout in the Clonmel plant would consist of an unit assembly area for each of the three product groups identified in the

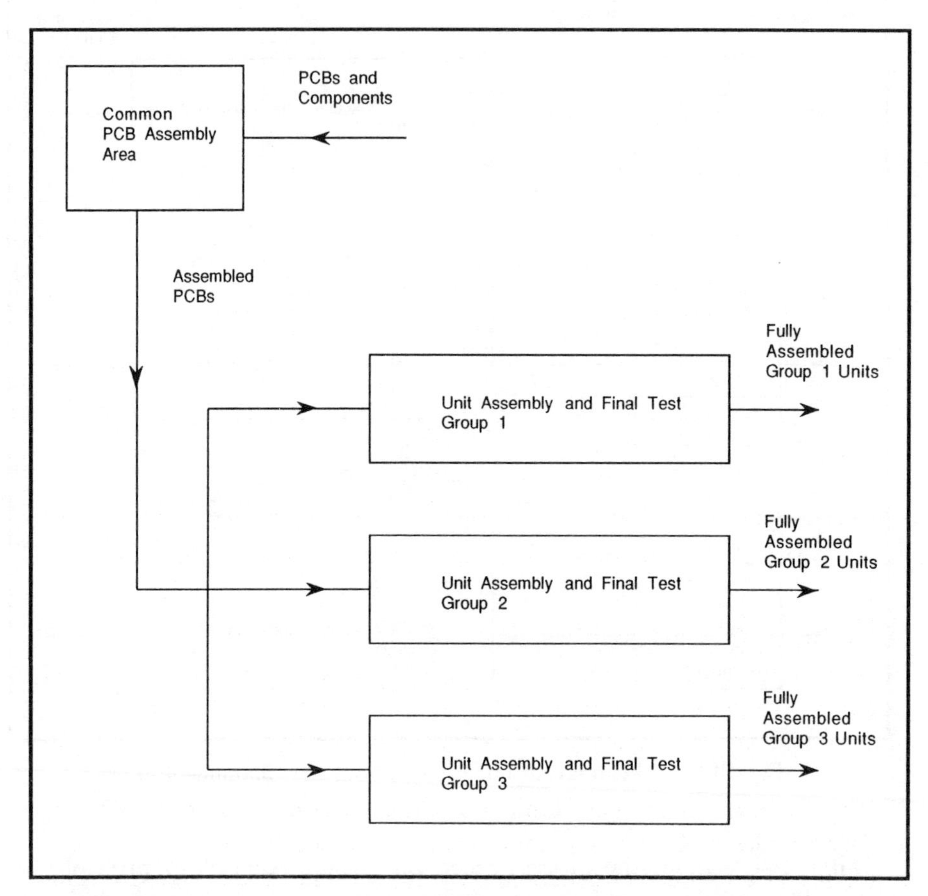

Fig. 10.4. The actual layout of the PCB assembly area and unit assembly cells.

initial redesign of the process, with each unit assembly area would be serviced by a PCB assembly cell (Fig. 10.3). As we discussed earlier, it may not be possible to define an ideal product based layout, because of the requirement for expensive equipment to be shared among the different cells. In the context of the Clonmel plant, this type of compromize had to be made in relation to the ideal layout . The main result of this compromize was the formation of one common PCB assembly cell for the three unit assembly cells (Fig. 10.4). The common PCB assembly cell consists of four workcells; surface mount, machine insertion, hand assembly and solder wave (Fig. 10.5). Workcell is a term used in the Clonmel plant to describe a group of resources. During the implementation

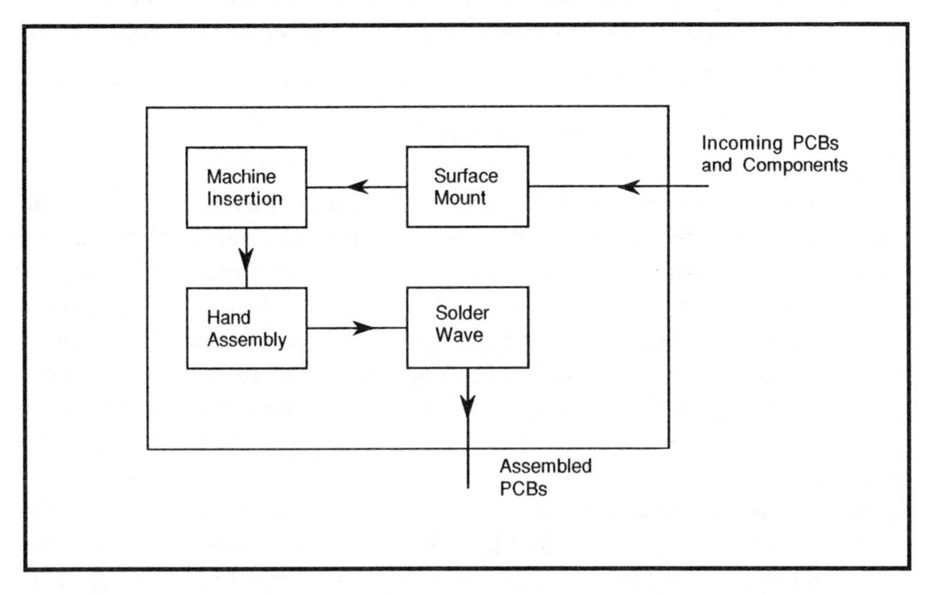

Fig. 10.5. The four workcells within the common PCB assembly cell.

of PAC in the Clonmel plant, we selected three of the workcells in the common PCB assembly cell; machine insertion, hand assembly and solder wave (see Chapter 11). The surface mount process was not selected, because it had just been introduced into the plant and was not fully integrated into the production process.

The technical sub-system within the Clonmel plant may be summarized as follows; the process has become a pull-type, continuous flow operation, consisting of a number of generic process stages and product groups. Manufacturing cycle time has been reduced from over 25 days, in 1983, to two days on average, today. Inventory management has also improved dramatically. In 1983 the plant turned over its entire inventory just twice; the current (1989) inventory performance has reached 5.8 turns. In a truly steady-state volume manufacturing environment, inventory turnover is likely to be considerably greater. Quality levels are now excellent, with a first time process yield of over 95 per cent. These improvements have all been made against a background of shrinking product life-cycles and rapid increase in both the number and volume of products supplied to the marketplace.

It should now be clear that such advancements are not attributable to the application of MRP II alone, but rely fundamentally on the design and organization of the production environment. However, the availability of a flexible, distributed computing environment, at Clonmel, has been (and remains) critical to the development of such an operational structure. But, perhaps the single greatest change in process organization has been the re-design of the social sub-system which supports the process. This, arguably, has also had the greatest impact.

10.4.1 The social sub-system

In the highly automated environment of modern discrete parts manufacturing, there is a temptation to disregard the personal values of production operators in the design of manufacturing systems. Yet, JIT philosophies suggest that we must capitalize on such virtues within the manufacturing organization. The sociotechnical design approach recognizes the importance of appropriate technology in the hands of highly motivated, skilled and responsible personnel. The organization and management of production operators at the Clonmel facility is an elegant example of the application of sociotechnical and JIT principles.

Development of the production social sub-system was central to the re-design of the manufacturing process at Clonmel, as outlined earlier in this chapter. Personnel were assembled into multi-skilled product based workgroups. Each group was assigned total responsibility for a specific range of products, from component delivery to the shop floor, to product shipment. Responsibility imparts ownership for the products (throughout the process), materials ownership and quality ownership. The group focus has necessitated the development of a flexible and well-skilled workforce. Where operators had previously been allocated to single functional tasks, they now move freely across a broad range of tasks and responsibilities. This, of course, has meant a continuous investment in personnel training and re-skilling. But the benefits are immediately evident in terms of worker satisfaction, morale and performance, as well as being measurable in terms of process cycle time and quality.

Workgroups, typically, consist of eight to ten operators and a production supervisor. They also have access to manufacturing and process engineers, who assist in the resolution of those problems which maybe outside the inherent capability of the group. Groups take full responsibility for all production related aspects of their assigned products, including scheduling, materials and quality. In essence, these function as self-managing mini-factories. They manage all of the process stages and interfaces for their specific products; they plan and coordinate the day to day operations and goals of the workgroups, and control their own development and training needs. So, the process technology and personnel resources are controlled directly by those who are closest to its operation. In this way, both technical and social resources are focused sharply on the core objectives of the manufacturing business. It should be noted that the workgroups at Clonmel have not, as yet, progressed to the fully autonomous concept of workgroups discussed in Chapter 7. But a considerable amount of self-management has already been achieved.

We have seen how workgroups and a product based layout contribute to effective production management at the Clonmel plant. However, there remains a need for improved coordination, planning and control at the operational level, to ensure greater efficiency in the collective efforts of discrete workgroups. Our pilot implementation of PAC-based operational control, discussed in the next chapter, was partly aimed at developing a greater understanding of this challenge. In Chapter 11, we shall see that the social sub-system itself poses a barrier to the effective implementation of PAC systems, unless adequate attention is given to the target social environment.

The installed information technology environment may also be a constraint in implementing PAC systems. We now deal specifically with the information technology environment at the Digital Equipment Corporation's Clonmel plant.

10.5 The information technology environment

As a modern computer industry production facility, the presence of a powerful information technology environment at Clonmel is, perhaps, taken for granted. However, as we have emphasized earlier, access to computer processing power does not itself guarantee effective or appropriate production management. Just as the technical and social sub-systems, which make-up the manufacturing process, are designed as a cohesive model, the underlying IT environment must directly support the manufacturing organization in achieving its production and business objectives.

Earlier in this chapter, we have seen that the production management thinking at Clonmel is based on the integration and automation of simplified business processes. In the broader context of the manufacturing business, we have also highlighted Digital Equipment Corporation's approach towards integrating diverse aspects of the enterprise (Fig. 10.1). In practice however, the integrated manufacturing model can only be realized through the provision of effective and timely communication between these business elements. Indeed, successful application of the hybrid PMS architecture (as described in Chapter 1) relies on communication throughout its various levels. At the operational planning and control level of manufacturing, the modular PAC structure functions within a real-time communication context. Availability of an efficient computer system and network is, thus, a prerequisite for operational management within the CIM-based PMS architecture.

At Clonmel, the information technology environment comprizes the installed computer hardware, Local and Wide Area Network (LAN and WAN) communications, and a wealth of application software. It is a truly distributed computing environment, consisting of an Ethernet based LAN, which connects all of the internal functions of the business, and WAN linkages to remote Digital Equipment Corporation sites throughout the world, in a private global network of more than 40,000 nodes. Public communication networks are also employed to connect non-Digital Equipment Corporation sites, such as suppliers, for electronic data exchange. The installed hardware consists of 3 main VAXclusters, which support the business, design engineering and systems development, activities respectively. These clusters include VAX 8800, VAX 11/785 and MicroVAX machines, all of which run the common, proprietary VMS operating system. Thus all applications and data are fully shared across the entire enterprise.

The large investment in networked computing at Clonmel is a recognition of the strategic value of information at all levels of manufacturing. The ability to transmit and share this asset quickly and accurately is central to the success of the plant's business. Management have estimated that over 60% of manufacturing overhead costs are information related. And, as advances in the process technologies and supporting organizations promote shorter manufacturing cycle times, the efficiency of related information processes becomes even more critical. The distributed computing environment enables the control of manufacturing operations to be performed locally, while planning functions are more centralized. Thus, individual process functions can be managed and monitored by dedicated device controllers, or producers, while these in turn are integrated to provide coordination of the complete process. While the horizontal integration and connection of production information is central to CIM, the vertical reporting of this data, to support the tactical and strategic executive decision levels, remains a challenge.

At Clonmel, information technology supports all of the individual functions, with application-specific systems for computer aided design, business planning, finance and asset management, materials and inventory control, supplier management, etc. Within the networked environment, these applications interact to enable, for instance, close cooperation between product and process design. Process equipment, such as DIP and VCD machines, automatically down-load patterns and insertion programs over the network, from CAD libraries stored centrally. Shop floor data collection and statistical process control are also integrated in this way.

This target information technology environment, thus, appears to represent an ideal host for our modular PAC and factory coordination systems. Much of the technology discussed in Chapters 4 and 5 is already installed at the plant, and this is managed by a well established MIS function. However, in implementing such systems, we are constrained by those PMS applications already in use, particularly by the MRP II and shop floor data collection systems. The integration of PAC with these systems is discussed in the next chapter.

10.6 Conclusions

In this chapter we have examined the target environment for our PAC implementation. We have attempted to develop a contextual understanding of the implementation in terms of business pressures, current PMS methods, the existing operational planning and control environment, and finally the plant's information technology base. This discussion has highlighted the importance of complementary development of the manufacturing environment and operational systems, and has provided insight into the effective application of sociotechnical and production environment design principles. In the next chapter, we describe the design, development and pilot implementation of PAC systems at the Clonmel plant.

11

Implementation of a PAC system

11.1 Introduction

In Chapter 10, we presented the host environment for the case study. We now proceed to the application of the architecture, design tools and implementation models in a real manufacturing environment. The PAC development life cycle (outlined in Chapter 9), provides an excellent opportunity to experiment with possible PAC solutions. But, it may not adequately emphasize the difficulties of the transition from PAC simulation to full implementation in a manufacturing site. This migration path is paved with an array of site-specific technical and social parameters within the target environment, including the integration with existing applications (such as MRP and data collection systems), and the promotion of PAC, as a viable operational production control method, among those currently responsible for production operations.

A requirement of the ESPRIT 477 COSIMA project, upon which our knowledge of PAC is based, was the validation of PAC concepts by means of a series of pilot implementations of PAC systems in real manufacturing environments. The implementation described here took place over a three week period, during February/March 1989. This not only validated the PAC architecture, but also highlighted the importance of the pilot implementation itself in easing the transition from simulation to full-scale implementation.

The principal objective in conducting our pilot implementation at Clonmel was, thus, to test the viability of our production planning and control concepts under real manufacturing conditions. This environment would represent the 'acid test' of our PAC systems, as normal production operations could not be disrupted for the sake of experimentation. Furthermore, as these systems were developed as part of a generic research project, and not in response to a specific requirement at the Clonmel facility, production personnel were likely to feel detached from our main objective. A considerable amount of preliminary justification, promotion and training was thus required before installing our PAC prototype. In the final implementation of PAC systems, where a pre-defined requirement exists at a site, we would expect the preliminary educational effort to be confined to user training.

Our concept of a PAC design tool, discussed in Chapter 9, advances the idea that much of the development required in building a PAC implementation system, is not site specific. It would seem reasonable then to make available to developers a set of re-usable building blocks, from which a manufacturing site's implementation team could build the final PAC system. This should considerably reduce the on-site development and customization effort needed, and lend portability (to other sites) to the applications thus developed. The Clonmel pilot implementation followed directly on our application of such a design tool in the design and simulation stages of the development life cycle, as described in detail in Chapter 9.

Apart from our principal aim of validating the PAC architecture and concepts, the pilot implementation at the Clonmel facility also provided an opportunity to:

1. Evaluate the PAC development life cycle, and especially the feasibility of the design tool concept as an aid to the construction of realizable site-specific PAC solutions;
2. experiment with a prototype PAC system and thus help to specify more clearly the needs for operational level production planning and control within the host environment of electronics manufacturing;
3. communicate our PAC ideas with potential end-users, so that the appropriateness and potential difficulties of this approach could be identified.

We now believe that pilot implementation is an essential step in the final phase (i.e. the implementation phase) of the PAC life cycle, outlined in Chapter 9. As in the application of good software engineering methodologies, the prototype and pilot implementation phase should give rise to a more stable and durable final implementation model, requiring less customization and enhancement, and subsequent maintenance, as the site-specific requirements have been met in the design of the final implementation system.

11.2 Description of the the pilot implementation

11.2.1 Approach to the pilot implementation

Before embarking on the pilot implementation, we had successfully experimented with the design tool (described in Chapter 9) in simulating the application of PAC within the Clonmel manufacturing process. We were thus confident that our modular PAC architecture adequately represented the production activities being performed on the shop floor. This simulation phase also provided us with an insight into the site-specific requirements for the scheduler, dispatcher and monitor building blocks which were to be implemented. In constructing the simulation model, we had also populated the manufacturing database with the raw material, product and process and order requirements data, which would subsequently be needed for the live implementation. However, the design tool could not take account of the many site-specific, and often intangible, demands imposed by the real manufacturing environment, and its use was thus restricted to simulation and experimentation, as outlined earlier in Chapter 9.

The site implementation team thus set about the task of specifying and developing implementable PAC applications, based on our experiments with the design tool and a wealth of experience gained at the plant, over a period of six months or so. During the six months preceding the implementation, members of the team (consisting of three engineers and a project manager) met regularly with shop floor operators, supervisors, manufacturing and process engineers, and production managers, in order to refine the PAC implementation proposals.

During this period we developed, in particular, a respect and understanding of the work group organization of production operators (as described in Chapter 10). We were also faced with the difficulties of introducing coordinated planning and control systems in an environment where the level of individual and group autonomy is high. The challenge then was to develop an appropriate PAC solution, which would enable centralized planning (i.e. scheduling) and distributed control, without infringing upon the product ownership and responsibility values held by these product focused work groups. A considerable amount of previous work had also been devoted to the development of factory scheduling techniques at the plant, although the manually created build plan was still in use for daily scheduling of the shop floor.

In preparing for the implementation, a large amount of effort was undoubtedly devoted to the promotion of PAC concepts, refinement of the prototype system and user training on the implementation software. A recurring need in the specification and refinement of this software was the demand for simplicity and transparency in the real time applications being developed for the shop floor. This deviated somewhat from the relative sophistication of the functionality made available in the design tool simulator. Final development of the implemented software was completed in a relatively short period (about six person-months of effort), once the simulation phase of the project was completed. Installation and testing of the software on the plant's main computer cluster was performed in a few hours on the first morning of the pilot implementation.

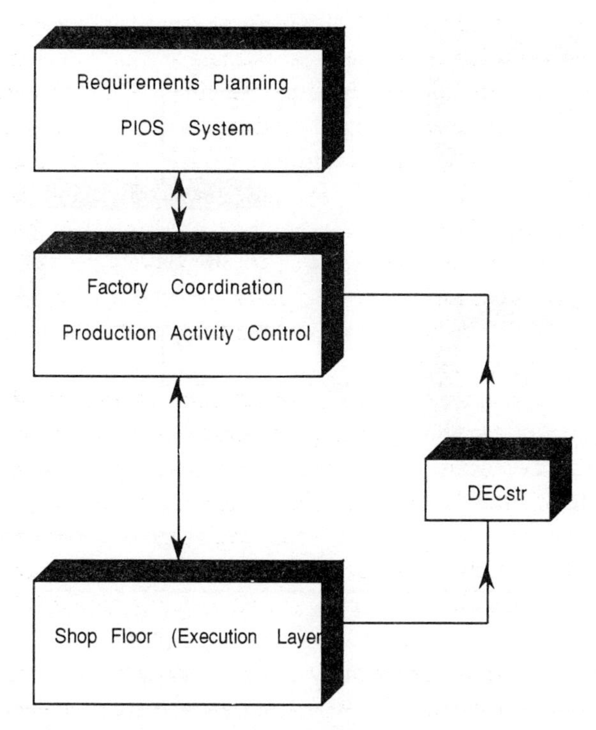

Fig. 11.1. The hierarchy of PIM systems in Clonmel.

11.2.2 Pilot implementation test area and functional overview

The Clonmel pilot implementation addressed production planning and control at a level below the plant's MRP II system, PIOS, and above the execution layer on the shop floor (Fig. 11.1). As can be seen from the diagram, the role of such PAC systems is to bridge the gap between requirements planning and shop floor activity. The elements of the PAC architecture which were implemented include: the scheduler, dispatcher and monitor, as well as the supporting application network and manufacturing database.

The live implementation took place in a section of the manufacturing process bounded by three discrete workcells: machine insertion, hand assembly and solder wave. The physical layout of these and a description of the processes performed within the workcells is given in Chapter 10. The mapping of the PAC architecture onto these workcells will be described in the next section. A number of products from the entire Clonmel product range was used for the pilot implementation. These were selected on the basis of having a reasonably high volume and stable demand throughout the implementation period.

In developing the implementation software, the need to physically integrate the PAC applications with both the MRP II planning and shop-floor data

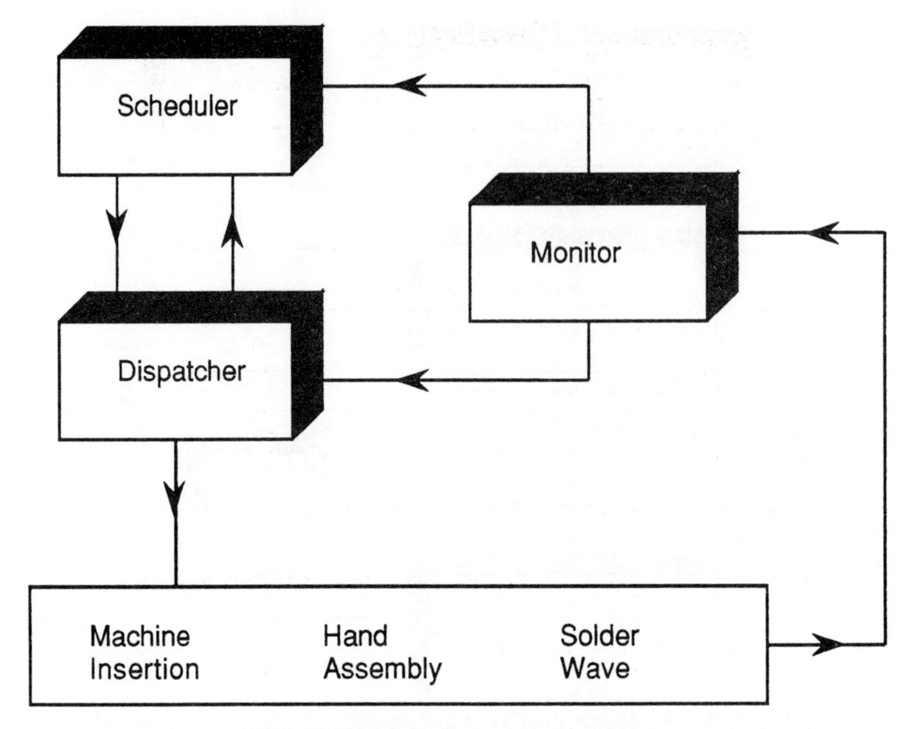

Fig. 11.2. PAC in the test environment.

collection systems was identified. By interfacing with the PIOS system, Bill of Materials (BOM) and production requirements data could automatically be made available to the PAC system, obviating the need for additional data entry. An interface was thus constructed which allowed regular up-dating of the PAC manufacturing database, in response to changes in the relevant data held in the PIOS database. This provided effective upwards integration of PAC with the business planning function of MRP II in the hierarchical hybrid model of production management systems, outlined in Chapter 1.

At the shop floor level, an interface to the plant's data collection system (DECstr) was also constructed, to allow the existing Work In Progress (WIP) tracking facilities to be used as part of the PAC system's monitor. This minimized the impact of the new PAC system on the shop floor, as much of the monitoring function was performed through existing DECstr terminals. However, the DECstr terminals facilitated the collection of data alone, and could not be configured to perform the user interface functions of the dispatcher, in issuing the scheduled commands. A workcell reporter was thus developed to perform the dual user interface functions of dispatcher and monitor; this will be further described in the next section.

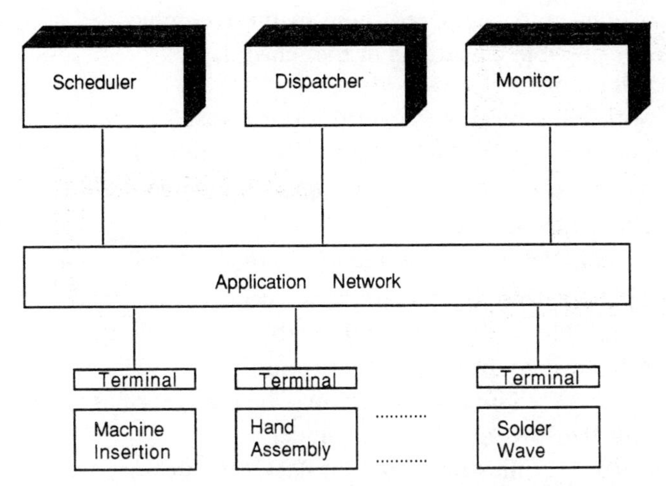

Fig. 11.3. PAC implementation model in the Clonmel environment.

11.3 Architectural mapping and implementation software

11.3.1 Mapping the PAC architecture onto the test area

The PAC architecture applied to the Clonmel environment was shown in Fig. 11.2. It consists of the scheduler, dispatcher and monitor building blocks which interact with the workcells on the shop floor. The PAC implementation model is shown in Fig. 11.3. Here the software blocks are shown as well as their interactions through the application network. In physical terms, the three workcells were connected into the PAC system by a user friendly application (later identified as the workcell reporter), which was displayed on a terminal located at each of the workcells. The implemented functions of each of the PAC building blocks are now outlined.

- Scheduler
  The role of the scheduler was to provide a schedule which would coordinate the flow of work through each of the workcells. It was not intended to produce a detailed schedule for each of the workstations within a workcell, but rather a preferred start time and finish time for each job at their respective workcells.
- Dispatcher
  The dispatcher ensured that the schedule was implemented by reporting to each of the workcells the required start and finish times of jobs.
- Monitor
  The monitor displayed the current status of jobs in the system. It received messages from each of the workcells indicating the actual start and finish

time of jobs, and compared these to the recommended start and finish times for possible deviations in the schedule.

- Workcell reporters
  The workcell reporters provided the means for the cell operators to communicate with the PAC system. These terminals allowed the following information to be passed to the application network:

  The actual start time of a job in the workcell;
  the completion time of the job in the workcell.

The use of the workcell reporters was crucial for the PAC dispatcher and monitor building blocks, as they both enabled the collection of data for the monitor and the display of information from the dispatcher. As these acted in the capacity of workcell interfaces, they may be thought of as virtual producers, consistent with the PAC architectural model.

Figs 11.4 and 11.5 illustrate how the PAC architecture was mapped onto the implementation area at the Clonmel plant. Fig. 11.4 shows the routine flow of work between the workcells* and the DECstr factory data collection point at the entry to the hand assembly cell. For the pilot implementation the scheduler, dispatcher, monitor, workcell reporter and application network were implemented in this environment, and integrated with existing and additional DECstr points, as shown in Fig. 11.5. The scheduler, dispatcher and monitor were displayed on a terminal in the workcell supervisor's area, for ease of access to the system. The DECstr and workcell reporter user terminals were distributed, as shown. All of the communication between (and among) the various building blocks and user terminals was achieved through the application network.

11.3.2 Pilot implementation application software

The software used for the implementation was based on the modules described in our discussion of the design tool in Chapter 9, and the complete software architecture used for the pilot implementation is shown in Fig. 11.6. It consists of: a manufacturing database with a user interface; the PAC scheduler, simulator, dispatcher and monitor modules; the workcell reporter and DECstr interface; and a communications system, the Application Network (AN). There was no requirement for a mover in the Clonmel site, as all parts are moved by the operators, and the workcell reporter is a virtual producer, since the pilot implementation did not need to control specific devices on the shop floor.

Database:
As described in Chapter 9, the manufacturing database is the data foundation

* The three workcells constitute a product based (i.e. PCB) assembly cell

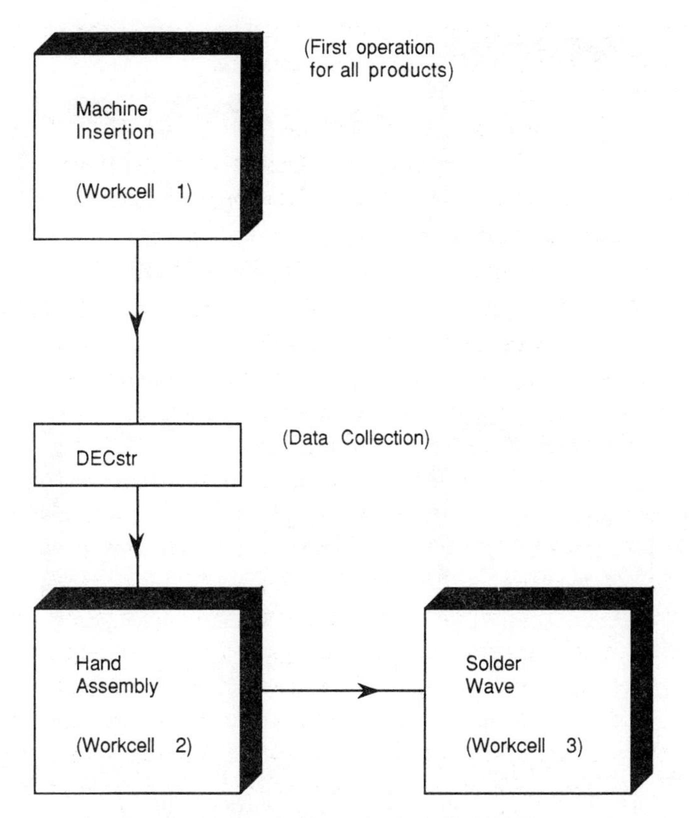

Fig. 11.4. Implementation area for pilot PAC system.

for the PAC system, and it holds all of the relevant manufacturing data needed to carry out the tasks of planning and control. The Clonmel manufacturing database used a software server which retrieved the monthly requirements figures from the PIOS database, obviating the need for manual intervention, thereby ensuring the integrity of data which was used for scheduling purposes.

Scheduler:
The scheduler had a number of options available which were used to develop schedules within the pilot implementation test area. These included production smoothing, bottleneck search, bottleneck analysis and scheduling heuristics (shortest processing time, earliest due date, backward schedule).

Dispatcher:
The dispatcher facilitated the implementation of a recommended schedule by relaying this schedule to the workcell operators; this was transmitted both as a hard copy output and by using the distributed message passing system provided by the application network to the workcell reporter.

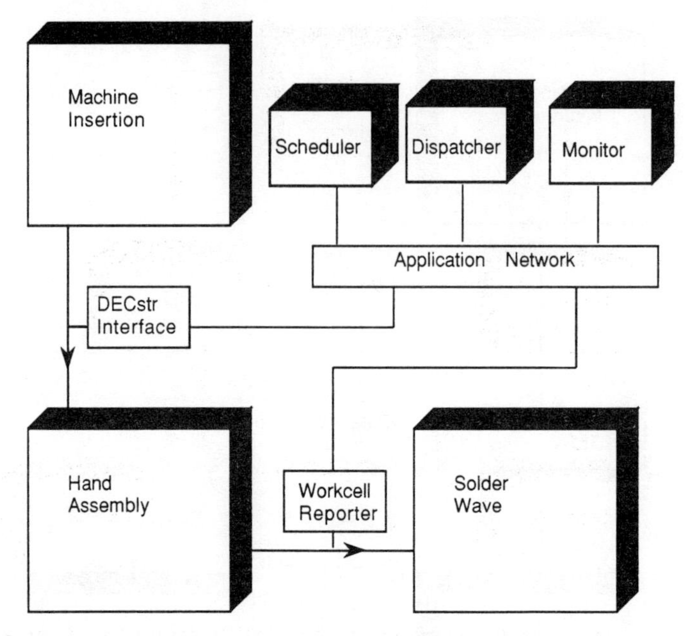

Fig. 11.5. Final PAC implementation.

Monitor:
The monitor provided feedback on the current status of work in progress on the shop floor. This data, when compared with the planned work data from the plant level scheduler, gave an accurate picture of the effectiveness of the implemented schedule. The monitor compared the planned schedule (received from the scheduler) and the actual implementation, based on data received from the DECstr data collection system and the workcell reporter. It consisted of a menu driven interface which allowed the user to monitor work in progress on a plant-wide or a workcell-wide basis. Information from the monitor was of the form:

- Workcell Name : Machine Insertion
- Part Name : DHQ11
- Transfer Batch Number : ES000121
- Scheduled Start Period : 10:45–12:30
- Actual Start Time : 10:56
- Status : On Schedule

Workcell reporter:
The workcell reporter enabled the recording of the start and finish times of jobs at a workcell. For the pilot implementation, the workcell reporter received the daily schedule from the dispatcher through the application network and then facilitated the recording of shop floor events. The workcell reporter

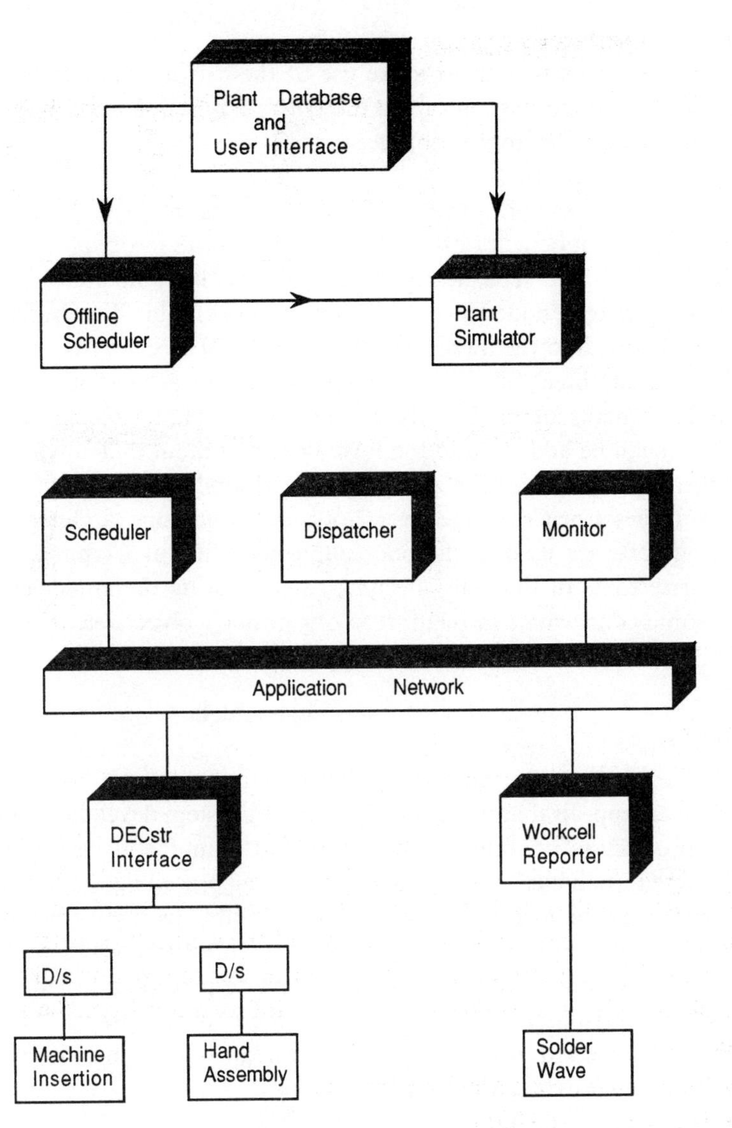

Fig. 11.6. The PAC implementation software.

operated in two modes: a supervisory mode and an operational mode. The supervisory mode showed a list of all the jobs to be completed at a workcell and highlighted started, completed and late jobs. The operational mode facilitated the recording of start and finish times for each transfer batch to pass through the workcell. These messages were then relayed through the application network to the monitor module to update the work in progress records.

DECstr interface:
This interface was written to make use of the existing data collection system in Clonmel. It took messages from the DECstr system and passed them to the monitor through the application network.

Application network:
The application network provides the message passing facility for the live PAC system. The main benefit of the AN is that it enables flexibility and modularity in the PAC system. The flexibility means that different building blocks can reside on different nodes in a local area network. This distributed feature of the AN is critical to the implementation of PAC. The Clonmel implementation software used three different local area nodes to achieve its goals. One of the main benefits of modularity is that more building blocks (e.g. workcell reporters) can be added on to the PAC system without difficulty, thus enabling incremental implementation and long-term flexibility.

Having described the objectives, approach, functions, architecture and application software used in the pilot implementation at Clonmel, we will now summarize some of the many insights gained during the implementation, and draw some conclusions in relation to our primary objectives.

11.4 Lessons and guidelines

In order to evaluate the PAC implementation, the pilot project was monitored by an appraisal group consisting of both system developers and potential users (production operators, supervisors and manufacturing engineers). This group acted as an advisory group in the months preceding the implementation, and provided a focus for comment, criticism and constructive feedback during (and after) the implementation period. Structured interviews with personnel from different functions were also conducted to provide a broader interpretation of the feasibility of PAC systems in the Clonmel plant. Our evaluation results are presented in summary and are categorized in line with the following system characteristics and criteria:

1 The PAC architecture;
2 the system behaviour;
3 the software and hardware requirements;
4 the target environment and existing systems infrastructure;
5 the system utility and benefits;
6 the user attitudes.

11.4.1 PAC architecture

This pilot implementation demonstrated the validity of the PAC architecture in the electronics assembly environment. The level of implementation and the type of environment places a different emphasis on the requirements of the individual building blocks. For example, in the Clonmel environment, all decision making in real-time (dispatching) is performed by the operators and supervisors, so the need for dispatcher functionality is not as great as it would be for a fully automated environment. It was indicted that PAC had a lot to offer in scheduling the workcell, particularly machine insertion, which is at the start of the process. This workcell could thus perform an important regulatory task in managing the flow of work, and in doing so, improve the coordination further down the manufacturing process.

11.4.2 System behaviour

A most positive aspect highlighted during the pilot implementation was the behaviour of the software system. The modularity of PAC, with the emphasis on different building blocks, allowed easy integration with existing software systems (PIOS and DECstr). The distributed facility provided by the application network allowed applications to run in a clustered environment, which increased reliability and flexibility of the PAC system. The user interfaces were designed in such a way that users found the system simple to use and easy to understand.

11.4.3 Software and hardware requirements

The main software and hardware requirements for the implementation were:

1. A relational database to hold the relevant manufacturing data, which was based on Digital's Rdb database system;
2. the application network, a communications bus to relay messages from the different building blocks in a distributed environment;
3. conventional third (Pascal) and fourth generation languages.

11.4.4 Host PMS environment

The manufacturing environment has a major influence on realisation of production planning and control systems. The main requirements at Clonmel were to integrate successfully with existing PMS applications. This integration was in two directions, upwards to the requirements planning system (PIOS), and downwards to the data collection system (DECstr). The technological challenge of integration was met successfully due mainly to the modular nature of the PAC system. Software-to-device integration may be needed for any future application of PAC which may be implemented at a lower level (e.g. device coordination level) to this pilot implementation.

11.4.5 System utility and benefits

Two significant benefits of implementing a PAC system at Clonmel were identified: one arising from scheduling, the other from monitoring. The main benefits from scheduling are a reduction in lead times, and an overall coordinated plan from which people can manage work. Effective monitoring gives users increased visibility of manufacturing activities. Some further detail on these specific aspects of our implementation is now provided.

Scheduling The implementation of scheduling at Clonmel involved regular meetings with the potential users of the system over a six month period. The rationale behind the drive towards scheduling was to reduce lead times through better coordination of the workflow. An internal project (continuous flow manufacturing) had significantly reduced lead times over a three year period, largely through reorganization of the process, and it was proposed that scheduling could reduce these even further.

Through these regular meetings, it was recognized that strict scheduling could not be applied in the present Clonmel environment (e.g.: start job number 100 at machine insertion at time 9:04), and the scheduling procedure was then expanded to offer a priority system. Users felt this was too flexible and would be of little benefit. The capability of the scheduler was then increased and a backward schedule, based on organizing work into a series of time buckets (to meet end-of-line due dates), was presented. This met with the most favourable reaction because it was not too strict and it mirrored the Clonmel social system environment, in that the time buckets recommended were natural break times in the working day (e.g.: start job 1 between 8:30–10:00 at machine insertion).

However, there was a general ambivalence to the proposed scheduling methods, as these were perceived to limit the individual control of supervisors over their areas of responsibilities. This highlights an important sociotechnical issue of the design and development of work practices in human intensive environments. The implementation of any IT based decision support system is likely to meet with similar barriers unless the design of such systems takes true cognisance of the responsibilities and needs of the end users. Pava (1983) discusses such difficulties in detail and specifically mentions the problems of coordinating self managed work groups similar to those found in the Clonmel plant. In a more traditional production line environment, or indeed within a fully automated plant, such issues should be of less significance. Fig. 11.7 illustates the broad functionality of the scheduler developed for the implementation, with the main menu options including bottleneck analysis and the selection of a particular strategy.

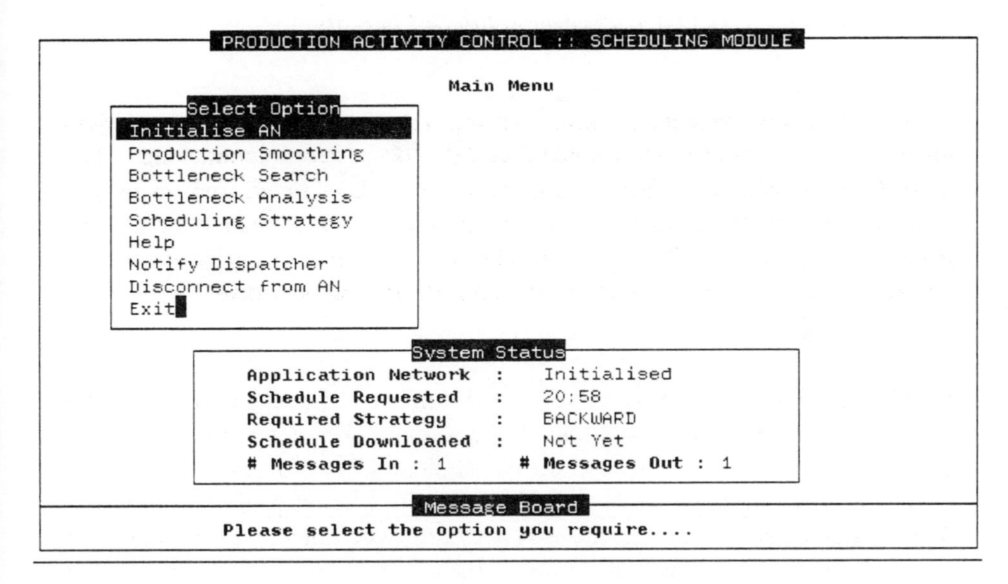

Fig. 11.7. The PAC scheduler.

Monitor The advantage of an effective monitor is that users may access relevant information in real time. This information includes data on machine availability, work in progress, actual performance data and deviations from schedules. The monitor, part of which is illustrated in Fig. 11.8, provides a window to the shop floor in a paperless environment, and a user friendly application means that the potential user can access important information at the touch of a button. A historical reporting facility was a vital part of the Clonmel monitor because it allowed different users to access up-to-date information from a terminal located at their own desk, or place of work. This meant that both plant management, supervisors and production operators could access the same source of data on manufacturing performance and generate reports as required. Fig. 11.8 illustrates the workcell monitor, which displays the expected start and finish times of jobs (transfer batch), with the actual start and finish times of jobs.

As the manufacturing cycle time shortens, the need for work in progress tracking will diminish, with a consequential change in the emphasis of the PAC monitor. The main requirements of a monitor in the future should be to ensure that:

1 The process is in control;

2 the schedule is operating according to plan;

3 quality levels are being maintained.

WORKCELL REPORTER ::: SOLDERWAVE

Options

<< Daily Schedule >>

Transfer Batch	Part Number	Job Status	Scheduled Start	Scheduled Finish	Actual Start	Actual Finish
2	CXA16	Started	15.00	17.00	10:43	-----
6	H7186-A	Started	15.00	17.00	10:46	-----
7	DESQA	Started	15.00	17.00	10:46	-----
8	DMB32	Started	15.00	17.00	10:44	-----
14	DZQ11	Started	15.00	17.00	10:44	-----
15	DZQ11	Started	15.00	17.00	10:44	-----
16	DZQ11	Started	15.00	17.00	10:44	-----

N - Next Screen, P- Previous Screen, Q- Quit >

Fig. 11.8. The workcell monitor.

11.4.6 User attitudes

The users' attitudes formed a valuable part of the evaluation of the pilot implementation, as ultimately, the success of PAC depends on how useful the potential users find the system. At Clonmel there was unanimous agreement on the effectiveness of a pilot implementation, as it gives system developers and users important feedback on the features of a prototype system.

The over-riding view among production operators and supervisors was that information technology applications, such as our PAC system, should advise rather than control the users, and that the applications must be clearly understood by potential users. This is a very important issue, especially in a human intensive environment such as Clonmel, which encourages product ownership and workcell autonomy, and where an implementation of a rigid 'black box' control system would erode the control and responsibility of the operations personnel.

11.5 Conclusions

The pilot implementation of PAC at the Clonmel facility realized, in practice, a concept which had developed from an architectural framework to a live manufacturing application during the course of the ESPRIT 477 research project. Our experience in implementing the PAC prototype at Clonmel confirms the validity of the PAC architecture and reaffirms its use as a viable model for production planning and control in discrete parts manufacturing. In particular, PAC systems can readily be configured to close the loop between business planning, as in MRP II, and operational control of production activities on the shop floor. The Clonmel implementation is a true example of this closed-loop planning and control at PAC, and exhibits the software integration of PAC with MRP II and shop floor data collection systems.

Our rapid development of site-specific implementation software at Clonmel would seem to give further support to the PAC Life Cycle methodology and design tool concept, discussed in Chapter 9. Undoubtedly the development time was greatly reduced by the availability of (and our previous experimentation with) a PAC simulation model. Apart from the insight gained through experimentation, this model greatly simplified the process of communication with users in specifying and developing the site specific applications.

We now believe that the PAC life-cycle should be thought of as an iterative procedure, where design, simulation and prototyping can be re-visited, as site-specific requirements for shop floor control systems become redefined. The pilot implementation itself is a critical stage in this cycle, during which substantial system refinement and user acceptance is generated. This approach is consistent with the most advanced software engineering methodologies, where incremental development is preferred to detailed blueprinting, as user needs may change frequently during long traditional development cycles.

In developing planning and control systems for semi-automated manufacturing environments, it is easy to overlook the human context within which these systems must operate. The Clonmel implementation greatly developed our understanding of the impacts of PAC systems on the social organization, and confirmed the sociotechnical principles of organization design, where both the technical and social sub-systems are designed as a unified model. The PAC architectural model certainly accommodates such principles and can be constructed to provide automated control of routine tasks, with real time feedback to support more effective decision making, planning and tactical management.

12

References and further reading

12.1 References

Ackoff, R.A. (1967) Management Misinformation Systems. *Management Science,* **14** (4), 147–156.

Ackoff, R. A. (1975) Does the Quality of Life have to be Quantified? *General Systems,* **20**.

Ackoff, R.A. Emery, F. (1972) *On Purposeful Systems,* Aldine-Atherton, Chicago, USA.

Ackoff, R.L. (1967) Management Misinformation Systems, *Management Science*, **14** (4), pp. 147–156.

Ackoff, R.L. (1975) *Does Quality of Life Have to be Quantified?*, General Systems, Volume XX.

Ackoff, R.L. (1977) Optimization + objectivity = opt out. *European Journal of Operational Research,* **1**.

Acock, M., Zemel, R. (1986) *DISPATCHER: AI Software for Automated Material Handling Systems.* Proceedings of the SME Artificial Intelligence in Manufacturing Conference, Long Beach, California September 23–25.

Adiga S., Li, R. (1987) *Research in The Application of Artificial Intelligence Techniques to Generative Process Planning, Modern Production Systems,* (ed. A. Kusiak), Elsevier Science Publishers, North Holland.

Albrecht, K. (1978) *Successful Management by Objectives.* Prentice-Hall, New Jersey, USA.

Amstead, B., Begeman, M. (1977) *Manufacturing Processes,* John Wiley and Sons, USA.

Anderberg, M. (1973) *Cluster Analysis for Applications*, Academic Press, New York.

Anon, (1960) Composite Components Give Long Production Runs, *Engineering*, August.

Ashby, W. (1956) *An Introduction to Cybernetics,* Chapman and Hall, London.

Baker, K.R. (1974) *Introduction to Sequencing and Scheduling.* John Wiley, New York.

Bals, P. (1989) *An Object Oriented Analysis and Design Methodology for Computer Integrated Manufacturing Systems*, Proceedings Tools Conference pp. 75–84.

Balzer, R., Cheatham, T.E., Green, C. (1990) Software Technology in the 1990's: Using a New Paradigm. *Computer,* November, pp. 39–45.

Bao, H. (1984) *Computer Aided Process Planning, Robotics and Factories of the Future* (ed. S. Dwivedi), Springer-Verlag, pp. 619–629.

Barkocy, B., Edelblick, W. (1984) *A Knowledge Based System for Machining Operation Planning*, Autofact 6 Conference Proceedings, USA, October.

Baron, N. (1986) *Computer Languages: a guide for the perplexed.* Penguin Books, London, UK.

Barr A., Feigenbaum (eds) (1981) *The Handbook of Artificial Intelligence*, Volume I, Addison-Wesley, Reading, MA, USA.

Beckhard, R. (1969) *Organisation Development: Strategies and Models*, Addison-Wesley, Mass. USA.

Bellman, R. (1957) *Dynamic Programming.* Princeton University Press, USA.

Birtwistle, G., Dahl, O., Myhrtag, B., Nygaard, K. (1973) *Simula Begin.* Auerbach Press, Philadelphia, USA.

Blackstone, J.H., Phillips, D.T., Hogg, G.L. (1982) A state-of-the-art survey of dispatching rules for manufacturing job shop operations, *Int. J. Prod. Res.*, **20** (1), 27–45.

Blumberg, P. (1969) *Industrial Democracy*. Schocken, New York.

Bobrowicz, V. (1975) *Coding and Classification Systems – A prerequisite for Group Technology*, SME Technical Paper CM76-273, Soc. of Mechanical Engineers, Michigan, USA.

Bowden, R. Browne, J. (1987) *Robex – An Artificial Intelligence Based Process Planning System for Robotic Assembly*, Ninth International Conference of Production Research, Cincinatti, USA.

Bowden, R., Browne J., Duggan, J. (1989) *The Design and Implementation of a Factory Coordination System*, Xth International Conference on Production Research, Nottingham, August.

Bowen, J., O'Grady P., Nuttle H., Terribile M. (1989) An Artificial Intelligence Approach to Loading Workstation Resources in a Distributed Job Shop Controller, *Computer Integrated Manufacturing Systems*, **2**, (1), February, 21–28.

Brankamp, A. (1969) *Outline of a Workpiece Classification and Coding program,* Group Technology: Proceedings of an International Seminar, International Centre for Advanced Technical and Vocational Training, Turin, Italy.

Brauer, W., Brauer, U. (1989) Better Tools – Less Education? *Information Processing 89*, G. Ritter, Elsevier Science Publishers, USA.

Bravco, R., Yadav, S. (1985) Requirement Definition Architecture – An Overview, *Computers in Industry*, **6**, 237–251.

Bray, O. (1988) *Computer Integrated Manufacturing – The Data Management Strategy*. Digital Press, USA.

Browne, J. (1988) Production Activity Control – a key aspect of production control. *Int. J. Prod. Res*, **26** (3), 415–427.

Browne, J. O'Gorman, P. (1985) Product Design for Small Parts Assembly. *Robotic Assembly : International Trends in Manufacturing Technology.* K. Rathmill, Springer-Verlag, Berlin.

Browne, J., Boon, J.E., Davies, B.J. (1981) Job Shop Control, *Int. J. Prod. Res.*, **19** (6), 663–643.

Browne, J., Dubois, D., Rathmill, K., Sethi, S., Stecke, K. (1984) Classification of flexible manufacturing systems. *The FMS Magazine,* April, pp. 114–117.

Browne, J., Duggan, J. (1988) *Esprit Project 477: Production Activity Control Design and Implementation.* Proceedings of ESPRIT Technical Week, November 14–17, North Holland.

Browne, J., Harhen, J., Shivnan, J. (1988) *Production Management Systems – A CIM Perspective.* Addison-Wesley, UK.

Browne, J., O'Gorman, P. (1985) Product Design for Small Parts Assembly, *Robotic Assembly: International Trends in Manufacturing Technology*, (ed. K. Rathmill), Springer-Verlag, Berlin.

Browne, J.A. (1954) *The Social Psychology of Industry.* Pelican, England.

Brownston, L., Farrell, R., Kant, E., Martin, N. (1985) *Programming Expert systems in OPS-5.* Addison-Wesely, Reading, Mass., USA.

Bruno, G., Marchetto, G. (1986) *Process Translatable Petri nets for the rapid prototyping of process control systems.* IEEE Trans. on software eng., Vol. SE-12 (2), February, 346–357.

Bucher, W. (1979) *An Integrated Group Technology program*, SME Technical paper MS79-977, Soc. of Manufacturing Engineers, Michigan.

Bunce, P. (1988) *CAM-I Intelligent Manufacturing Program: Accomplishments and Plans*, Proceedings of the Production Planning and Control Information Exchange between CAM-I and ESPRIT Projects, 25th May, Munich, West Germany, P-88-IMMP-01.

Burbidge, J. (1975) *The Introduction to Group Technology* , Heinemann, London.

Burbidge, J. (1985) Automated Production Control in *Modelling Production Management Systems* (eds P. Falster and R. Mazumber), Amsterdam, North-Holland.

Burbidge, J. (1989) *Production Flow Analysis for Planning Group Technology*, Oxford Science Publications, England.

Burbidge, J.L. (1971) Production Flow Analysis, *Production Engineer*, **50**, 39.

Burbidge, J.L. (1977) A manual method of Production Flow Analysis, *Production Engineer,* October.

Burbidge, J.L. (1985) Production Flow Analysis, Proc. of 8th ICPR Conference, Stuttgart.

Burbidge, J.L. (1989) *Production Flow Analysis for Planning Group Technology*, Oxford Science Publications, Clarendon Press, England.

Burns, T., Stalker, G.M. (1961) *The Management of Innovation.* Tavistock Publications, London.

Bylinsky, G. (1983) An Israeli shakes up US Factories, *Fortune* 5th September, pp 120–132.

COSIMA Project Team (1987) *Development Towards an Application Generator for Production Activity Control.* Proceedings of the 4th Annual ESPRIT Conference, Brussels, September 24–28, North Holland, pp. 1648–1661.

Cerf, V.G., and Kahn, R. (1974) A protocol for packet network interconnection, *IEEE Transactions on Communication*, Vol. COM-22, no.5, pp. 637-648.

Campbell, J., Dunette, M., Lawler, E. Weick, K. (1970) *Managing Behaviour, Performance, and Effectiveness.* McGraw-Hill, New York.

Chamberlin, D.D., Boyce, R.F. (1974) *SEQUEL: A Structured English Query Language.* Proc. 1974 ACM SIGMOD Workshop on Data Description, Access and Control. May.

Chang, T.C., Wysk, R.A. (1985) *An Introduction to Automated Process Planning Systems,* Prentice-Hall.

Cherns, A. (1976) The Principles of Sociotechnical Design. *Human Relations.* **29**.

Cherns, A. (1977) Can Behavioural Science help Design Organisations? *Organisational Dynamics.* Spring.

Choobineh, F. (1984) Optimum Loading for GT/MRP manufacturing systems, *Computers and Industrial Engineering*, **8**, 197–206.

Chryssolouris, G. (1987) MADEMA: An Approach to Intelligent Manufacturing Systems, CIM Review *Artificial Intelligence in Manufacturing*, Spring 1987, pp. 11–17.

Codd, E. (1970) A Relational Model of Data for Large Shared Data Banks. *Communication of the ACM,* **13**, (6), June.

Coll, A., Brennan, L., Browne, J. (1985) *Digital Systems Modeling of Production Systems*, Modeling Production Management Systems, North Holland.

Collinot, A., Le Pape, C., Pinoteau, G. (1988) SONIA: a knowledge based scheduling system, *Artificial Intelligence in Engineering*, **3** (2).

Connolly, R., Middle, G., Thonley, R. (1970) *Organising the manufacturing facilities in order to obtain a short and reliable manufacturing time*, 11th MTDR Conference, England.

Conway, R.W., Maxwell, W.L. (1962) Network dispatching by shortest operation discipline, *Operations Research*, **10**, 51.

Conway, R.W., Maxwell, W.L., Miller, L.W. (1967) *Theory of Scheduling*, Addison-Wesley, Reading, Mass, USA.

Copas, C., Browne, J. (1990) A rules-based scheduling system for flow type assembly, *Int. J. of Prod. Res*, (not yet published).

Corbett, J. (1988) *Strategic Options for CIM. Computer Integrated Manufacturing Systems.* **1** (2), May, pp. 75–81.

Cunningham, P. and Browne, J. (1986) A LISP-based heuristic scheduler for automatic insertion in electronics assembly, *Int. J. Prod. Res.*, **24** (6), 1395–1408.

D-Class User Manual (1985) *Computer Aided Manufacturing Lab.*, Brigham Young University, Provo, Utah, USA.

Daellenbach, H.G., George, J.A., McNickle, D.C. (1983) *Introduction to Operations Research Techniques*, 2nd edn, Allyn and Bacon, Inc, Newton, Massachusetts, USA.

Date, C.J. (1988) *Database Systems.* Volume 1, Addison-Wesley Publishing Company, USA.

Davies, B.J., Darbyshire, I.L., (1984) The Use of Expert Systems in Process Planning, *Annals of CIRP*, **33** (1), 303–306.

Davis, L.E. (1977) Evolving Alternative Organisation Design: the Sociotechnical Bases. *Human Relations*, **30** (3).

Davis, L.E. (1982) *Organisational Design.* Handbook of Industrial Engineering, G. Salvendy, Wiley & Sons, USA.

De Beer, C., De Witte, J. (1978) Production Flow Synthesis, *Annals of CIRP*, **27**.

DeVries, M. (1976) *Group Technology: An Overview and Bibliography*, The Machinability Data Center, Cincinnati Ohio.

Dekleva, J., Kusar, J., Menart, D., Sarbek, M., Zavbadlav, E. (1988) Extended Production Flow Analysis, *Robotics and Computer Integrated Manufacturing*, **4** (1/2), 63–68.

Deming, W.E. (1986) *Out of the Crisis*, 2nd edn. Massachusetts Institute of Technology, USA.

Descotte Y., Latcombe J.C. (1981) GARI: A Problem Solver that Plans How to Machine Mechanical Parts, *IJCAI*, pp 766–772.

Doumeingts, G. (1991) Private communications with the authors. (University of Bordeaux, France).

Drucker, P.F. (1969) *The Practise of Management*. Pan Books, London.

Duggan, J., Bowden, R., Browne, J. (1989) *A Simulation Tool to Evaluate Factory Level Schedules*. Proceedings of the Third International Conference on Expert Systems and the Leading Edge in Production and Operations Management, 21-24 May, Hilton Head Island, South Carolina, Management Science Department, University of South Carolina.

Duggan, J., Browne, J. (1988a) *ESPNET: Expert System Based Simulator of Petri Nets*. IEE Proceedings-D Control Theory and Applications, **135** (4), July.

Duggan, J., Browne, J. (1988b) *An AI Based Simulation for Production Activity Control Systems*. Proc. 4th Int. Conf. Simulation in Manufacturing, November, IFS Publications, pp. 177–194.

Dumaine, B. (1989) What the Leaders of Tomorrow See. *Fortune*. July, pp. 24–34.

ECMA (1987) STANDARD ECMA 127, *RPC Basic Remote Procedure Call using OSI Remote Operations*. European Computer Manufacturer's Association, Geneva Switzerland.

Edwards, J.N. (1983) *MRP and Kanban, American Style*, APICS 26th Annual International Conference Proceedings, pp. 586–603.

El-Essawy, I.F. (1971) *Component Flow Analysis*, Ph.D Thesis, Department of Management Sciences, UMIST, Manchester, England.

El-Essawy, I.F., Torrance, J. (1972) Component Flow Analysis: An effective Approach to Production System's Design, *Production Engineer,* **51**, 165.

Emery, F. (1967) The next Thirty Years: Concepts, Methods and Anticipations *Human Relations.* **20** (3), 199–237.

Emery, F. (1977) *Futures We are In.* Leiden, Nijhoff.

Emery, F., Thorsrud, E. (1976) *Democracy at Work: The Report of the Norwegian Industrial Democracy Program.* Leiden, Nijhoff.

Emery, F., Trist, E. (1960) Sociotechnical Systems. *Management Science: Models and Techniques.* C. Churchman and M. Verhulst, Pergamon, New York.

Emery, F., Trist, E. (1973) *Towards a Social Ecology.* Plenum, New York.

Erman, L.D., Hayes-Roth, F., Lesser, V.R., Reddy, D.R. (1980) The Hearsay-II speech understanding system: Integrating knowledge to resolve uncertainty. *Computing Surveys,* **12**, 213–253.

Eshel, G., Barash, M., Chang, TC. (1988) Generate and Test and Rectify - A Plan Synthesis Tactic for Automatic Process Planning, *Artificial Intelligence in Engineering,* **3**, 1.

Factrol (1986) *FACTOR: Modeler's Reference Manual*, FACTROL Inc., P.O. Box 2569, W. Lafayette, Indiana 47906, USA.

Fazakerley, G.M. (1974) Group Technology: Social Benefits and Social Problems. *The Production Engineer.* October, pp. 383–386.

Fazakerly, (1974). The Human Problems of Group Technology, *European Business*, Summer.

Fordyce, K., Sullivan, G. (1989) *Logistics Management System (LMS): Integrating Decision Support and Knowledge Based Expert Systems to Monitor and Control Manufacturing Flow*, Proceedings of the Third International Conference on Expert Systems and the Leading Edge in Production and Operations Management, May 21-24, 1989, Hilton Head Island, South Carolina, Management Science Department, College of Business Administration, University of South Carolina, USA.

Fox, M. (1983) *Constraint Directed Search: A Case Study of Job Shop Scheduling*, PhD Thesis, Carnegie Mellon University, December.

Fox, M.S., Allen, B.P., Smith, S.F., Strohm, G.A. (1983) *ISIS: A Constraint-Directed Reasoning Approach to Job Shop Scheduling*, Proceedings of IEEE Conference on Trends and Applications '83, National Bureau of Standards, Gaithersburg, Maryland, USA.

Fox, R.E. (1982) OPT An Answer for America Part II, *Inventories and Production Magazine* 2 (6).

French, S. (1982) *Sequencing and Scheduling: An Introduction to the Mathematics of the Job-Shop*, Ellis Horwood Ltd, West Sussex, England.

Galbraith, J. (1973) *Designing Complex Organisations*, Addison-Wesley, Mass., USA.

Galbraith, J. (1977), *Organisation Design*, Addison-Wesley, USA.

Gallagher, C., Knight, W. (1973) *Group Technology*, Butterworth, London.

Genrich, H.J., Lautenbach, K. (1981) System modelling with high level Petri nets. *Theoretical Compt. Sci.*, **13** 109–136.

Goldberg, A., Robson, D. (1983) *SMALLTALK 80 – The Language and its Implementation.* Addison-Wesley Publishing Company, USA.

Goldratt E., Cox, J. (1986) *The Goal.* Creative Output Books, USA.

Gombinski, J. (1969) *The Birsch Classification system and Group Technology*, Group Technology: Proceedings of an Inter. Seminar, Inter. Centre for Advanced Technical and Vocational Training, Turin, Italy, pp 53–66.

Grant, H. (1987) *Controlling the Shop Floor*, CIM Technology, November.

Graves, S. C. (1981) A Review of Production Scheduling, *Operations Research*, **29** (4), July–August.

Greene, T., Sadowski, R. (1983) Cellular Manufacturing Control, *J. Manufacturing Control*, **2**, 2.

Greene, T., Sadowski, R. (1984) A Review of Cellular Manufacturing Assumptions, Advantages and Design Techniques, *J. Operations Mgt*, **4** (2), February.

Groover, W.R. (1980) *Automation, Production Systems, and Computer Aided Manufacturing*, Prentice-Hall, USA.

Grzyb, G.J. (1981) Decollectivation and Recollectivization in the Workplace: The Impact of Technology on Informal Work Groups and Work Culture. *Economic and Industrial Democracy*. **2**, pp. 455–482.

Gunasingh, K., Lashkari R. (1989) The Cell Formation Problem in Cellular Manufacturing Systems – A Sequential Modeling Approach, *Computers and Industrial Engineering*, **16** (4), 469–476.

Gunn, T. (1982) The Mechanisation of Design and Manufacturing, *Scientific American*, **247** (3), 87–110.

Gunn, T. (1987) *Manufacturing for Competitive Advantage*. Ballinger Publishing, Mass., USA.

Hackman, R. Oldham, G. (1980) *Work Redesign*. Addison-Wesley, USA.

Hadavi, K. (1987) *An Integrated Planning and Scheduling Environment*, Proceedings of the conference on Artificial Intelligence in Manufacturing, Long Beach, California, Published by the Society of Manufacturing Engineers.

Halberstam, D. (1986) *The Reckoning*, William Morrow and Company Inc., New York.

Hall, R.W. (1983) *Zero Inventories*. Dow Jones-Irwin, Illinois, USA.

Hall, R.W. (1987) *Attaining Manufacturing Excellence*. Dow Jones-Irwin, Illinois, USA.

Handy, C. (1985) *Understanding Organisations*. Penguin, England.

Haray, F. (1971) *Sparse Matrices and Graph Theory in Large Sparse set of linear equations* (ed. J. Reid), Academic Press, London, pp. 139–167.

Haverman, H. (1991) *Optimization of Production Schedules using A1 Techniques*, 9th Annual IEOR Symposium and 4th Annual WATMINS Workshop on Applied Scheduling, Columbia University, New York, October.

Hayes, C., Fonda, H. (1986) *People in Business: The Chief Executive's Role*. NEDO/Prospect Centre.

Hayes-Roth, B. (1985) A blackboard architecture for control, *Artificial Intelligence*, **26**, 251–321.

Heard, E., Plossl, G. (1984) Lead Times Revisited, *Production and Inventory Management,* Third Quarter, pp. 32–47.

Herbst, P.G. (1974) *Alternatives to Hierarchies.* Leiden.

Hewitt, C.E. (1977) Control structure as patterns of passing messages, *Artificial Intelligence*, **8**, 323–363.

Higgins, P., Browne, J. (1989) The Monitor in Production Activity Control Systems. *Int. J. Production Planning and Control,* (not yet published).

Higgins, P.D. and Browne, J. (1989) The Monitor in Production Activity Control Systems, *Production Planning and Control*, **1** (1), January–March.

Hill, P. (1971) *Towards a New Philosophy of Management*. Gower, London.

Hinds, S. (1982) The Spirit of Materials Requirement Planning. *Production and Inventory Management*. **23** (4), pp. 35–50.

Hoefer, H. (1985) Theory and Practice of Decentralised Production Mgt. Systems – a Philosophy rather than Methods, *Computers In Industry*, **6**, 515–527.

Holstein, W.K., Berry, W.L. (1970) Work Flow Structure – An Analysis for Planning and Control. *Management Science.* **16** (6), February.

Holtz, R. (1978) GT and CAPP cut work in time 80%, *Assembly Engineering*, **21**, pp. 16–19.

Houtzeel, A. (1975) *MICLASS a classification system based on group technology*, SME Technical Paper MS75-721, Society of Manufacturing Engineers, Dearborn, Michigan, USA.

Hyer, N.L., Wemmerlov, U. (1985) Group Technology Oriented Coding Systems: Structures, Applications and Implementation, *J. Irish Production and Inventory Control Society,* Second Quarter, **26** (2).

ISO (1984) *International Standard 7498: Information Processing Systems – Open systems Interconnection Basic Reference Model.* UDC 681.3.01, Ref No. ISO 7498-1984(E).

ISO (1986) *The Ottawa Report on Reference Models for Automated Manufacturing Systems ISO*, Geneva, Switzerland.

Imai, M. (1986) *Kaizen: the Key to Japan's Competitive Success.* Random House, New York.

Ingram, F. (1982) *Group Technology, Production and Inventory Mgt.*, 4th Quarter, pp. 19–34.

Iri, M. (1968) On the Synthesis of Loop and Cutset Matrices and the Related Problems, *SAAG Memoirs*, **4**, A-XIII, 376.

Jackson, S., Browne, J. (1989) An Interactive Scheduler for Production Activity Control. *Int. J. Computer Integrated Manufacturing*, **2**, (1), 2–15.

Jayaran, G. (1978) *Open Systems Planning. Sociotechnical Systems.* W. Pasmore and J. Sherwood, University Associates, USA.

Jensen (1981) Coloured Petri nets and the invarient method. *Theoretical Compt. Sci.*, **14**, 317–336.

Jones, A., McLean, C. (1986) A Proposed Hierarchical Control Model for Automated Manufacturing Systems, *Journal of Manufacturing Systems*, **5** (1).

Jones, B., Scott, P. (1986) Working the system: a comparison of the management of work roles in American and British flexible manufacturing systems. *Managing Advanced Manufacturing Technology.* C.A. Voss, IFS Publications, New York, USA.

Joyce, R. (1986) *Functional Specification of a Production Activity Control System in a CIM Environment*, Masters of Engineering Science Thesis, Industrial Engineering Department, University College Galway, Ireland.

Keen, P., Morton, M. (1978) *Decision Support Systems: An Organisational Perspective.* Addision-Wesley, Reading, Mass., USA.

Kempf, K.G. (1985) Manufacturing and Artificial Intelligence, *Robotics*, 1, Elsevier Publishers, New York, pp. 13–25.

Kempf, K.G. (1988) *Artificially Intelligent Tools for Manufacturing Process Planners*, Proceedings from the First International Conference on Expert Systems and the Leading Edge in Production Planning and Control, (ed. M. Oliff), Benjamin Cummings Publishing Company, USA, pp. 131–163.

Kerr, C., Rosow, J. (eds) (1979) *Work in America: The Next Decade.* Van Nostrand Reinhold, New York, USA.

King, J.R., Nakornchai, V. (1982) Machine Component Group Formation in Group technology: Review and Extension, *Int. J. Prod. Res.,* **20** (2), 117–133.

Kiran, A.S. and Smith, M.L. (1984a) Simulation studies in job shop scheduling – I: A survey, *Comput. & Indus. Engng,* **8** (2), 87–93.

Kiran, A.S., Smith, M.L. (1984b) Simulation studies in job shop scheduling – II: Performance of priority rules, *Comput. & Indus. Engng,* **8** (2), 95–105.

Knight, W.A. (1974) The Economic Benefits of Group Technology. *The Production Engineer,* May, pp. 145–151.

Kreppel, H. (1988) *CNMA - Progress in Industrial Communication.* Proceedings of the 5th Annual ESPRIT Conference, North Holland, Amsterdam.

Kuhn, T.S. (1970) *The Structure of Scientific Revolutions,* 2nd edn, The University of Chicago Press, USA.

Kukla, D.E. (1984) *Successful Manufacturing Systems Through MRP. Production and Inventory Management.* **4** (8), pp. 1–17.

Kusiak, A., Chow, W. (1987) Efficient Solving of the Group Technology Problem, *J. Manufacturing Systems,* **6** (2).

Kusterer, K. (1978) *Know-how on the Job: The Important Working Knowledge of Unskilled Workers.* Macmillan, London.

Lange, K. (1980) Computer Aided Process Planning in Cold and Forging, *Annals of CIRP.*

Latham, D. (1981) Are You among MRP's walking wounded? *Production and Inventory Management.* **22** (3), pp. 33–41.

Law, A.M., Haider, S.W. (1989) Selecting Simulation Software for Manufacturing Applications: Practical Guidelines & Software Survey, *Industrial Engineering,* May.

Lawrence, A. (1986) Are CAPM Systems Just Too Complex?, *Industrial Computing,* September, p. 5.

Lawrence, P. and Lorsch, J. (1967) ***Organisation and Environment.*** Harvard University Press, Cambridge Mass., USA.

Lawrence, P., Lorsch, J. (1967) *Organisation and Environment*, Harvard Business School, USA.

Leonard, R., Koenigsberger, F. (1972) *Conditions for Introducing Group Technology*, Proc. of 13th Int. MTDR Conf., England.

Lewis, F.A. (1986) Statistics aid planning for JIT production. *Chartered Mechanical Engineer.* June, pp. 27–30.

Link, C.H. (1976) *CAPP – CAM-I Automated Process Planning System*, Proc 13th Numerical Control Soc. Annual Meeting and Technical Conference, Cincinnati, USA, November.

Lodge, G. (1975) *The New American Ideology.* A. Knopf, New York, USA.

Long, M.W. (1984) *Guidelines for the Collection and Use of Shop Floor Data*, Execution and Control Systems, Auerbach Publishers Inc.

Lubben, R.T. (1988) *Just in Time Manufacturing: An Aggressive Manufacturing Strategy.* McGraw-Hill, USA.

Lundrigan, R. (1986) What is this thing called OPT? *Production and Inventory Management* **27** (2), pp 2–12.

Marklew, J.J. (1970) An example of the Cell System of Manufacture at Ferranti, *Machinery and Production Engineering,* August.

Mason, R. (1969) A Dialectical Approach to Strategic Planning, *Management Science,* **15** (8).

McAuley, J. (1972) Machine Grouping for Efficient Production, *Production Engineer,* **51** (53).

McEwen, N., Carmichael, C., Short, D., Steel, A. (1988) *Managing Organisational Change – A Strategic Approach.* Long Range Planning, **21** (6), 71–78.

McKay, K.N., Buzacott F.A., Safayeni, F.R. (1989). *The Schedulers Knowledge of Uncertainty: The Missing Link*, Knowledge Based Production Management Systems (ed. Jim Browne) North Holland, Amsterdam.

Meyer, W., Isenberg, R., Huebner, M. (1988) Knowledge-based factory supervision – The CIM Shell, *Int. J. Computer Integrated Manufacturing*, **1** (1), 31–43.

Mill, F., Spraggett, S. (1984) Artificial Intelligence for Production Planning, *Computer Aided Engineering*, December.

Mitchell, D. (1979) *Control Without Bureaucracy*. McGraw-Hill, New York, USA.

Mitrofanov, S.P. (1966) *Scientific Principles of Group Technology*, National Lending Library, England.

Monden, Y. (1983) *Toyota Production System: Practical Approach to Production Management*, American Institute of Industrial Engineers.

Monden, Y. (1983) *Toyota Production System: Practical Approach to Production Management*, American Institute of Industrial Engineers.

Nau, D.S., Chang, T.C. (1985) *A Knowledge Based Approach to Generative Process Planning*, Winter Meeting of Americian Soc. of Mechanical Engineers, November, pp. 65–71.

Nevins, J., Whitney, D. (1978) Computer Controlled Assembly, *Scientific American*, **238**, February, pp. 62–74

Nii, P.H. (1986a) Blackboard systems: The blackboard model of problem solving and the evolution of blackboard architectures (Part One). *The AI Magazine*, Summer, pp. 38–53.

Nii, P.H. (1986b) Blackboard systems: The blackboard model of problem solving and the evolution of blackboard architectures (Part Two). *The AI Magazine*, August, pp. 82–106.

Nof, S., Suranjan, D., Whinston, A. (1985) Decision Support in Computer Integrated Manufacturing. *Decision Support Systems* **1**, 37–56, Elsevier Science Publishers.

Nolen, J. (1989) *Computer-Automated Process Planning for World Class Manufacturing*, Marcel Dekker Inc., New York.

O'Grady, P., Lee, K.H. (1988) An intelligent cell control system for automated manufacturing, *International Journal of Production Research*, **26** (5), 845–861.

O'Toole, J. (1973) *Work in America.* MIT Press, USA.

OIR (1983) *Group Technology CAD/CAM*, Organisation for Industrial Research, Waltham, Massachusetts.

OSH (1991) Digital Equipment Corporation, *Open Systems Handbook, a guide to building open systems.*

Opitz, H. (1964) *Workpiece Statistics and Manufacture of Parts Families, VDI Zeitschrift*, **106** (26), September, MTIRA Translation T146.

Opitz, H. (1970) *A Classification System to Describe Workpieces*, Pergamon Press, Oxford.

Opitz, H. Wiendahl H. (1971) Group technology and Manufacturing Systems, *I.J. Production Research,* **9** (1), 181–203.

Orlicky, J. (1975) *Materials Requirement Planning*, McGraw-Hill, USA.

Orlicky, J. (1975) *Materials Requirements Planning: The New Way of Life in Production and Inventory Management*, McGraw-Hill, New York.

Owen, T. (1985) *Assembly with Robots*, New Technology Modular Series, Kogan Page, London.

POSIX (1988). IEEE *Standard Portable Operating System Interface for Computer Environments*, IEEE Std 1003.1-1988, The Institute of Electrical and Electronics Engineers Inc, New York, USA.

Panwalker, S.S., Iskander, W. (1977) A survey of scheduling rules, *Operations Research*, **25**, 48.

Pava, C. (1983) *Managing New Office Technology: An Organisational Strategy.* The Free Press, New York.Ω

Peterson, J.L. (1977). *Petri Net Theory and the Modeling of Systems.* Prentice Hall Inc., NJ USA.

Pritsker, A.B. (1984) *An Introduction to Simulation and SLAM II*, System Publishing Corporation, West Lafayette, Indiana, USA.

Pugh, D.S., Hickson, D.J., Hinings, C. (1971) *Writers on Organisations.* Penguin, England.

Purcheck, G.F., Oliva-Lopez, E. (1978) *Suitability Testing for Group Technology*, 19th MTDR Conference, September, Birmingham, England.

Rajagopalan, R., Batra, J.L. (1975) Design of Cellular Production Systems: A Graph-Theoretic Approach, *Int. J. Prod. Res.*, **13** (6), 567–579.

Rathmill, K., Leonard, R. (1977) Group Technology Myths, *Management Today*, January.

Requicha, A. (1980) Representations for rigid solids: theory, methods, and systems, *ACM Computer Surveys*, **12** (4), 437–464.

Rickel, J. (1988) *Issues in the Design of Scheduling Systems, Expert Systems and Intelligent Manufacturing*, (ed. M.D. Oliff) Elsevier Science Publishing Company, pp. 70–89.

Rinnoy Kan, A.H.G. (1976) *Machine Scheduling Problems: Classification, Complexity and Computations*. Martinus Nijhoff, The Hague, Holland.

Rodammer, F.A., White, K. P. (1988) A Recent Survey of Production Scheduling, *IEEE Transactions on Systems, Man and Cybernetics*, **18** (6), November/December.

Ross, D.T. (1985) Applications and Extensions of SADT, *Computer*, April, pp. 25–34.

Sacerdoti, E D. (1974) Planning in a Hierarchy of Abstraction Spaces, *Artificial Intelligence*, **5**, 115–135.

Sacerdoti, E D. (1977) *A Structure for Plans and Behavior*, Elsevier Publishers, USA.

Safizadeh, M., Raafat, F. (1986). Formal/Informal Systems and MRP Implementation. *Production and Inventory Management*, **27** (1), 115–120.

Schaffer, G. (1981) Implementing CIM, *American Machinist*, **125** (8), 151–174.

Schein, E.H. (1980) *Organisational Psychology*. Prentice-Hall.

Schonberger, R. (1982) *Japanese Manufacturing Techniques : Nine Hidden Lessons*. Free Press, New York.

Schonberger, R. (1986) *World Class Manufacturing: The Lessons of Simplicity Applied*. Free Press, New York.

Shambu, G. (1989) *A Rule Based System for Scheduling in a Hybrid Group Technology Environment*, Proceedings 3rd Int. Conf. on Expert Systems and the Leading Edge in Prod. & Op. Mgt, May, S. Carolina, USA.

Shingo, S. (1985) *A Revolution in Manufacturing: The SMED System*, The Productivity Press, USA.

Skinner, W. (1984) *The Focused Factory*, Harvard Business Review, May–June.

Skinner, W. (1985) *Manufacturing: The Formidable Competitive Weapon*, John Wiley and Sons, Inc., New York.

Smith, A. (1776) *The Wealth Of Nations.* Penguin, London, (Latest Edition 1970).

Smith, P.D., Barnes, G.M. (1987) *Files and Databases: an introduction.* Addison-Wesley Publishing Company, USA.

Smith, S., Fox, M. (1984) ISIS: A Knowledge Based System for Factory Scheduling, *Expert Systems*, 1, 25–49.

Smith, W.E. (1956) Various optimizers for single-state production, *Naval Re. Logist. Quart.*, **3**, 59–66.

Sneath, P.H., Sokal, R.R. (1973) *Numerical Taxonomy*, Freeman and Co., USA.

Softtech: Application of SADT, *SADT Structured Analysis and Design Technique: Author Guide*, Volumes 1 and 2.

Solberg, J.J. (1989) *Production Planning and Scheduling in CIM*, *Information Processing '89*, (ed. G.X. Ritter), Elsevier Science Publishers B.V., North Holland, pp. 919–925.

Stincombe, A. (1959) Bureaucratic and Craft Administration of Production : A Comparative Study. *Administrative Science Quarterly,* September, pp. 168–187.

Susman, G. (1976) *Autonomy at Work.* Praeger, New York.

Susman, G.I., Chase, R. (1976) A Sociotechnical Analysis of the Integrated Factory. *J. Applied Behavioural Science.* **22** (3), 257–270.

TNO (1981) *Introduction to MIPLAN,* Organisation for Industrial Research Waltham, MA, USA.

Taha, H. (1982) *Operations Research: An Introduction*, Macmillan Publishing Co., New York, USA.

Taylor, F.W. (1911) *The Principles of Scientific Management*. Harper and Row, New York.

Tempelhof, K.H. (1980) A System of Computer aided Process Planning for Machine parts. *Advanced Manufacturing Technology*, (ed. P. Blake), North-Holland, pp. 141–150.

Thompson, G. (1967) Managerial Work Roles and Relationships. *J. Management Studies*. **3** (3), 270–284.

Thorsrud, E. (1976) *European Contributions to Organisational Theory*. G. Hofstede and M. Kessen, Van Gorcum, Assem.

Tiemersa, J.J. (1988) *The Development of a Production Control System for Small Batch Part Manufacturing*, Proceedings of the 5th Annual ESPRIT Conference, Brussels, pp. 1528–1544.

Tierney, K., Bowden, R., Browne, J. (1987) Robex – An Artificial Intelligence Based Process Planning System for Robotic Assembly, *Recent Developments in Production Research* (ed. A. Mital), Elsevier Science Publishers, The Netherlands.

Toffler, A. (1970) *Future Shock*. Pan Books, England.

Trist, E. (1982) *The Sociotechnical Perspective*. Perspectives on Organisation Design and Behaviour, A. Van de Ven and W. Joyce, Wiley, New York.

Trist, E., Emery, F. (1965) The Casual Texture of Organisation Environments. *Human Relations*, **18**.

Trist, E., Higgins, G., Murray, H., Pollock, A. (1963) *Organisational Choice*. Tavistock Publications, London.

Tulkoff, J. (1981) *Lockheed's GENPLAN*, Proc. 18th Numerical Control Society General Meeting and Technical Conference, USA, May, pp. 417–421.

Turano, T., Bauman, R. Production based language simulation of Petri nets. *Simulation*, **47** (5), 191–198.

Tzafestas, S. (1989) *Petri Net and Knowledge Based Methodologies in Manufacturing Systems Modeling Simulation and Control*, Proc. CIM Europe Conference, May, pp. 39–50.

Vere, S. (1983) Planning in Time: Windows and Durations for Activities and Goals, *IEEE Transactions on Pattern Analysis and Machine Intelligence,* PAMI-5(3), May, pp. 246–266.

Vickers, G. (1965) *The Art of Judgement.* Basic Books, New York.

Vollmann, T., Berry, W., Whybark, D., (1988) *Manufacturing Planning and Control Systems,* Dow Jones-Irwin, USA.

Voss, K. (1983) Using Predicate/Transition Net to model and analyze distributed data base systems. *IEEE Trans. on sofware engineering,* **6**, Nov, 733–745.

Wadhwa, S., Browne, J. (1989) Modeling FMS with Decision Petri nets. *Int. J. Flexible Manufacturing Systems,* **1**, 225–280.

Waghodekar, P., Sahu, S. (1984) Machine component cell formation in group technology, *Int. J. Production Research,* **22**, 937–948.

Walsh, M., Tierney, K., Browne, J. (1989) Process Planning as a Two Stage Procedure Between Product Design and Manufacture, *IJPR* (not yet published).

Walton, R.E. (1980) Establishing and Maintaining High Commitment Work Systems. *The Organisational Life Cycle,* J.R. Kimberly and R.H. Miles, Jossey-Bass, USA.

Weber, M. (1964) *The Theory of Social and Economic Organisation.* A. Henderson and T. Parsons, Free Press, New York.

Weill, R., Spur, G., Eversheim W. (1982) Survey of Computer Aided Process Planning Systems, *Annals of CIRP,* **31** (2).

Weinberg, G. (1975) *An Introduction to General Systems Thinking,* Wiley, New York.

Whale, R., Mills R. (1988) *The Application of Finite Element Methods to Manufacturing Process Modeling,* Proc. 4th CIM Europe Conference, May.

White, E. (1980) Implementing an MRP System Using the Lewin-Schein Theory of Change. *J. American Production and Inventory Control Soc.,* **21** (1).

Whyte, W.F. (1969) *Organisational Behaviour.* Irwin, USA.

Wiener, R.S., Pinson, L.J. (1988) *An Introduction to Object-Oriented Programming and C++*. Addison-Wesley Publishing Company, USA.

Wight, O. (1981) *MRP II: Unlocking America's Productivity Potential*, CBI Publishing, Boston MA.

Willenborg, J. (1987) *FMS and Organisation*. Second International Conference on Robotics and Factories of the Future, San Diego, USA.

Willenborg, J. and Krabbendam, J. (1987) Industrial Automation requires Organisational Adaptations. *Int. J. Production Research,* **25** (11).

Winston, P.H. (1977) *Artificial Intelligence*, Addison-Wesley Publishing Company, Reading, Mass. USA.

X/OPEN Group Members (1987) *XOPEN Portability Guide*. Elsevier Science Publisher B.V., Amsterdam, Netherlands.

Yau, S., Caglayan, M. (1983) Distributed software systems design representation using modified Petri nets. *IEEE Trans. on Soft. Eng.*, **6**, November 733–745.

Zander, A. (1983) *Making Groups Effective*. Jossey-Basy, USA.

12.2 Further reading

Carlsson, M., Trygg, L. (1987) Assembly philosophies: some implications on engineering organisation, design work, manufacturing systems and the product itself. *Int. J. Production Research,* **25**, (11).

Cochran, W.D. (1984) *An approach to CAPP Cost Justification*. Proceedings of Autofact 6, pp. 4-1–4-11.

Dunn, M.S., Bertelsen, J.D., Rothauser, C.H., Strickland, W.S., Milsop, A.C. (1981) *Implementation of Computerised Production Process Planning*. Report R81-945220-14, United Technologies Center, East Hartford CT, USA, June.

Hartson, H., Hix, D. (1989) Human Computer Interface Development Concepts and Systems for its management. *Computing Surveys ACM*, **21** (1), March.

IEEE Standard 1003.1-1988, (1988) *Portable Operating System Interface for*

Computer Environments. The Institute of Electrical and Electronics Engineers, Inc New York, USA.

ISO 7498-1984 (1984) *Information processing systems. Open Systems Interconnection – Basic Reference Model.* International Standards Organisation.

ISO TC4 N38 (1989). *Draft standard for the Exchange of Product Model Data.* International Standards Organisation.

King, J.R. (1980) Machine Component Grouping in production Flow Analysis: An Approach using a Rank Order Clustering Technique. *Int. J. Prod. Res*, **18**, 213.

McCormick, W.T., Schweitzer, P.J., White, T.E. (1972) Problem Decomposition and Data Recognition by Clustering Technique. *Opns Res* **20** (993).

Schaffert, C. (1986) *An introduction to Trellis/Owl.* OOPSLA '86 Proceedings, September, pp. 9–16

Skinner, W. (1969) Manufacturing – The Missing Link in Corporate Strategy. *Harvard Business Review,* May/June.Ω

INDEX

Application generator 244
  application network 255, 265
  dispatcher 255, 258
  emulator 255, 260
  information flow 267
  interactive scheduler 257
  manufacturing database 246, 248
  model building 248
  monitor 255, 263
  operational modes 246
  order requirements model 252
  PAC simulation model 246, 255
  Petri nets and simulation 259
  physical model 249
  product and process model 249
  raw materials model 249
  rulesbase 246, 252
  sample study 269
  scheduler 255, 256
ANSI 135
Autonomation 12
Architecture
  for FC entities and core services 101–8
  for PAC entities and core services 101–8
  for production planning and control 26

Batch size 21
Bill of Materials 14, 15, 17

Bill of Materials Processor 13
BITBUS 128
Blackboard/actor based systems 170–1
Bottlenecks 20, 21

Capacity Requirements Planning (CRP) 13, 17
Closed Loop MRP 17
CNMA 124, 126
Combinatorial explosion 157
Computer Aided Manufacturing International (CAM–I) 173
Control task
  definition 54
  dispatcher 59–60
  monitor 60–2
  scheduler 57–9
Coordinate factory (SADT model) 71
  analyse manufacturing system 81
  context diagram 71
  control cells 87–8
  coordinate product flow 83
  design production environment 75
  develop process plans 77
  dispatch
    cell 88, 90
    factory 86
  maintain product based layout 79
  monitor
    cell 89, 91
    factory 92–3
  move 88
    between cells 87
  produce 88
  schedule
    cell 88, 89
    factory 84
COSIMA 84
Cost accounting (and OPT) 22

Database systems 128–37
  definition 128
  databases
    comparison of 134–5

hierarchical 131
network 131–2
relational 130–6
Digital Equipment Corporation, Clonmel (Irl.) 279
business
environment 280
integration 283
philosophy 284
continuous flow manufacturing 289
Group Technology 291
information technology environment 295
manufactured products 280
manufacturing challenges 281
marketplace pressures 280, 281
MRP II experience 285–6
technology 286–8
PAC implementation 297
product life cycles 281
production management system environment 283
social subsystem 288, 294
technical subsytem 288, 289, 293
Discrete-event simulation 161
classification of 162
DISPATCHER 166
Dispatching
blackboard 169
implementation 169, 303
rules 157

EDI 124, 127–8
EEC (European Economic Community) 168
EFTA (European Free Trade Association) 168
ESPRIT 125, 168, 244
project number 809 169
project number 932 170
Ethernet 126

FACTOR 167
Factory coordination (FC)
definition 46
product based layout vs. process based layout 48–9
social subsystem 231–4

technical subsystem 235–9
Fieldbus 128
Flexibility (in software) 120
definition of 121

Gantt chart 241, 257
General computing model (for PAC and FC) 120, 123–4
and communication systems 123–8
data management systems 128–37
object oriented approach 144–7
processing systems 137–40
user interfaces 141–4
GKS 142, 143
Group Technology 7, 19, 25, 29, 181–3
implementation of 289

Hand assembly 271, 290

IBM 176
Information Resource Data Dictionary 130, 136–7
Inspection time 3
Inter-operability (in software) 120, 136, 143
definition of 121–2
ISIS 174
ISO 27, 98, 99, 124

JIT 3, 4, 5–12, 19
and suppliers 7
manufacturing planning techniques 8–9
quality 11
compared to MRP and OPT 24
Johnson's Algorithm 153

Kanban 5, 6, 8, 11, 12

Lead time 3, 7, 22, 30
Leitstand Approach 177
LMS (Logistics Management System) 176
Lot size, *see* Batch size

Machine insertion
DIP machine 271, 289
VCD machine 271, 289

MADEMA 173–4
Manufacturing systems analysis
  layout of the production process 204–5
  line balancing operations standard 206
  product design 205
  set-up reduction 206–7
  use of flexible resources 205–6
Manufacturing lead time, *see* Lead time
MAP 124, 126–7
Master Production Schedule (MPS) 9, 14, 15, 18
Mixed model assembly 9
MMS 124, 127
Monitor
  blackboard 172
  implementation of 170–1, 304
  user attitudes 312
MRP 4, 12–19
  bottom up replanning 16
  compared to JIT and OPT 24
  criticisms of 19
  top down planning 16
MRP II 12, 13, 17–19
  experience of 285–6
  implementation of 285
  technology 287–8
Multiskilled operators 10

NBS 27
  hierachy of production planning and control 28
NP complete 155

Object oriented approach (to programming) 144–8
  see also General computing model
Operation time 10
Operation types
  combinative 250
  disjunctive 251
  inspection 251
  sequential 250
OPT 4, 19–24, 256
  compared to JIT and MRP 24
  rules of OPT 20–3

Optimization 164
OSI 98
  and the OSI model 124–5
Open Systems Foundation (OSF) 143
  OSF/Motif 143

Paradigm 164
PERT/CPM 160
Petri nets 98
  descriptive summary 108–13
  marking of a Petri net 109
  model of the information flow between a dispatcher and a producer 113-19
  Petri net timing 111
  state of a Petri net 111
PLATO-Z 170
Portability (of software) 97
  definition of 122
POSIX 137, 141
Priority rules 157
Process
  time 3
  batch 21
Process planning
  computer aided 197
  generative and variant 200–3
  strategies 199–200
Product
  based manufacturing 180–1
  design 6–7, 8
  families, *see* Group Technology
Product family formation
  classification and coding systems 184–92
  composite product 193
  production flow analysis 192–3
  visual inspection 184
Production Activity Control (PAC) 13, 17, 31
  definition 36
  dispatcher 39–41
  implementation of 298–312
  life cycle model 245
  monitor 41–4
  mover 44–5

Production Activity Control (PAC) (*cont'd*)
  producer 45
  scheduler 38–9
  subsystem
    social 231–4
    technical 235–9
Production Activity Control (implementation) 298
  aims of 299
  approach adopted 299
  background preparation 300
  computing requirements 308
  database 304
    dispatcher 303–5
  interface to
    data collection 301
    MRP II 301
  lessons and guidelines 308–12
  monitor 303, 306
  PAC life cycle 298
  role of the manufacturing environment 309
  scheduler 303, 305
  system
    behaviour 309
    utility and benefits 310
  test area 301
  use of simulation 299
  user attitudes 308, 312
  validity of PAC 308
  workcell reporter 306
Production Environment Design task (PED)
  analysis of a manufacturing system 52–3
  definition 48
  maintenance of a product based layout 51–2
  process planning 50–1
Production
  lead time, *see* Lead time
  smoothing 9, 12, 30
Programming languages classification 138–40
Protocol (software) 99

Queuing time 3, 10, 22

Remote Procedure Call (RPC) 127, 137

Rough Cut Capacity Planning (RCCP) 13, 17

SADT (Structured Analysis and Design Technique) 65
  activity diagrams 68–9
  dual aspects of a system 67
  functions versus implementation modelling 67
  graphical format 69
  input, output, control and mechanism 68–9
  model building 65
  top down decomposition 65
Scheduling
  algorithmic solutions 153
  artificial intelligence approaches 162
  blackboard 170–1
  branch and bound 155–6
  control
    paradigm 164
    theory 160
  data processing paradigm 164
  discrete event simulation 161
  drawbacks of the traditional approach 163
  due date rules 159
  dynamic
    programming 155
    rules 158
  enumeration methods 155
  FC and PAC 177
  implementation 171, 303
  mathematical notation 152
  modern approaches 164
  new approach characteristics 166
  optimization paradigm 164
  processing time rules 158
  project scheduling 160
  simple rules 159
  single machine solutions 153
  traditional approaches 152
  user
    attitudes 308, 312
    requests 298
Set-up time 3, 10–11, 19, 21
Slack time 158

SMED 11, 26
Sociotechnical design
  basic principles 213–20
  best fit of social and technical subsystems 239–41
  contribution towards implementation of PMS 226–8
  definition 212
  sociotechnical analysis procedure 223–6
  subsystem
    social 294
    technical 288
  technical and social subsystems 220–2
Surface mount 289
Solder wave 271, 290
SONIA 175
SQL 130, 133, 135, 136, 137, 138, 140
Standards, development of 169

TCP/IP 125–6
Throughput time, *see* Lead time
Transfer batch 21, 26
Transport time 3

User interfaces, classification of 141
  see also General computing model
U-shaped layout 9, 10, 26

Xwindows 141–2, 143